Environment and Society in Florida

Environment and Society in Florida

Howard T. Odum
Elisabeth C. Odum
Mark T. Brown

A CRC Press Company
Boca Raton London New York Washington, D.C.

Library of Congress Cataloging-in-Publication Data

Odum, Howard T., 1924-
Environment and Society in Florida / by Howard T. Odum, Elisabeth C. Odum, Mark T. Brown
p. cm.
Includes bibliographical references and index.
ISBN 1-57444-080-2 (alk. paper)
1. Environmental management—Florida. 2. Florida—Environmental conditions. 3. Ecosystem management—Florida. I. Odum, Elisabeth C. II. Brown, Mark T. (Mark Theodore), 1945- . III. Title.
GE315.F6028 1997
333.7′15′09759—dc21 97-41621
CIP

Direct all inquiries to CRC Press LLC, 2000 N.W. Corporate Blvd., Boca Raton, Florida 33431.

Lewis Publishers is an imprint of CRC Press LLC

No claim to original U.S. Government works
International Standard Book Number 1-57444-080-2
Library of Congress Card Number 97-41621
Printed in the United States of America 2 3 4 5 6 7 8 9 0
Printed on acid-free paper

Preface

Increasingly, there is a call for a unified view of the system of nature and humans in which we live. We need to understand the interdependence of the economy with resources and the environment, and the way economic vitality and quality of life depend on a symbiotic system. In the renewed efforts to educate our young people in basic sciences, none is more important than the unified science of environment and society. This text presents Florida as a system, with its subsystems, and the ways they work. Although it is written as a text, this book may be suitable for the general reader interested in Florida's ecology and economics.

This book is an introduction to the emerging field of *Environmental Science*. Since the first Earth Day in 1966, schools and colleges have experimented with approaches to environment. All too often the policy adopted was to put a smattering of ecology, chemistry, and pollution discussion in every grade, kindergarten through twelfth grade, with "environmental studies classes" introducing environmental problems in college. Environment was regarded as bits of various disciplines. That approach did not provide an adequate understanding of the environmental system and its economics. This book contains the principles for introducing the new discipline of environmental science.

Whereas many subjects such as physics, chemistry, and cellular biology use similar examples wherever they are taught, environmental science deals with the larger scale around us. It is best taught with state, regional, and local examples. Thus, *Environment and Society in Florida* presents principles that apply everywhere but uses examples from Florida.

More and more educators and curriculum designers are seeking a unified course that enables students to use their training to evaluate the relationship between humans and nature — between human systems and natural systems. This change of emphasis arises out of a recognition that the impact of human systems on nature has been badly managed, and that this mismanagement has important implications for us all. The next generation will have to cope with a complex of problems stemming from interactions among population, resources, economics, and environment. The first step towards coping is a clear understanding of the problems.

The contents include materials normally scattered in biology, geography, earth science, economics, ecology, and computer science, here combined in a whole systems approach to the environment. The text is organized for a course to teach unified environmental science, economics, and public policy.

The concept of energy systems is used to introduce basic principles. Systems modeling really comes alive when combined with computer simulations, which show the consequences of facts and mechanisms in graphs with time. The microcomputer is optional. Where schools have the equipment, students are provided with sufficient guidance to run some simple computer models. The teacher can run the model on one computer, with the students suggesting "what if" experiments, or, in a computer room, groups of three students can work together at their own pace. Where no computer is available, the exercises can be completed by hand. The short programs in BASIC are in the Appendix and easily typed into an available computer, with versions easily adapted for the Apple II, Macintosh, PC, and Windows.

Although we have calculated energy units in joules, as standardized in international measurements, we sometimes refer to kilocalories also, because they are more familiar. Common units are used with metric quantities in parentheses.

Part I starts with the environmental problems in Florida and then introduces the basic principles of a language of symbols. The student is taught to program computers, enough to simulate simple models that show what a system does over time. Setting up microcosms is suggested as a critical-thinking hands-on project.

In Part II we introduce the main kinds of ecological systems (biomes) in Florida, explaining their special characteristics, and problems in their relationships to the human economy.

In Part III are systems that use the environment for economic purpose. Ecological microeconomics relate resources to human uses with chapters on agriculture, forestry, fisheries, mining, parks, cities, and information centers.

In Part IV we consider the whole system of Florida and its spatial organization, relating the state to resources, past and present populations, alternate energy sources, waste processing, national and international exchanges and possibilities for the future.

We hope this unified approach to education can lead to a new generation that understands better what shapes their lives so that they can play a better role in their communities, developing sound policies for economic and environmental prosperity in the future.

Finally, we recognize that the text offers something of a "voyage of discovery" for most teachers. The new discipline drawing from many old disciplines, the potential for computer-assisted teaching, and the use of overview systems approach may be new and challenging.

Teachers involved in classroom testing of previous similar texts have reported that it also functions well as a self-instruction text, with class discussion of suggested questions and activities.

Acknowledgments

This introduction was prepared by combining examples from Florida with models of systems, energy, environment, and economics that have been developed for teaching over two decades. Opportunities to develop, teach, and test content and organization were provided by: Joint Centre for Environment, University of Canterbury, Christchurch, New Zealand, in 1976; Lyndon B. Johnson School of Public Affairs, The University of Texas, Austin, in 1982; Department of Environmental Engineering Sciences, The University of Florida, and Santa Fe Community College, Gainesville, 1970 to present.

College presentations were developed as part of the traveling Chatauqua program of the American Association for the Advancement of Science. Testing in high schools was aided from 1985 to 1987 by workshops for teachers sponsored by the Curriculum Development Division of the National Science Foundation.

Many people have provided materials, taught earlier drafts, and made suggestions including: A.A. Anderson, C. Bersok, J.L. Bogart, K.S. Braun, J.A. Dickinson, Z.P. Efird, S.B. Everett, J. Ewel, B.L. Freeman, E. Gavin, M.J. Gingras-Kuhn, D. Guerin, C. Houck, J.E. Knight, D. LaHart, S. Lan, G.P. Marshall, C.L. Mau, R.H. Meyers, C.L. Montague, J.A. Moxley, N. Meith, L.J. Norton, R.J. O'Connell, J. Peppler, F. Savage, J. Sendzimir, S.D. Stark, D. Tilley, Q. Trimmer-Smith, K.M. Turner, J.C. Van de Venter, and I. Winarsky. Joan Breeze, secretary, University of Florida, facilitated preparation of manuscripts, figures, updates, and revisions.

The Authors

Howard T. Odum is Graduate Research Professor, Emeritus, in Environmental Engineering Sciences, University of Florida, Gainesville. He received his A.B. degree from the University of North Carolina and his Ph.D. from Yale University. With 300 publications and nine books, he has pioneered new fields of systems ecology, ecological economics, and ecological engineering and started the Center for Wetlands in 1973. He received the Crafoord Prize of the Swedish Royal Society and Honorary Doctor of Science degree from Ohio State University.

Elisabeth C. Odum, an instructor at Santa Fe Community College in Gainesville, Florida, since 1986, teaches biology, ecology, and honors courses in the department of natural sciences. She has a B.A. in Zoology from Swarthmore College and an M.Ed. from the University of Florida. Active in public affairs, she is especially concerned with environmental education in Florida and gives lectures, workshops, and short courses for high school and college teachers. Her publications include textbooks, environmental computer simulation programs, and journal articles.

Mark T. Brown is Assistant Professor in the Department of Environmental Engineering Sciences, University of Florida, where he teaches and conducts research in the fields of systems ecology, ecological engineering, energy analysis, and environmental planning. Raised in south Florida, he has traveled throughout the world studying systems of energy and environment toward achieving a broader understanding of humanity's role in the biosphere.

Contents

Part I. Florida and systems concepts

Chapter 1
Overview of Florida 3
1.1 Northern and southern Florida 3
1.2 Colonization and the culture of development 5
1.3 Modern Florida 7
1.4 One system 8
Questions and activities for chapter one 10

Chapter 2
Florida's environmental problems 11
2.1 Dirty waters 11
2.2 Scarred land 12
2.3 Trashy dumps 12
2.4 Acid rain and ozone 12
2.5 Climate changes 13
2.6 Beach erosion 13
2.7 Stressed reefs 14
2.8 Fewer fish 14
2.9 Population explosion 14
2.10 Decaying image 15
2.11 Expensive electricity 15
2.12 Education for hope 16
Questions and activities for chapter two 16

Chapter 3
Ecosystems and symbols 17
3.1 Ecosystem processes 17
3.2 Symbols 18
3.3 Model of a forest ecosystem 22
3.4 The money transaction symbol 23
Questions and activities for chapter three 25

Chapter 4
Flows of energy 27
4.1 Metabolism in ecosystems 29
4.2 Showing the quantities of energy flow 30
4.3 Fire and a symbol for switching 31
4.4 Energy flow in a forest 31
Questions and activities for chapter four 33

Chapter 5
Cycles of materials 37
5.1 The water cycle in the forest 37
5.2 Principle of conservation of matter 38
5.3 The carbon cycle 38
5.4 The phosphorus cycle 39
5.5 Phosphorus in Florida 40
5.6 The nitrogen cycle 41
Questions and activities for chapter five 43

Chapter 6
Production and limiting factors 45
6.1 Production 45
6.2 Gross and net production 45
6.3 Limiting factors 46
6.4 Soil 48
6.5 Carrying capacity 48
Questions and activities for chapter six 49

Chapter 7
Energy webs and transformations 51
7.1 Structure of a food web 51
7.2 Energy chain 51
7.3 Energy relationships in modern society 57
7.4 Emergy 57
7.5 Solar transformity 59
7.6 Energy hierarchy 59
Questions and activities for chapter seven 61

Chapter 8
Self-organization and succession 63
8.1 Self-organization in ecosystems: succession 63
8.1.1 Early succession 64
8.1.2 Later states of succession 65
8.1.3 Diversity 65
8.1.4 Reorganization and oscillation 67
8.2 Maximum power principle 67

8.2.2 Good uses and feedback ... 69
8.2.3 Feedback control ... 70
8.2.4 Converging and dispersing of material in cycles ... 71
Questions and activities for chapter eight ... 72
Directions for making a pond-type microcosm ... 72
Directions for making a terrestrial microcosm ... 73
Directions for making a stream-type microcosm ... 73

Chapter 9
Simulation of quantitative models ... 75
9.1 TANK, a sample model ... 75
9.2 Simulate by hand ... 77
9.3 How you make the computer simulate your model ... 78
9.4 Spreadsheet calculation ... 79
9.5 "What if" experiments ... 79
Questions and activities for chapter nine ... 81

Chapter 10
Growth models ... 83
10.1 Model 1: exponential growth ... 83
10.2 Model 2: logistic growth ... 84
10.3 Model 3: growth on a constant-flow source ... 85
10.4 Model 4: storage loss ... 87
10.5 Model 5: growth on a nonrenewable source ... 87
10.6 Model 6: growth on two sources ... 88
Questions and activities for chapter ten ... 90

Chapter 11
Oscillating systems ... 91
11.1 An oscillating model ... 92
11.2 Two-population plot ... 92
11.3 Switching model ... 93
11.4 Pulse model ... 94
11.5 Chaotic oscillations ... 96
Questions and activities for chapter eleven ... 97

Part II. Environmental systems

Chapter 12
Weather and climate ... 101
12.1 Energy from sunlight ... 103
12.2 Global atmospheric circulation ... 104
12.2.1 Water vapor and rain ... 106
12.2.2 Air pressure and winds ... 107
12.2.3 Fronts in storms ... 109

12.2.4 Florida's winter weather ... 109
12.2.5 Florida's summer weather ... 110
12.3 Severe storms ... 111
12.3.1 Thunderstorms ... 111
12.3.2 Severe thunderstorms and tornadoes ... 111
12.3.3 Tropical storms and hurricanes ... 112
12.4 Climate and biomes ... 114
12.4.1 Microclimate ... 114
12.4.2 Climatic zones of the Earth ... 115
12.4.3 Land biomes ... 115
Questions and activities for chapter twelve ... 116

Chapter 13
Geology and hydrology ... 119
13.1 Soil formation and weathering ... 119
13.2 Chemical processes ... 120
13.3 Geologic structure of Florida ... 121
13.4 Hydrologic cycle in Florida ... 123
13.4.1 Variations in sea level ... 126
13.4.2 Beaches ... 127
Questions and activities for chapter thirteen ... 129

Chapter 14
Oceanography ... 131
14.1 Physical processes ... 131
14.1.1 Tide ... 131
14.1.2 Wind waves and currents ... 134
14.1.3 Gulf Stream and shelf currents ... 135
14.1.4 Seawater salinity and density currents ... 136
14.1.5 Rivers and coastal currents ... 137
14.1.6 Storms ... 138
14.2 Chemistry of seawater ... 138
14.3 Ecological self-organization with coastal eutrophication ... 139
Questions and activities for chapter fourteen ... 139

Chapter 15
Marine ecosystems ... 141
15.1 Blue-water ecosystem ... 141
15.2 Continental shelf ecosystem ... 143
15.3 Shallow-water grass flats ... 145
15.4 Coral reefs ... 145
15.5 Beach ecosystems ... 150
15.5.1 Ecosystem of the wave zone ... 150
15.5.2 Sand dune ecosystems ... 152
Questions and activities for chapter fifteen ... 152

Chapter 16
Estuaries 155
16.1 Salinity 155
16.2 Circulation 156
16.3 Food-rich estuarine nursery 157
16.4 Systems overview 158
16.5 Estuarine ecosystems 160
16.5.1 Plankton-benthos 160
16.5.2 Grass flats 161
16.5.3 Oyster reefs 162
16.5.4 Marshes and mangroves 164
16.6 Florida estuaries 166
16.6.1 St. Johns River estuary 166
16.6.2 Indian River and Cape Canaveral estuaries 166
16.6.3 Miami estuaries 167
16.6.4 Florida Bay 167
16.6.5 Charlotte Harbor and Tampa Bay 167
16.6.6 Apalachicola Bay 168
Questions and activities for chapter sixteen 168

Chapter 17
Freshwater ecosystems 171
17.1 Ponds and lakes 171
17.1.1 Eutrophic and oligotrophic waters 174
17.1.2 Productivity 174
17.1.3 Problems with eutrophic ponds and lakes 176
17.1.4 Lake Okeechobee: a eutrophic lake 177
17.2 Stream ecosystems 178
17.2.1 Ecosystem of Silver Springs 180
Questions and activities for chapter seventeen 183

Chapter 18
Wetland ecosystems 185
18.1 Water regime and wetlands 186
18.2 Types of wetland ecosystems 190
18.2.1 Freshwater marshes 190
18.2.2 Lake margin wetlands 191
18.2.3 Floodplain forest 191
18.2.4 Saltmarsh and mangroves 191
18.3 Kissimmee River and the Everglades 192
18.3.1 Everglades canals, dikes, and highway fill 193
18.3.2 Peatland oxidation 196
18.4 Wetlands management 197
18.4.1 Constructed wetlands and self-organization 197
Questions and activities for chapter eighteen 198

Chapter 19
Upland ecosystems 201
19.1 Forest succession 201
19.2 Soil formation 203
19.3 Pine forests and fire climax 204
19.3.1 Type of pine forest and groundwater 206
19.4 Hardwood forests 208
19.4.1 Temperature and type of hardwood forests 208
19.4.1.1 Deciduous hardwood forest 208
19.4.1.2 Subtropical evergreen hardwood forest 209
19.4.1.3 Tropical evergreen hardwood hammocks 211
19.5 Forest conservation 211
19.5.1 Role of diversity in protecting forests 211
19.5.2 Pine forests and epidemics 212
19.5.3 Endangered species 212
19.5.4 Sustainable management 212
Questions and activities for chapter nineteen 213

Chapter 20
Landscape mosaic of ecosystems 215
20.1 Regional maps of Florida ecosystems 215
20.2 Interdependence 218
20.3 Soils of Florida 219
20.4 Ecosystem overview 221
Questions and activities for chapter twenty 222

Part III. Systems that use the environment

Chapter 21
Economic use and development 225
21.1 Typical economic use interface 225
21.2 Money and real wealth 226
21.2.1 Emergy evaluation 227
21.3 Costs, prices, and scarcity 227
21.3.1 Market supply and demand 228
21.3.2 Real wealth and market price 228
21.4 Economic development 229
21.4.1 Predicting economic success with the emergy investment ratio 229
21.4.2 Wildlife habitat 230
21.4.3 Water and Florida's economic development 230
21.4.4 Planning and designing with nature 231
Questions and activities for chapter twenty-one 233

Chapter 22
Forestry 235
22.1 Pine plantation system 235
22.2 The money part of the business 237
22.3 Comparison of plantations and mature forest 237
22.4 Natural tracts and diversity 239
Questions and activities for chapter twenty-two 240

Chapter 23
Agricultural systems 243
23.1 Agroecosystems of Florida 244
23.1.1 Citrus groves 244
23.1.2 "Truck farming" of winter vegetables 245
23.1.3 Sugarcane 247
23.1.4 Beef cattle system 248
23.2 Energy and food requirements 250
23.2.1 Migrant labor 250
23.2.2 Energy and vegetarian diet 250
23.2.3 Self-sufficient agriculture 250
Questions and activities for chapter twenty-three 251

Chapter 24
Fisheries 253
24.1 Yield and overfishing 253
24.2 Competition between sports and commercial fishermen 257
24.3 Aquaculture 259
Questions and activities for chapter twenty-four 260

Chapter 25
Mining 261
25.1 Phosphate mining 261
25.1.1 Environmental impacts and restoration 264
25.2 Peat 265
25.3 Sand 265
25.3.1 Titanium mineral sands 265
25.4 Limestone 266
25.5 Fossil fuels 266
25.6 Net benefits of mining 266
Questions and activities for chapter twenty-five 267

Chapter 26
Parks, tourists, and biodiversity 269
26.1 A measure of environmental use 269
26.2 Parks 270

26.2.1 Everglades National Park 270
26.2.2 National forests 271
26.2.3 Aquatic parks 272
26.3 Biodiversity 273
26.3.1 Measuring biodiversity 274
26.3.2 Speciation 274
26.3.3 Endangered species and subspecies 274
26.3.4 Wildlife corridors and greenways 275
26.3.5 Ecosystem reserves for automatic reseeding for restoration 276
26.3.6 Land rotation 276
26.3.7 Residential biodiversity 276
26.3.8 Hunting 276
Questions and activities for chapter twenty-six 277

Chapter 27
Cities 279
27.1 Urbanization with fossil fuels 279
27.1.1 The agrarian city 279
27.1.2 The fuel-based city 281
27.2 Overview of the city system 281
27.3 Spatial organization of the city 284
27.3.1 Residential neighborhood 285
27.4 City waste 286
27.5 City malfunction 287
Questions and activities for chapter twenty-seven 287

Chapter 28
Industry, technology, and information 289
28.1 Industry 289
28.2 Information 290
28.2.1 Information system in the economy 291
28.2.2 Global information processing 292
28.2.3 Space futures in Florida 294
Questions and activities for chapter twenty-eight 294

Part IV. The Florida state system

Chapter 29
Resource basis for the Florida economy 297
29.1 Renewable resources of Florida 297
29.1.1 Purchased fuels 299
29.1.2 Goods and services 299
29.1.3 Population and immigration 300
29.1.4 Information 300

29.2 Stored resources reserves in Florida 300
29.3 Resources and the circulation of money 300
29.3.1 Emdollars 301
Questions and activities for chapter twenty-nine 301

Chapter 30
Early human society in Florida 303
30.1 The Indian landscape 303
30.1.1 Hunting and gathering society 303
30.1.2 Agricultural society 305
30.2 Colonial economy of Florida 306
30.2.1 Systems view of cultural competition 307
30.2.2 Nonrenewable economy 307
Questions and activities for chapter thirty 308

Chapter 31
Overview of the Florida economy 311
31.1 Environmental-economic system of Florida 311
31.1.1 More detailed view of Florida 313
31.1.2 Carrying capacity 315
31.1.3 Attraction and image 315
31.1.4 Emergy/money ratio 316
31.2 Comparisons between Florida and other areas 316
31.3 Florida's trade 317
31.3.1 International influences 318
31.3.2 Resources for the future 318
Questions and activities for chapter thirty-one 319

Chapter 32
Florida's networks 321
32.1 Hierarchy of population centers 321
32.2 Transportation 321
32.2.1 Ships and barges 323
32.2.2 Transportation alternatives 324
32.2.3 Ecological engineering of transportation 325
32.2.4 Florida and international networks 325
32.2.5 Energy supply networks in Florida 325
32.2.6 Information networks 326
Questions and activities for chapter thirty-two 326

Chapter 33
Evaluation of energy sources 329
33.1 Power from heat engines 329
33.1.1 Optimum efficiency for maximum power 329
33.1.2 Temperature measurements 331

33.1.3 Efficiency of heat engines 331
33.2 Net emergy evaluation of fuels 332
33.2.1 Emergy yield ratio 332
33.2.2 Effect of transportation 333
33.2.3 Emergy yield ratios of foreign oil 334
33.2.4 Comparison of fuels 334
33.2.5 Transportation fuels 335
33.3 Emergy yield ratios of electric power production 336
33.3.1 Nuclear power 336
33.4 Emergy evaluation of renewable energy sources 337
33.4.1 Solar energy use through biomass 338
33.5 Solar technology 339
33.5.1 Solar voltaic cells for electric power 339
33.5.2 Solar energy hot water heaters 340
33.6 Other renewable energy flows 340
33.6.1 Wind 340
33.6.2 Geothermal power and ocean thermal electrical conversion 341
33.6.3 Hydroelectric power 341
33.6.4 Waves and tides 341
33.6.5 Using an appropriate energy quality 342
33.6.6 Future sources for Florida 342
Questions and activities for chapter thirty-three 342

Chapter 34
Wastes and recycle 345
34.1 Background 345
34.2 Principles of reuse and recycle 346
34.2.1 Closing broken cycles 346
34.2.2 Converging products and dispersing recycle 346
34.2.3 Economics and emergy evaluation of waste alternatives 347
34.2.4 Reuse and environmental recycle 348
34.2.5 Environmental technology and industrial ecology 349
34.2.6 Self-organization underway 349
34.3 Wastewaters 349
34.3.1 City wastewaters 349
34.3.2 Paper mill wastes 352
34.4 Air wastes 352
34.4.1 Ozone 353
34.4.2 Greenhouse gases and climate change 354
34.5 Solid wastes 354
Questions and activities for chapter thirty-four 356

Chapter 35
Population and carrying capacity 357
35.1 Population and standard of living 357
35.2 Carrying capacity 358
35.3 The effect of declining resources on population 360
Questions and activities for chapter thirty-five 361

Chapter 36
Simulating the future 363
36.1 Simulation of a world mini-model 363
36.2 Simulating a Florida mini-model without exchange 365
36.3 Simulation of Florida assets with world trade 366
36.3.1 Simulation of state growth with steady increase in world assets 367
36.3.2 Simulation of Florida growth driven by the world pulse of growth 368
Questions and activities for chapter thirty-six 368

Chapter 37
The future 371
37.1 A scenario of continued growth 371
37.2 A scenario of sustaining the present economy 372
37.3 A scenario of descent 373
37.4 A scenario of steady but smaller economy 374
Questions and activities for chapter thirty-seven 377

Appendix A. Programs for computer simulations 381
Appendix B. Emergy evaluation of Florida 395
Appendix C. Useful conversions 399
Appendix D. Types of wetlands 401
Glossary 405
Suggested readings 421
References 423
Index 429

part one

Florida and systems concepts

Part I introduces environmental systems concepts that help us see wholes, parts, and connections and what these relationships cause over time. To understand the complex world, human minds first use words to make simple overviews, which are called models. We use the method of systems diagrams and ecological examples to make the models clear, concrete, quantitative, and amenable to simple computer simulation. With emphasis on whole systems, simplified models are used to present the principles of matter, energy, information, and money. Chapters 1 and 2 begin with a verbal overview of the Florida state system and its parts and problems. The symbol diagrams are introduced in Chapter 3; energy and material cycles in Chapters 4 and 5; production and limiting factors in Chapter 6; networks and hierarchy in Chapter 7; simple computer simulation of storage and growth models in Chapters 9 and 10; and pulsing in Chapter 11. It is hoped that you learn from this section how to aggregate and simplify your view of environments and their problems so as to see consequences and solutions.

chapter one

Overview of Florida

Let's begin by examining Florida a thousand years ago. Imagine you are a high-flying bald eagle looking down on a green landscape on a perfectly clear day. Imagine the altitude to be high enough so that the entire landscape is visible at one glance (Figure 1.1).

You can see a peninsula jutting out into the sea that is 150 miles (240 km) wide, 500 miles (833 km) long, and only 345 feet at its highest elevation. The coastline is very long, over 1350 miles. Including all the undulations, islands, and small pieces of land that penetrate into the sea, the coast is an amazing 8500 miles (13,600 km) long. Due to the pounding surf of the ocean, the east coast has "smooth" beaches backed by dunes, while the south and west coasts, with less wave and wind energy, develop an irregular coastline of mangrove swamps and saltwater marshes.

Because the land is nearly surrounded with ocean, the climate is warm and humid with abundant rainfall for 6 months of the year, and cooler with much less humidity and rainfall for the remaining 6 months. It is covered from coast to coast with a rich diversity of plant and animal life adapted to large seasonal rainfall. The natural vegetation of the landscape is organized according to topography, availability of water, geology, sunlight, and tide and wave energy to form a mosaic of pinelands, hardwood forests, freshwater swamps, marshes, prairies, and mangrove swamps.

1.1 Northern and southern Florida

The peninsula is divided at its middle into a northern and southern region, each with special characteristics visible from above. The north has upland forest of tall straight pines and stately oaks growing so thick and tall that sunlight barely penetrates to the ground. On sandy ridges of ancient beaches are scrub pines and other species unique to Florida. Thousands of lakes collect rain water that ever so slowly flows toward the sea. Swamps of enormous, buttressed cypress trees dot the landscape and line hundreds of small streams. These freshwater marshes and lakes, some thousands of acres in size, are host to millions of water fowl.

Figure 1.1 "Bird's-eye" view of the Florida peninsula before European colonization.

There are many small rivers, and, while the water itself is not visible, they are recognized by the thin green floodplain swamps that line them. Rivers like the Suwannee, Apalachicola, Oklawaha, Withlacoochee, Santa Fe, and Peace, to name but a few, resemble thin green cracks in the landscape through which the wet season rains flow slowly to the Atlantic Ocean on the east and to the Gulf of Mexico on the west (Figure 1.2).

In the middle of the southern region is a great lake, called Okeechobee, fed by the Kissimmee River, a river flowing southward out of the central part of the peninsula. Water flows south out of the great lake throughout the year in an almost continuous sheet across a vast sawgrass marsh, exiting through saltwater mangroves and carrying organic matter and nutrients to the estuaries. In some places, only a few inches deep, the sheet of water through the sawgrass resembles a broad ill-defined river — thus, the name, Everglades, "river of grass".

Just to the east of the Kissimmee River, another river is visible — the St. Johns River, which flows north, finally exiting to the Atlantic Ocean on the east coast about 300 miles from its beginning.

Where the land meets the sea, mangrove swamps and saltwater marshes line the coast, filtering the waters that run off the land. Each tidal cycle

Figure 1.2 Rivers of Florida.

flushes tons upon tons of organic matter to the estuaries, feeding incredibly rich marine life and fisheries.

All over this land are the Indians, who migrated to Florida at least 10,000 years ago. They depend for survival on hunting, fishing, and the gathering of fresh foodstuffs, with some cultivation of corn, beans, and squash. A balance between Indians hunting and gathering food and the landscape providing that food was maintained for centuries. Generation after generation, the seasonal cycles continued, the Indian cultures adapting as needed.

1.2 Colonization and the culture of development

But, colonists with different ways then appear on the land. They bring a culture whose way is not to live with the environment, but rather to consume it. This culture draws a distinction between humanity and nature.

Powered by energy and technology not seen before, this new culture changes the face of the land. Vegetation is cleared, crops of vegetables and citrus are planted, cattle are introduced, and cities seem to spring up overnight. Railroads soon follow, connecting the cities, crisscrossing the landscape, and interrupting long-established drainage patterns. With the railroads, new lands are "opened up" for "development", and more people come, needing more food and shelter. In the north, forests recede under the assault of this new culture, converted into pasture and field crops. The vast sawgrass marshes of the south are drained, and large areas are cultivated in sugar cane.

Everywhere, north and south alike, the landscape is altered to fit this new culture. The abundant rainfall of this land is viewed as "too much water", so swamps are drained, rivers straightened, and levees built. Even in the upland areas where high groundwater tables ensured the productivity of vegetation during the dry season, drainage ditches are constructed to lower water tables, making the land more "suitable" for housing and agriculture. The net result is too much water downstream, as the wet season's rains run off far more quickly, and not enough water during the dry season, since it has been shunted to the sea. More ditches are dug to "fix" the downstream flooding, more rivers are straightened, more levees built. The landscape that once acted to smooth the differences between wet and dry season now alternates between flood and drought. Wildlife habitat is diminished.

This landscape that once resembled a patchwork mosaic of forests, swamps, and marshes is now a mosaic of forests, agriculture, and cities, the proportions of each changing almost daily as the population grows. Where there were once 2 Indians for each square mile, at this stage there are 20 people per square mile. Where there once stood a semi-permanent settlement of Indians, there now stands a permanent city. The density where people have clustered in cities is as much as 500 people per square mile. Still the people come. Where once there was plentiful water, now wells must be drilled and water withdrawn from the deep storages below ground. And still the population grows.

Energy and technological products, imported from elsewhere, permit population growth. The new culture is not tied directly to the productive capabilities of the land. As long as increasing supplies of food, materials, and energy can be imported, the population can grow, and grow it does.

Imagine, in a mere 200 years, the population of this landscape growing from 100,000 Indians to almost 10,000,000 people, more than a 100-fold increase. The changes, as a consequence of this incredible growth, are everywhere, so extensive there are hardly any places within this landscape that have not been altered.

Exports are very important to this new economy. Remaining forests are cut, and the wood is burned to make steam to power the machinery that alters the land. Some is used to build cities and some exported. Large deposits of the mineral phosphate are discovered in central Florida, and mining begins on a grand scale. Most of the phosphate is exported and utilized as fertilizer for agricultural crops elsewhere. Fishing becomes a major industry,

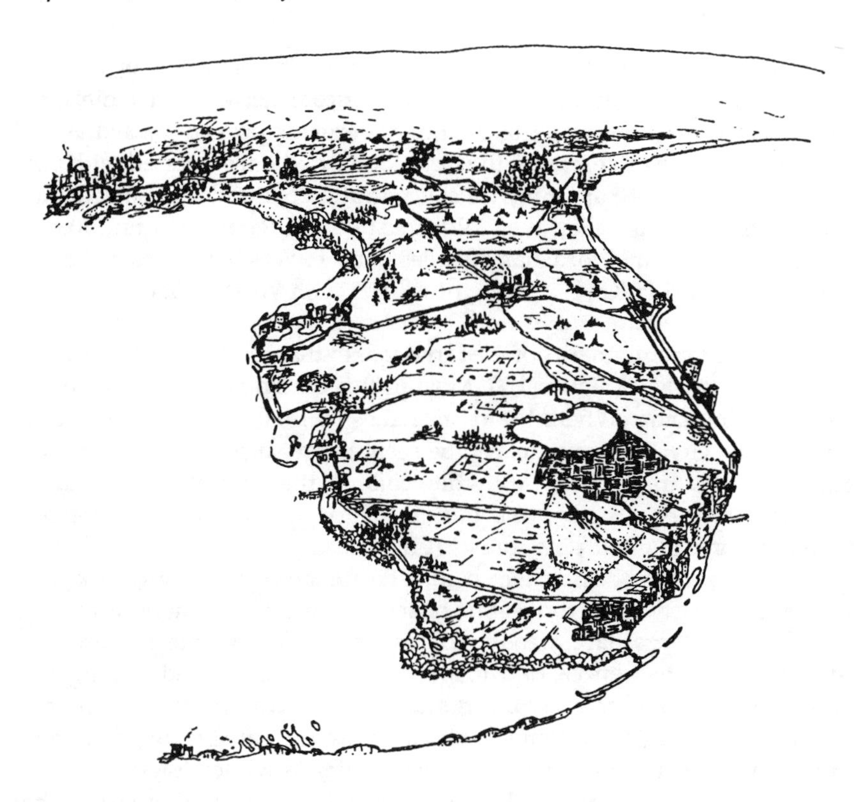

Figure 1.3 "Bird's-eye" view of the Florida peninsula in the 1990s after economic development.

but the stocks are overfished and diminish. Specialized agricultural develops with citrus and winter vegetables.

So, through the sale of citrus, vegetables, wood, phosphate, and the like, machinery, tools, vehicles, and fossil fuels are bought that enable further growth and development of the culture. These developments use up the centuries-old storehouses of soil, minerals, and woods, which are mostly sold to pay for growth.

1.3 Modern Florida

Imagine flying over this landscape today (Figure 1.3). It has been given a name — "Florida" — and divided, subdivided, and divided again into smaller and smaller parcels of land under differing political control and individual ownership. In south Florida, the "river of grass" no longer flows freely. Shown in Figure 1.3 are various dikes, ditches, and roads cutting across the Everglades from east to west and north to south. Each interrupts the flows of water, impounds them in some areas, and diverts them into other areas. One fourth of the original sawgrass marsh has been converted to the cultivation of sugar cane, where an intricate network of canals and pumps has lowered the water to expose the organic peat now used as soil.

To the east of the Everglades the limestone ridge, once covered with pine, has been replaced with an immense concentration of buildings, roads, and 3 million people that is called the Gold Coast. The ridge is a continuous city stretching from Miami more than 60 miles north to West Palm Beach. Farther south are the Florida Keys, a string of islands with West Indian forests and coral reefs.

Northward are remnants of the large marsh that formed the headwaters of the St. Johns River (Figure 1.2). Most of this marsh has been drained for pasture and housing. Where the river makes its final turn to flow into the ocean, a large city, Jacksonville, can be seen. Large ships come and go with commodities for trade. The stark white color of the concrete and buildings makes it very easy to trace the outline of the city against the green of agricultural and planted pine forests (Figure 1.4).

Extending across the northern peninsula are the very smooth, dark green pine plantations covering thousands of acres. Scattered throughout the pine plantations are large bare areas ready for replanting, where recently harvested pines have been carried to paper mills. Interspersed among pine plantations are thousands and thousands of small swamps of pond cypress. Easily recognized by their almost circular forms, they are bright green in the spring, orange in autumn, and a gray bark color in winter (Figure 1.4).

On the west coast, about half way up the peninsula, another large urban center (Tampa-St. Petersburg) can be seen with 1.5 million people surrounding a saltwater bay. Inland and northeast of Tampa-St. Petersburg is the stark white concrete of Orlando, a tourist city at a crossroads, almost equidistant from both coasts. Here, people from all over the world come to see the antics of imaginary creatures, such as Mickey Mouse, from children's stories.

Toward the north is a patchwork of agricultural lands with hundreds of squares, some bright green with new growth, some plowed and brown, and some dull green to yellow, ready for harvest (Figure 1.4).

1.4 One system

Viewed as a whole, the landscape now has a truly amazing organization. Something of the original network of streams and swamps remains. The cities and towns are connected by roads, railroads, electric power lines, and natural gas pipelines. Rural areas exchange products and services with smaller towns, which in turn exchange with the large cities. Each item viewed separately seems to make sense, but when related to the environment, there appear to be conflicts that threaten the overall system of people and the natural world.

So often, in everyday life, traveling through the landscape between school and home, or driving to the beach, or running errands, only part of the

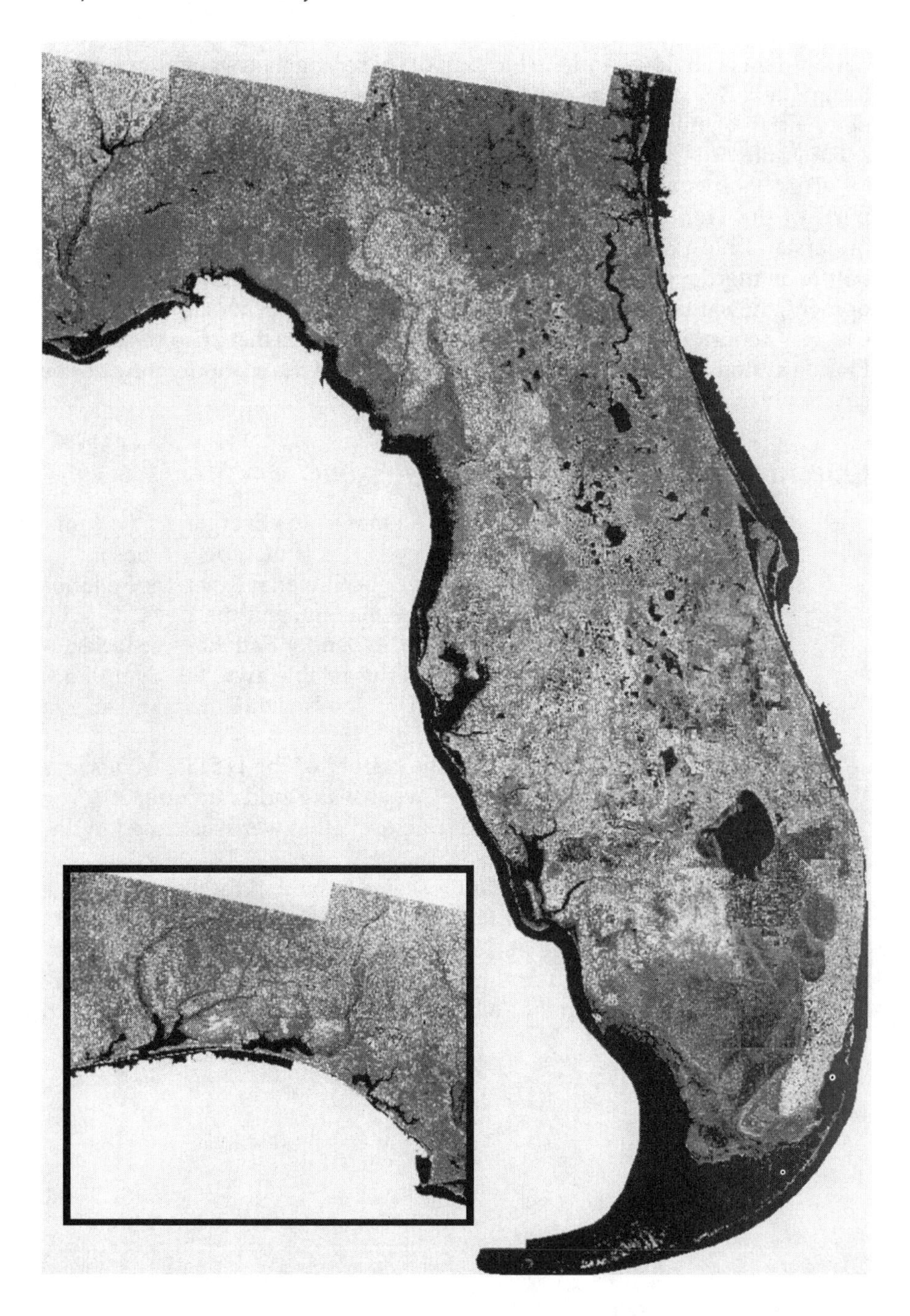

Figure 1.4 Satellite view of Florida. Image is a composite of 15 Landsat scenes. The image mosaic was created by the South Florida Water Management District and is included here with permission.

new system is obvious, and the life support of nature that is not seen is taken for granted. But viewed as one system, it is easy to recognize many interdependencies of humanity and nature. In 1995, there were 300 people per square mile.

This book is about the overall system of humanity and nature. While parts of the Florida system are discussed separately, the focus here is to understand how Florida works as one large system, including people and nature in interdependent activities. In the process of human work and development, the natural sector may be damaged and the partnership threatened. Chapter 2 considers some of the conflicts and problems that arise because the Florida system is not in environmental balance and the economy may not be sustainable.

Questions and activities for chapter one

1. Define: (a) peninsula, (b) swamp, (c) marsh, (d) Everglades, (e) hunting and gathering, (f) density, (g) cycle, (h) landscape, (i) mosaic.
2. In Figure 1.1, label: Lake Okeechobee, the Suwanee River, the St. Johns River, Tampa Bay, Gold Coast, Biscayne Bay, and the Keys.
3. In Figure 1.3, label: Miami, Tampa, Orlando, Tallahassee, Jacksonville, the sugar cane area, the winter vegetable area, the Everglades National Park, and highways I-75, I-95, the Florida Turnpike, I-4, and I-10.
4. Relate the history of Florida to the history of the U.S. In what ways were they the same, and in what ways was Florida unique?
5. Describe five problems of modern Florida that were discussed in this chapter.
6. The point-of-view of the chapter seems to be that the Florida Indian culture fit into their environment better than our modern culture does. Discuss why you agree or do not agree.
7. If there are any second or third generation Floridians in the class, ask them to find out what the "olden days" were like. Pictures would add reality.

chapter two

Florida's environmental problems

A new mission for humans on this earth is to use the great power of our information and collective social action to make a better biosphere that includes an appropriate role for each person. The challenge is to develop a symbiosis between the human economy and our environmental life-support system. To do this requires better public policies on environment. Everyone needs to try to understand how to solve local and global environmental problems by fitting nature and society together.

Many books considering environment are organized according to the problems such as those with water, climate, air pollution, mining, solid waste, energy, population, agriculture, urban growth, radioactivity, and biodiversity. Florida has its share of problems in these areas, as identified in this chapter. Later chapters consider how the Florida systems of environment and economy work so that we can see better how to solve the problems.

2.1 Dirty waters

In many places, water is becoming short in supply or poor in quality. The nutrient-rich sewage of Florida's 15 million people is dumped into streams and the ocean where it is too much for the natural ecosystems. Wastes pumped underground threaten the use of groundwater for drinking. The nutrients could be turned into a resource, valuable to agriculture and forests in appropriate quantities, stimulating the economy.

Because wetlands are nature's way of filtering and re-absorbing wastes, they can be used to process human sewage and many toxic wastes. Instead of draining or paving over our wetlands, they can be used for conservation of water quality and quantity and for recycling wastes.

Many people in Florida love to recreate in natural areas, fish in lakes and estuaries, and take their families boating. Yet, the suitable areas are decreasing, the fish are scarce, and high-powered motor boats are killing endangered manatees.

In the Everglades of south Florida, canals were dug to channel water out of peat lands to grow vegetables and sugar cane. The dried-out peat oxidizes, losing a centimeter (almost a half inch) each year. Nutrients released from the farming make the canals too rich with chemicals, weedy plants fill the canals, and pesticides decrease the natural life.

Highways built across the Everglades blocked the gradual flow of water going south, and cars killed the wildlife trying to cross at night. For one of the new highways, 14 underpasses were built to allow water, panthers, and other wildlife to flow more naturally.

Dikes and canals changed the broad flows of water which nourished the Everglades, the "river of grass". The Everglades National Park receives too much water in wet years, drowning deer, and too little in other years. The earlier pattern saw the growth of fish and frogs during wet periods, while in dry seasons they were concentrated into smaller pools, providing food for wading birds such as the wood storks. Now, changes in the quantities and timing of water have decreased the number of birds.

2.2 *Scarred land*

Florida's richest nonrenewable resource is phosphate rock, found in central Florida. It is processed into fertilizer and sold around the world. The process results in large holding ponds for the by-product clays that take years to settle. The mining and processing have scarred the land. Although some restoration work is being done, more of the mined areas could be productive ecosystems again. Some of the older mined areas with irregular topography have developed into good wildlife habitat.

2.3 *Trashy dumps*

Products made and not reused accumulate in dumps that each year take up more and more land, draining wastes into groundwaters which are the source of our drinking water. Think of all the plastic bags, styrofoam cups, and plastic picnic forks thrown away. Some of it can be reused, and recycling for reuse is increasing in most cities. The remainder needs to go back to nature in some way that is compatible with good landscape maintenance. In one experiment in Gainesville, shredded solid wastes were added to pine plantations and stimulated their growth.

2.4 *Acid rain and ozone*

Acids and heavy metals emitted from smoke stacks are damaging to the health of humans and ecosystems. Sulfur dioxide from the combustion of high-sulfur coal combines with water vapor to produce acid rain. Acid rain retards plant and tree growth, corrodes monuments, and makes sink holes by dissolving the underground limestone (Figure 2.1); however, some forests

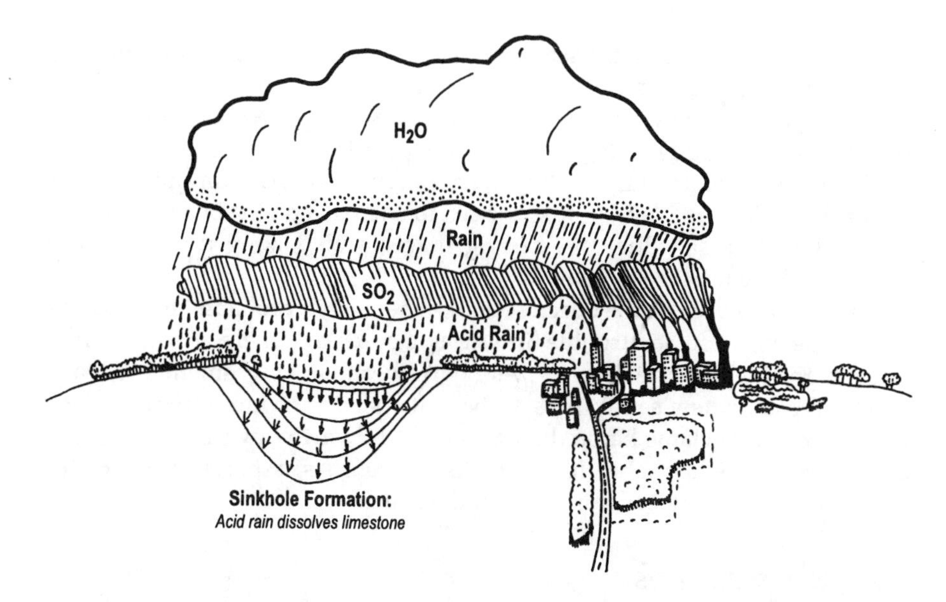

Figure 2.1 Acid rain pollution. SO_2 = sulfur dioxide from smoke stacks.

and crops require small amounts of sulfur. Sunlight acting on exhausts from cars and industries creates ozone, which is a stress on plants as well as humans.

2.5 Climate changes

In Florida, as in the rest of the world, the burning of fossil fuels and other natural resources by industries and transportation is producing carbon dioxide and other "greenhouse" gases that are changing the global climate. Scientists disagree as to whether water levels will rise and Florida will become smaller or water levels will fall, making a larger Florida.

Chlorofluorocarbons (CFCs) are depleting the ozone shield in the high atmosphere. This lets more ultraviolet light through to humans on Florida's beaches, causing more skin cancers.

Cutting down forests, draining wetlands, and building houses in central and south Florida may have affected the weather. Drying the lands changed the rain patterns, causing local showers to be less frequent. Because water changes temperature more slowly than does air, wetlands slow the rate of freezing. As wetlands are drained, the freeze line moves south. This has caused farmers to move orange groves farther south to avoid killing frosts.

2.6 Beach erosion

Originally, rains inland percolated into the groundwaters and flowed out to the sea in springs and by underground seepage. Now, however, cities and industries, by pumping up groundwaters, have reversed the flow, causing saltwater to come in from the sea or from levels deep underground. Saltwater replaces the fresh groundwater used for drinking.

The sea has only risen a few inches worldwide in the last century, but seawater is eroding Florida's beaches. It is possible that coastal zones have sunk, as observed in Italy and Taiwan, because we are pumping out the water. These lowered lands cause the sea to encroach on the beaches, diminishing their tourist value, threatening hotels, and making it necessary to pump sand back from offshore.

If you have been to the beach recently, there were probably more people than in the past, you had trouble finding a place to park, there was more trash at the high-tide line, the beaches were narrower than you remembered, and there were just a few sandpipers at the water's edge trying to pick up sand crabs or coquina clams before being forced to fly away by people, dogs, or cars. There were probably fewer shells for tourists, especially where the shells are crushed by cars allowed on the beach.

2.7 *Stressed reefs*

Off the Florida keys are coral reefs which support beautiful corals, many kinds of fish, and delicious seafood such as lobsters. Blue waters are becoming turbid with microbes from pollution and loose sediments from dredging. The coral reefs are stressed because they require clear water and bright sunlight. The balance among species maintains the biodiversity of the reef, but this is disturbed by too many spear fishermen taking fish and lobsters.

2.8 *Fewer fish*

Pollution by sewage in estuaries, where the rivers come into the saltwater, is becoming such a serious problem that oysters are contaminated and carry hepatitis and other diseases that they filter from the water. Cutting channels for boats allows high-salinity water to come into estuaries, interfering with the fluctuations in salinity required for productive oyster reefs.

Over-fishing in the waters of Florida and in international waters around Florida is reducing the quantity and diversity of fish and other seafood. Local fish are becoming scarce in Florida restaurants, and imported fish are expensive and discourage tourists. Sports fishermen and commercial fishermen compete for the few fish left, causing further loss of stocks. The 1996 fishnet ban amendment is an experiment to limit net fishing and allow the inshore fish populations to recover, including sports fish such as redfish and sea trout and the food chain fish such as mullet and menhaden on which larger fish depend.

2.9 *Population explosion*

More and more, as an exploding population is taking more agricultural space for cities, soil erosion and nutrient depletion are also decreasing productivity of good agricultural land. Not only is soil being depleted, but the diversity of wild plants and animals is also decreasing. Natural ecosystems have the capability of using their biodiversities to restore soil, but it takes time. We need policies so that use of land is rotated to give it enough time to renew itself.

Some people see the number of people increasing faster than available natural resources. Think how your town has changed as more people have moved in. It takes more schools, roads, libraries, jails, and taxes to provide services to the new people. Cost of living is higher in densely populated areas (for example, Miami) than in rural areas.

Those providing housing developments encourage people to immigrate into Florida. Some are attracted because Florida has a mild climate, others because there is less tax on income and inheritance. Many are retired. Some are affluent and bring their savings and buy property. Others come without much savings and may require many public services, but, with lower taxes, there is a problem providing this help. As more people come to Florida, some are finding it hard to find jobs. Some turn to crime.

2.10 Decaying image

Tourism brings in much money to Florida. It is dependent on people in the rest of the country having money for a Florida vacation and on Florida's image as a beautiful state with lovely beaches, clean campsites, wildlife, tourist facilities, and open spaces. With more urban development, costs rise, and the rest of the country may not be able to afford such trips. As attractive environmental areas are displaced by developments, we may not be able to maintain our vacation-land image.

2.11 Expensive electricity

Florida has five large nuclear power plants that provide much of the basis for high technology, cities, and comfortable life styles. Electric power is used most in summer due to air-conditioning. Nuclear power plants do not last more than about 40 years, and public opinion does not favor replacing them with new power plants. When they are decommissioned, the highly toxic radioactive wastes may have to be sealed up inside the concrete shells.

Florida imports almost all of its fossil fuels (coal, oil, natural gas, gasoline) and uranium for the production of electricity, heating and air-conditioning, and running cars. Over the world, fossil fuels are being used up faster than natural systems are producing them. When costs of fuels rise we will have to decrease our standard of living, decrease our population or both. Understanding and planning our environment means understanding the way the whole economy of county, state, nation, and the Earth is based on diminishing resources. As for the Earth as a whole, many of the social problems in our society are rooted in global limitations of environment resources, overpopulation, and economic policies.

2.12 *Education for hope*

Up to now, most environmental education has concentrated on pieces of systems, such as air pollution or solid waste. Studying whole systems is a new emphasis. Environmental systems education is essential for understanding Florida's ecosystems and dealing with Florida's environmental problems.

Questions and activities for chapter two

1. Define: (a) fossil fuels, (b) tourism, (c) wetland, (d) saltwater intrusion, (e) CFCs (chlorofluorocarbons), (f) keys (geographical), (g) pollution, (h) acid rain.
2. In the first class meeting, as the students are introducing themselves, ask each one to pose an environmental question. Write the questions on the board and ask one student to copy them to be printed and distributed for later discussions.
3. Discuss Florida's environmental problems with fossil fuels, tourism, population growth, water, mining, and waste. Explain how Florida's problems fit with national and global problems.
4. Choose a local problem to investigate. Organize task groups to find out as much as you can about it, including what has been proposed to fix it. Come up with your own suggestions. After you write it up, put it aside to bring out at the end of the course — to see if you would change your suggestions.
5. Find an environmental problem for which some action by your class could make a difference. It might be the one from Question 4. Perhaps you could go as a group to learn and interact with those most concerned.
6. List aspects of Florida that give it a negative image for tourists. Discuss how they could be improved.
7. Research and discuss your area's water supply problems.
8. Discuss the waste problems closest to you. Your local newspaper is a good resource.
9. Visit a local park. Discuss its problems. Perhaps the class can put in a workday to help.
10. Take a walk around your school grounds, identify ecosystems, biodiversity, natural areas, water runoff channels, ways of management.
11. Start an environmental improvement club, which is an effective way to combine action and sociability.

chapter three

Ecosystems and symbols

A *system* is a group of parts which are connected and work together. The Earth is covered with living and nonliving things that interact to form systems. Systems with living and nonliving parts are called *ecosystems* (which is short for ecological systems). A typical ecosystem contains living things such as trees and nonliving things such as nutrient substances and the air.

The surface part of the Earth where living things exist is called the *biosphere*. It contains many smaller ecosystems. Examples are forests, fields, ponds, and estuaries.

All the individuals of one species of organism in an ecosystem are a *population*. Every ecosystem contains many different populations. For example, an ecosystem may contain a population of oak trees, a population of squirrels, and a population of grasshoppers.

3.1 Ecosystem processes

Some organisms (algae and green plants) make their own food from chemicals using the energy of sunlight in the process of *photosynthesis*. The organisms that make the food products are called *producers*. The food produced is used by living cells to make more cells and to form *organic matter* such as wood and fats. Organic products of living organisms are sometimes called *biomass*.

The organisms that consume products made by producers are called *consumers*. Consumers can be plant eaters (*herbivores*), meat eaters (*carnivores*), or digesters of dead organic matter (*decomposers*). Cows and grasshoppers are herbivores, coyotes and spiders are carnivores, and mushrooms, bacteria, and vultures are decomposers.

After a consumer has digested and used its food, a few chemical waste products remain. Those waste products that are needed as fertilizer for plant growth are called *nutrients*. When nutrients are released by consumers and go back to the soil for reuse by the plants, we say that the nutrients have been *recycled*.

The forest is an example of a typical ecosystem. Its trees and other plant producers use the sun's energy and nutrient chemicals to make organic

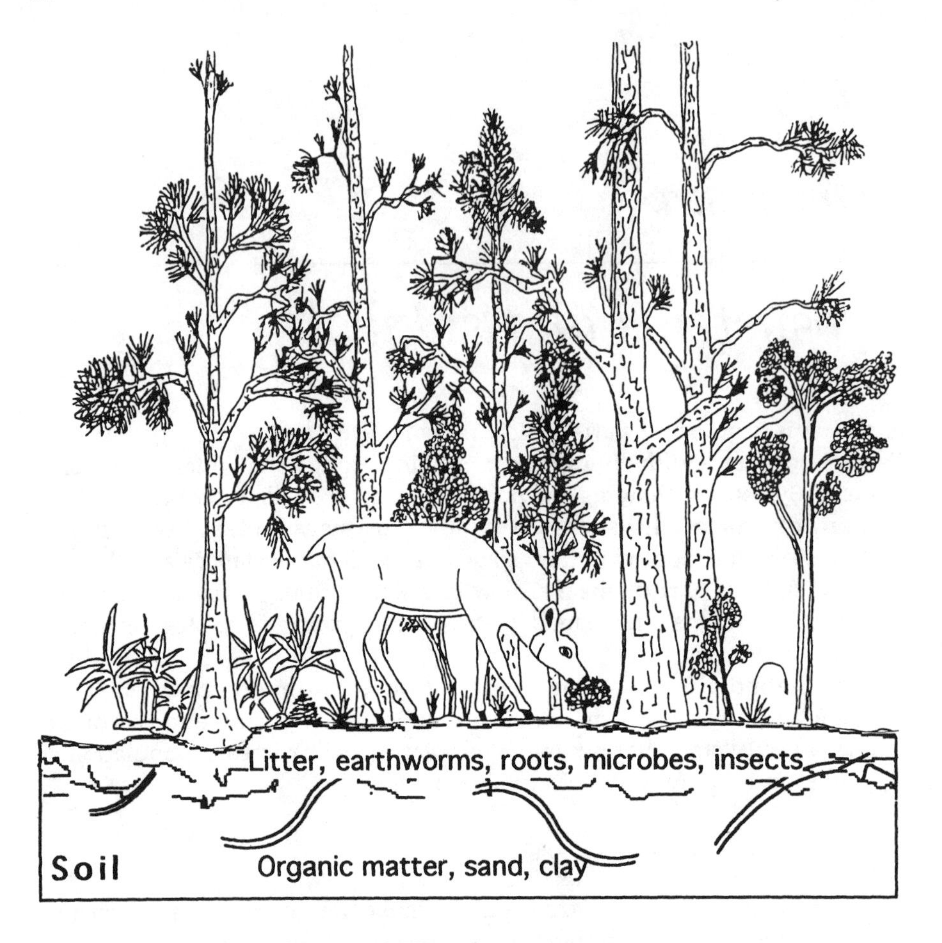

Figure 3.1 Pine forest ecosystem.

products. These are eaten by the consumers, which release nutrients back to the plant roots in the soil.

Soil is an essential part of the forest ecosystem. It is a combination of rock particles, chemical nutrients, dead organic matter from plants and animals, and living components such as roots and worms.

Figure 3.1 is a picture of the forest. Figure 3.2 depicts parts of a forest system with the arrows showing the way energy, food, and nutrients are flowing.

3.2 *Symbols*

Symbols show general system relationships. The first groups of symbols which need to be learned are given in Figure 3.3. Figure 3.4 shows the forest system drawn with these symbols. The units and pathways are the same as

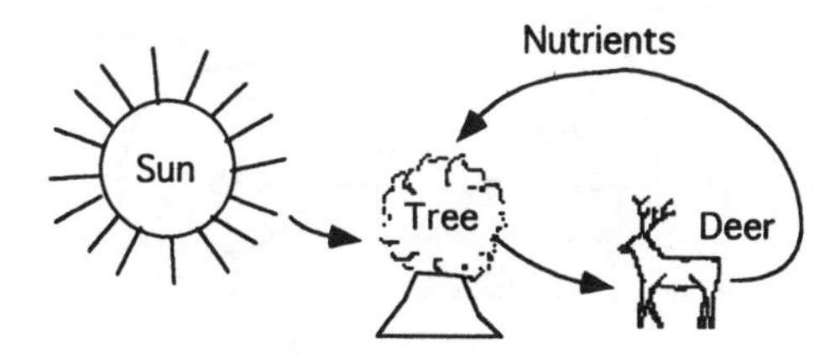

Figure 3.2 Some parts of a forest depicted by icons.

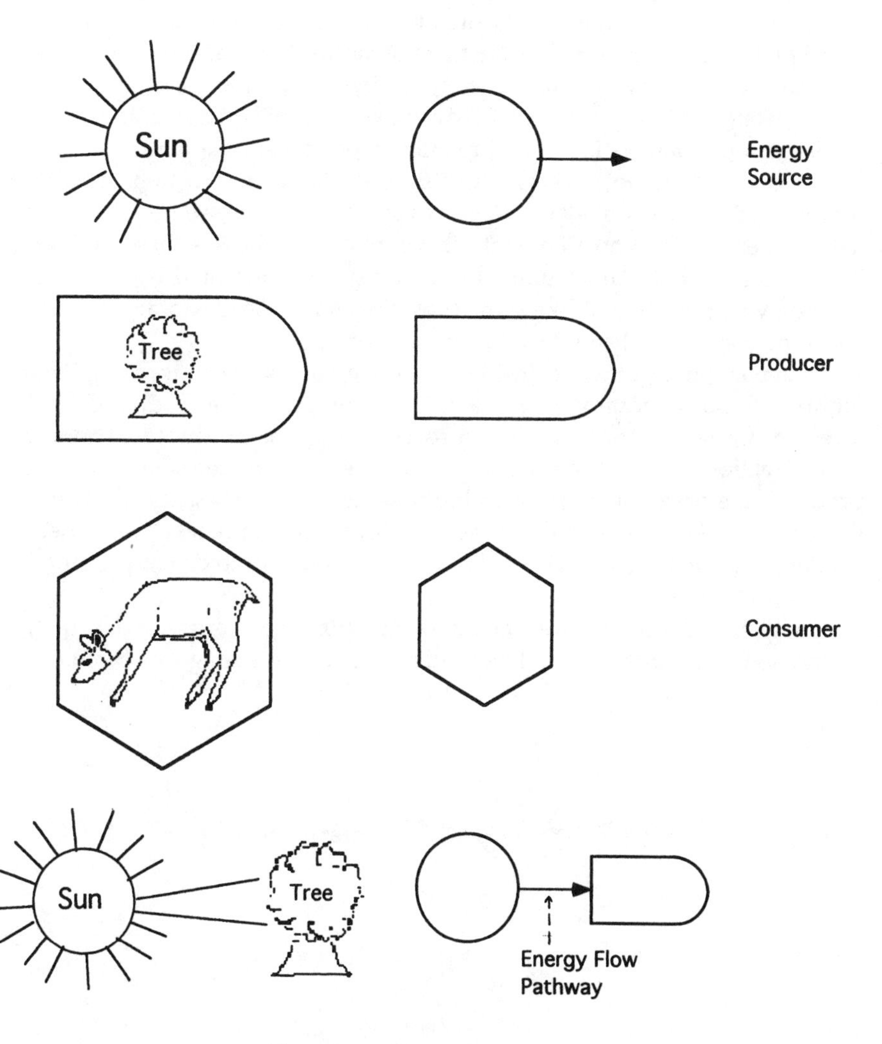

Figure 3.3 Icons and symbols.

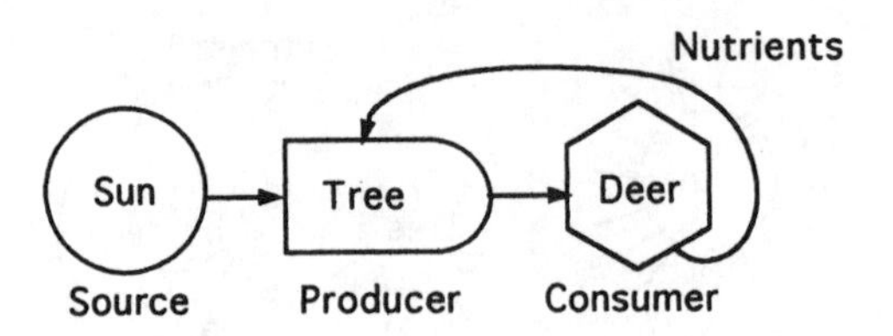

Figure 3.4 Use of symbols to represent parts of the forest.

in Figure 3.2, except that symbols have been substituted for the pictures of trees, deer, etc. The sun is represented by the symbol for energy source; the green plants are represented by the symbol for producer; and the animals are represented by the symbol for consumers. The arrows represent the flows of energy from one unit to the next. Many pathways carry materials and energy. A diagram that shows important relationships in a simple way is a *model*.

Two more symbols are given in Figure 3.5. An interaction process is represented in energy systems diagrams by an *interaction symbol*. The combination of energy and nutrient materials in photosynthesis is shown with the interaction symbol. An accumulated quantity is represented by the *storage* symbol, which is shaped like some towns' water tanks. A storage of acorns and pine cones is indicated by the storage symbol.

The parts and pathways inside a producer and a consumer are given in Figure 3.6. Both contain an interaction, a storage, and a feedback. In the forest example, the photosynthetic process within the producer is shown as an interaction that combines energy and the plant biomass. This process produces a storage of more plant biomass, which is necessary to continue the process. For the deer consumer, the interaction process is the deer's consumption of the green plants (interaction of plants and deer). The storage is the biomass of deer tissue.

In Figure 3.6, there are lines flowing from the storages back to the interaction processes. These lines indicate that the storage of biomass is

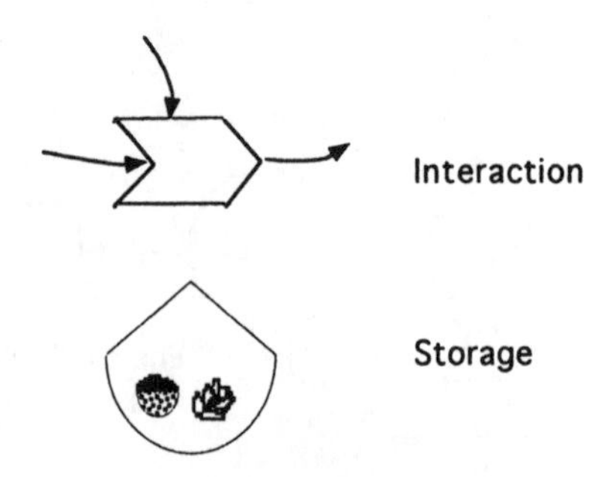

Figure 3.5 Interaction and storage symbols.

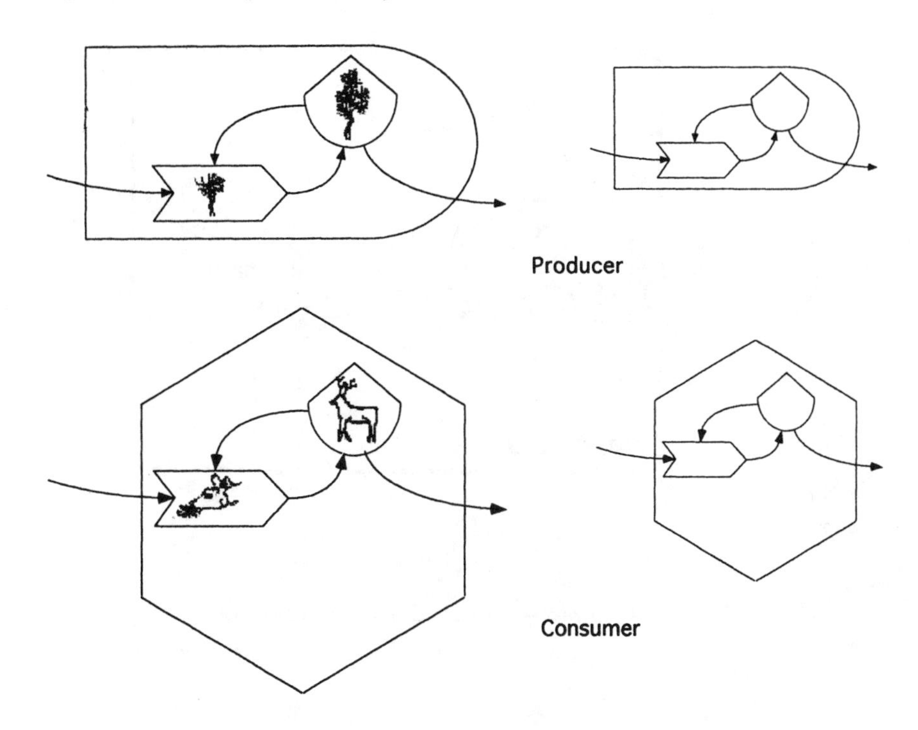

Figure 3.6 Parts inside producer and consumer symbols.

involved in producing more biomass. A line going back to the left is called a *feedback.*

The producer and consumer symbols can be used with or without the inside details, but the interaction, storage, and feedback are always implied. You can draw the symbols either way, depending on whether or not you want to emphasize the internal processes.

Energy is available to do work only when it is relatively concentrated and is said to be dispersed when it spreads out, losing its concentration and ability to do useful work. Some energy is always being dispersed from a storage of concentrated energy and from symbols representing processes, such as the interaction. The dispersal of energy from all storages and processes is shown with the *heat sink* symbol in Figure 3.7. Dispersed energy cannot do work again.

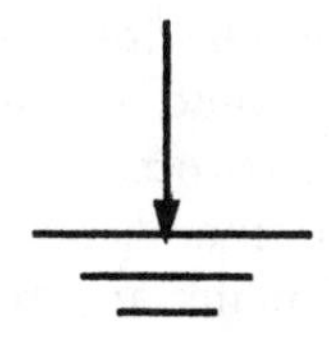

Figure 3.7 Heat sink symbol.

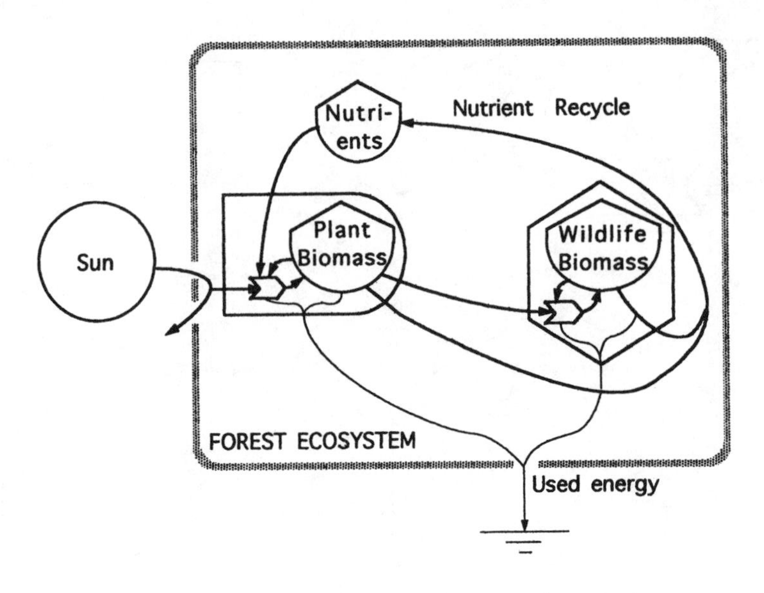

Figure 3.8 Forest ecosystem drawn with symbols.

For example, most of the solar energy used in the production process is dispersed by its use. In the case of a producer, it is necessary to disperse most of the input solar energy in order to make a small storage of biomass energy. When a consumer animal eats a plant, most of the food energy is dispersed as part of the biological processes keeping the animal alive.

3.3 *Model of a forest ecosystem*

The parts of the forest given in previous figures are combined to show a whole forest system in a diagram (Figure 3.8). In Figure 3.8, the box drawn around the symbols marks the boundary selected for this model. Only the source and the heat sink are shown outside the boundary lines. The energy source symbol is outside because the energy is provided from an external source. The heat sink is outside the box because this energy is dispersed from the system to the surroundings.

Because some of the sunlight energy flows through the forest without being used, the sunlight line is shown with one branch coming out again. The nutrients released by the producers and consumers are shown recycling back to the left into the plant production process. In summary, the energy symbols show the way in which the producer and consumer parts of an ecosystem are connected, use energy, recycle materials, and use their storages to help interaction processes.

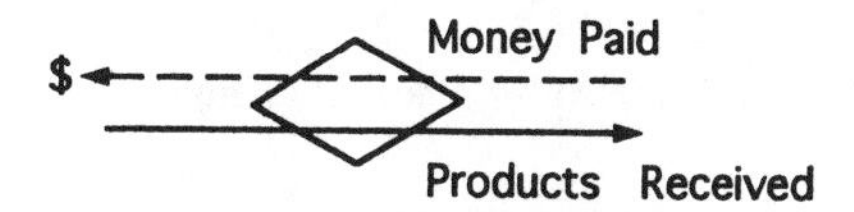

Figure 3.9 The money transaction symbol.

3.4 The money transaction symbol

The *money transaction symbol* (Figure 3.9) shows energy and materials flowing one way and money flowing in the opposite direction to pay for them. The diamond ties the money to the energy and the materials for which it pays.

In an economic system, the money is used to pay for goods and services. On the right in the energy diagram in Figure 3.10, energy and materials flow in one direction (the solid lines) and money flows in the opposite direction (the dashed lines). Wood goes from the production of the forest to the city market, and dollars go back to pay for the goods and services required to cut the trees. The interaction is the cutting and processing of the wood.

Ten symbols with their usual pathway connections are given in Figure 3.11. Nine are used in this chapter; the switch is used in Chapter 4.

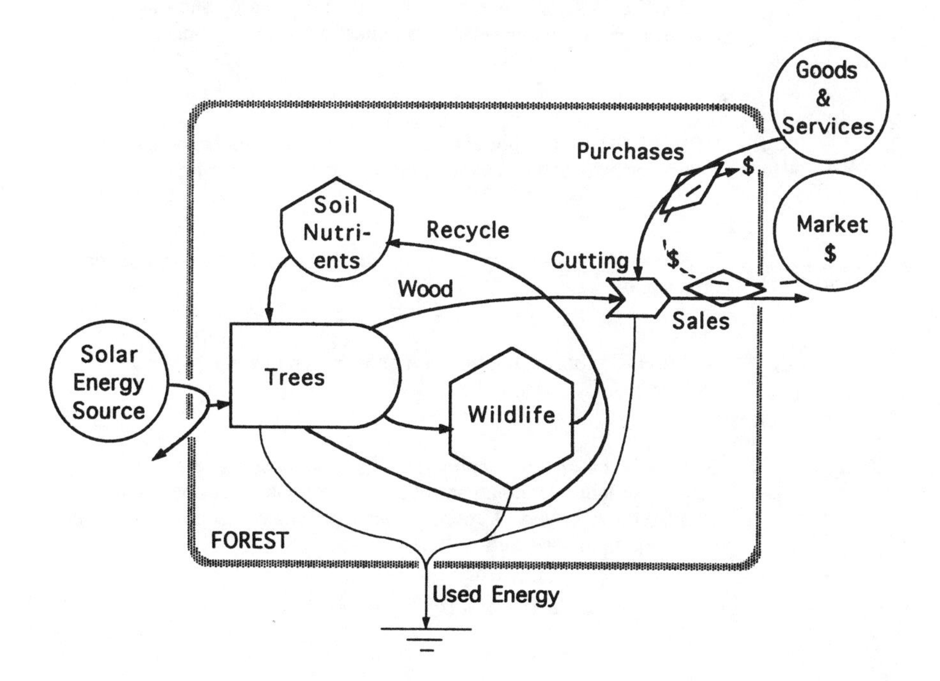

Figure 3.10 The forest economic system showing that energy and money flow in opposite directions.

ENERGY PATHWAY — a flow of energy, often with a flow of materials.

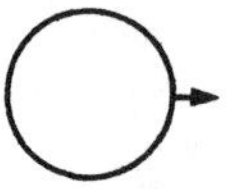

ENERGY SOURCE — energy that accompanies each of the resources used by ecosystems, such as sun, winds, tidal exchanges, waves on the beaches, rains, seeds brought in by wind and birds.

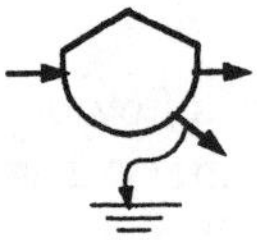

STORAGE — a place where energy is stored. Examples are resources such as forest biomass, soil, organic matter, groundwater, and sands in beach dunes.

HEAT SINK — energy that is dispersed and no longer usable, such as the energy in sunlight after it is used in photosynthesis, or the metabolic heat passing out of animals. Heat sinks are attached to storage tanks, interactions, producers, consumers, and switching symbols.

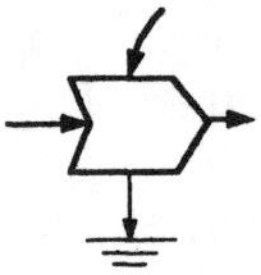

INTERACTION — process that combines different types of energy flows or flows of material. In photosynthesis, sunlight, water, and nutrients interact to produce organic matter.

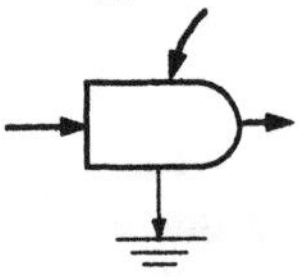

PRODUCER — unit that makes products from energy and raw materials. Examples include trees, grass, crops, and factories.

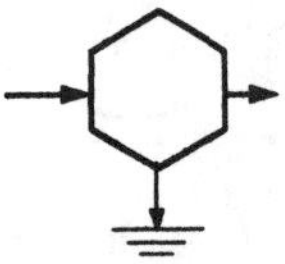

CONSUMER — unit that uses the products from producers. Examples include insects, cattle, microorganisms, humans, and cities.

$ TRANSACTION — business exchange of money for energy, materials, or services.

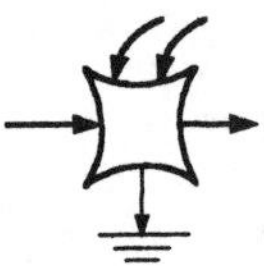

SWITCH — process that turns on and off, such as starting and stopping fire and the pollination of flowers.

BOX — miscellaneous symbol for subsystems such as the soil subsystem found in a diagram of a forest, or fishing business found in a diagram of an estuary. A rounded rectangle is also used for the system boundary that is selected.

Figure 3.11 Energy symbols.

Questions and activities for chapter three

1. Define: (a) system, (b) ecosystem, (c) consumer, (d) producer, (e) decomposer, (f) photosynthesis, (g) population, (h) nutrients, (i) energy dispersal, (j) interaction process, (k) storage, (l) production, (m) source, (n) biosphere.
2. Draw the nine symbols used in this chapter and explain what each stands for.
3. Trace energy through the forest diagram in Figure 3.8. Trace nutrients through the same diagram.
4. Use Figure 3.8 to explain the forest ecosystem and Figure 3.10 to explain the logging industry.
5. Draw a diagram of the process of you selling four pencils to a friend for $1.
6. Make a systems diagram of your school grounds. First list the outside energy sources coming in, the storages inside, and the flows going out. Group these into about five symbols and arrange them into a diagram. If you include money, draw it last, as a dashed line.
7. List four systems of which you are a part. Make an energy diagram of one (use no more than six symbols).
8. Draw picture icons in each symbol of your diagram in Question 7.
9. Debate the question of whether deer hunting should be allowed in Florida. Look at the question from all levels — from the larger system view down to the hunter and the deer.

chapter four

Flows of energy

Energy goes with everything. It goes with all processes and is stored in all things. Even the transmission of information requires some energy. There are many kinds of energy: solar, chemical, mechanical, electrical, etc., but they can all be recognized and measured by converting them into heat. In the energy systems diagrams, such as the figures in Chapter 3, there is energy flowing in all the pathways and energy stored in all the storage tanks. Figure 4.1 is a picture of many forms of energy.

Heat is the molecular motion of molecules and is one form of energy. When an object is heated, the molecules vibrate more. The increased motion of molecules of the fluid in a thermometer causes it to expand, and we use that expansion to measure the temperature. A high temperature means that the concentration of heat energy is greater. If you touch a hot pan, your nervous system senses this concentration.

A practical way to define energy is in terms of heat. Since all kinds of energy can be turned into heat, energy may be defined as a property of something that can be converted into heat. The energy of motion in a fast car appears as heat when the car crashes into a wall. The energy from the power plant comes through wires to run a motor and ends up in heat. Small amounts of energy come with sound waves to the ear and end up heating the head slightly.

Energy is "conserved". This means that the energy that goes into something does not disappear but either is stored there or flows out. This is called the *first energy law*. In the energy systems diagrams, the energy comes into the system boundaries from outside sources (circular symbols) and goes out through flows and the heat sink.

When energy in one place is in a concentrated form, it can cause *transformation* processes which we call *work*. Concentrated energy is said to be "available" to do work. Energy available to do work is sometimes called potential energy. As the work is done, the energy loses the concentration that made work possible. The energy is dispersed, degraded, often converted into heat. Although the energy has not been lost, it no longer can support work. The used energy cannot be reused, as it has lost its concentration. For

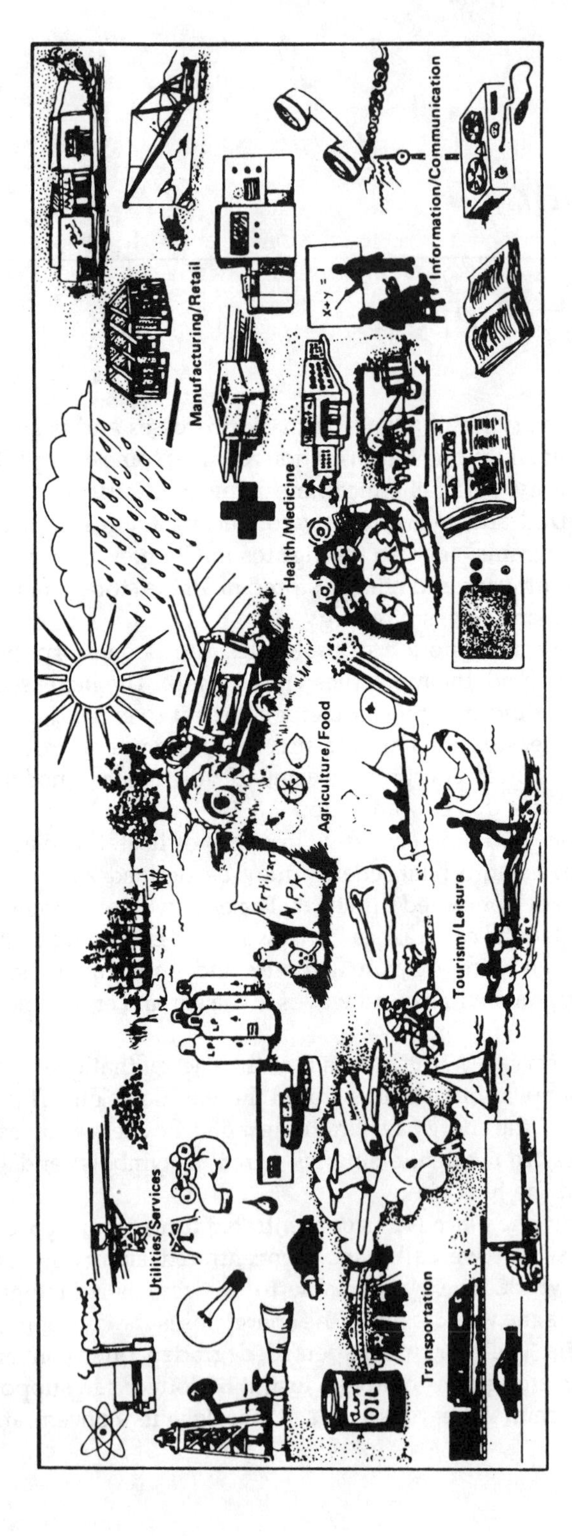

Figure 4.1 Energy — it comes in many forms. (From Browder, J. et al., *South Florida: Seeking a Balance of Man and Nature*, Bureau of Comprehensive Planning, Division of State Planning, Tallahassee, 1977.)

processes to take place and work to be accomplished, potential energy must be degraded. The fact that energy tends to be degraded is called the *second energy law*. When using the energy systems diagrams, we recognize the second law by drawing the heat sinks (Figure 3.7) down from all processes.

Any storage of matter, fuel, or information is also a concentration of energy which tends to disperse, to diffuse, to depreciate. In our systems diagrams we also show the heat sink to represent the spontaneous dispersal of stored concentrations. When your car rusts, when the sugar in your coffee diffuses, when batteries lose the ability to operate your flashlight, when buildings fall apart, depreciation has occurred. These are all examples of the second energy law.

Systems of nature and humanity have no exceptions to the energy laws. They have flows of energy on all their pathways and stored in all the components. The work of nature and of humans is accompanied by the conversion of energy concentrations to a degraded form, but in the process something special is done that is useful to the system. Trace the energy flows in the forest diagram in Figure 3.8. The processes of energy flow generate plant biomass and wildlife biomass. The main system of the biosphere also consists of plants undergoing photosynthesis and consumers, including humans, animals, microbes, and power plants, utilizing the products of photosynthesis.

4.1 Metabolism in ecosystems

In ecosystems, the plant producers may support the consumers with their products, and the consumers return materials and services to help plant production. The main processes of plants making food and consumers using it are called *metabolism*. These processes can be expressed as summary chemical reactions.

The process of photosynthetic production by green plants with the help of sunlight energy (for example, forest leaves) can be written as follows:

$$\text{carbon dioxide} + \text{water} + \text{nutrients} \rightarrow \text{organic matter} + \text{oxygen}$$

An equation for the process of organic consumption by consumers (including fire and industrial use of fuels) is the reverse:

$$\text{organic matter} + \text{oxygen} \rightarrow \text{carbon dioxide} + \text{water} + \text{nutrients}$$

These main processes of photosynthetic production and consumer consumption in an aquarium are shown in Figures 4.2 and 4.3. The sun's energy comes to the plants and is carried with the organic matter and oxygen to the consumers; then, when they have done their work, the by-products (carbon dioxide, water, and nutrients) circulate back for use by the plants again. The available energy in the sunlight is used up in the processes of these transformations.

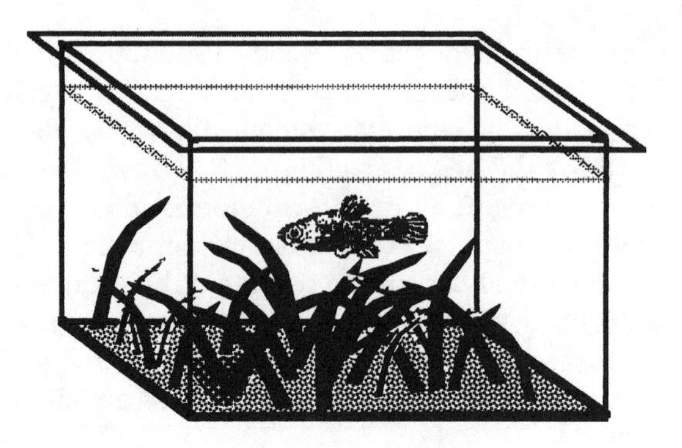

Figure 4.2 Picture of an aquarium with producers and consumers.

4.2 *Showing the quantities of energy flow*

Amounts of energy can be measured by the heat released in a process. There are two units commonly used to measure energy. The *calorie* is the heat required to raise the temperature of a gram of water one degree on the Celsius (centigrade) scale of temperature. A *kilocalorie* is a thousand calories. A human body releases about 2500 kilocalories per day which come from the energy in the food consumed.

By international agreement, a different unit of energy is increasingly being used: the *joule* (abbreviated J). A kilocalorie (often written Calorie, with a capital C) is equivalent to 4186 joules.

Energy is required for all of the processes in an ecosystem, which uses energy from the sun (solar energy) and small amounts from other sources. A good way to see how materials, energy, or money are flowing in a system is to write numbers on the pathways of the diagram. For example, numbers on

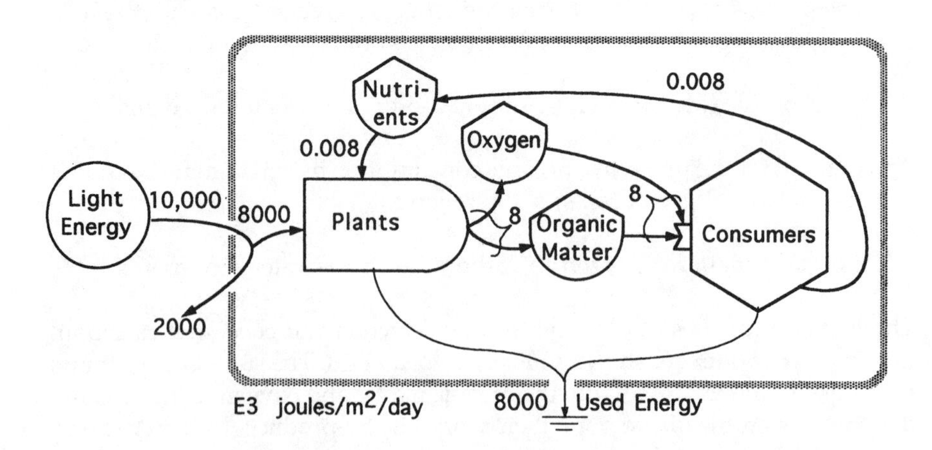

Figure 4.3 Energy flows in the metabolic processes in an aquarium.

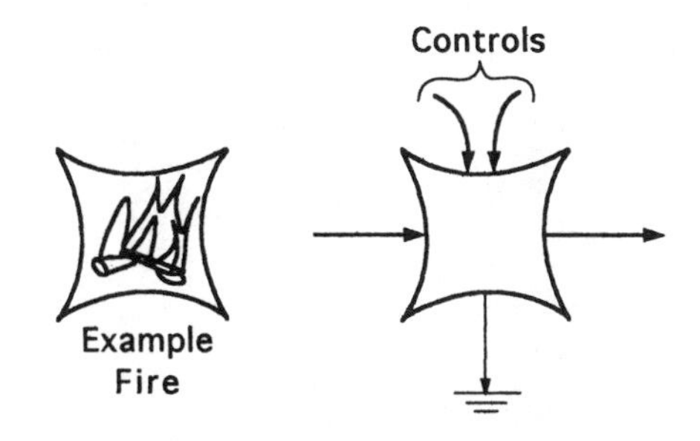

Figure 4.4 Switch symbol for a switching pathway that turns on and off.

the flow lines in Figures 4.3 and 5.6 are the rates of flows of energy in joules per year. In other diagrams, we keep track of weights of materials, such as the grams of water in Figure 5.2 and of phosphorus in Figure 5.4.

The diagrams include some large numbers. Large numbers with many zeros can be represented as the product of the front part of the number multiplied by 10 for each zero. For example, 627,000 can be represented as:

$$6.27 \times 10^5$$

The numeral 10 combined with the exponent 5 means "10 multiplied by 10 five times". Or, you can use the way of writing large numbers used by microcomputers:

6.27 E5

where E5 (meaning exponent 5) means to multiply by 10 five times. For a whole number, this is the same as adding five zeros (6 E5 = 600,000).

4.3 Fire and a symbol for switching

In Figure 4.4 is a new symbol which represents a switching action. This *switch* symbol is used to indicate that a pathway turns on and off according to some controlling conditions. For example, fire is a switching action. It turns on when the biomass is high enough and something starts a flame. The fire consumes biomass, and nutrients are released. It turns off when the biomass fuel is too small. The switching symbol is used to represent forest fire in Figure 4.5.

4.4 Energy flow in a forest

The systems diagram of forest processes in Figure 3.8 has been expanded to show more detail in Figure 4.5. In addition to the energy of the sunlight, the energies of wind, rain, and rocks are used up in operating the plants and

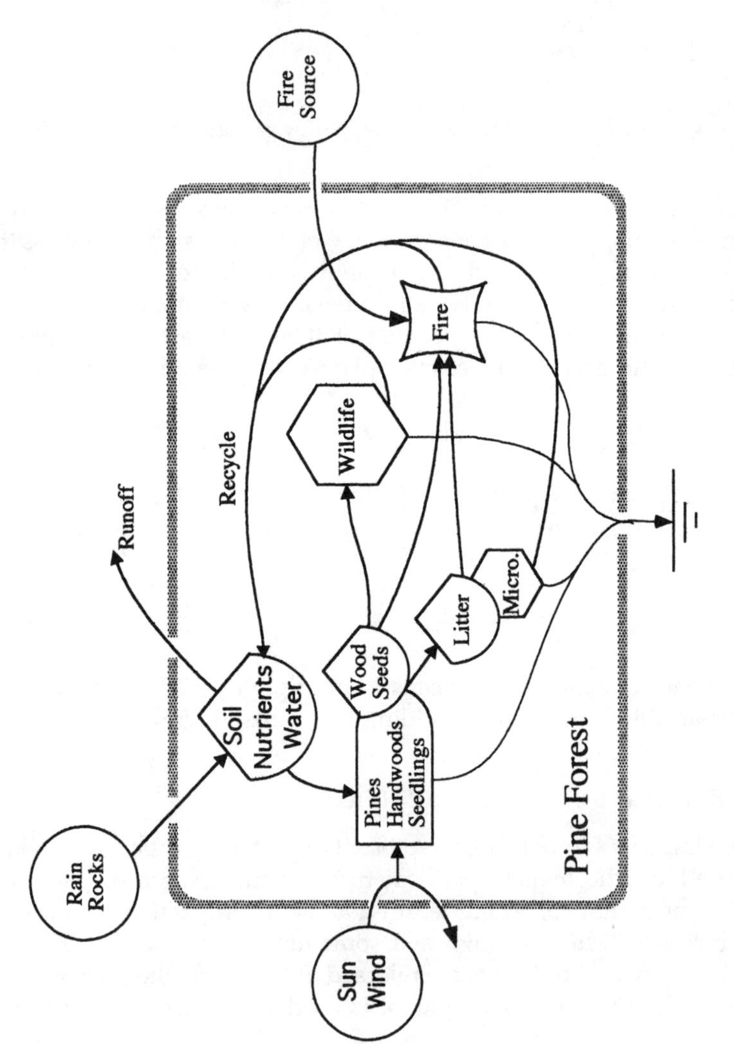

Figure 4.5 Systems model of a pine forest with fire; Micro. = microbes.

animals. Wind moves carbon dioxide and oxygen around and breaks limbs in storms. Rocks are slowly weathered into soil.

Trees and seedlings use energy from the sun and water and nutrients from the soil to grow and reproduce. Parts of them are eaten by the wildlife. As plants and animals give off wastes or die, their remains fall to the forest floor as dead organic matter. This material is called *litter*. Included in the litter are dead leaves, sticks, trunks, animal droppings, and feathers. Seeds fall to the ground to become part of the litter, ready to germinate if conditions are right. Many kinds of soil animals, including a large biomass of earthworms, feed on litter, shredding it into small particles. Fungi, bacteria, and other microorganisms (*microbes*) use the remaining organic matter as food. These decomposers break down complex organic molecules into nutrients (such as phosphates, nitrates, potassium, and many other chemical substances) available for uptake by roots again.

When the forest has developed a thick layer of litter, it may burn in dry periods if fire is set by people or by lightning. Fire is a good example of the conversion of energy concentrations into work. The fire generates turbulent eddies in the atmosphere and makes many changes in structure of soils.

After some years, a forest ecosystem may come into a steady-state balance. Water flows in and out. Nutrients move from the soil into living organisms and back to the soil again. Organisms grow, die, and decompose, and their nutrients are returned to the system.

Questions and activities for chapter four

1. Define: (a) energy, (b) transformation, (c) work, (d) biomass, (e) switch, (f) respiration, (g) kilocalorie, (h) joule, (i) steady state, (j) microbes.
2. Describe three ways wind is important in the forest ecosystem.
3. Describe two energy sources (other than the sun) in the forest ecosystem.
4. Describe two consumers in the forest ecosystem.
5. Use Figure 4.3 to explain the first and second energy laws.
6. Explain how Figure 4.6 illustrates the second energy law. List the materials produced by the decomposition of the leaf.
7. Write the equations for photosynthetic production and organic consumption.
8. Connect the symbols of the diagram of a pond ecosystem in Figure 4.7.
9. Draw the symbols for each of the terms below: soil, leaves, hawks, rain, fire, production.

10.1 Figure 4.8 is an energy systems model of a Florida pine forest ecosystem. Use the following to label the parts of the model: (a) sun, (b) rain, (c) long-leaf pines, (d) squirrels, (e) decomposers, (f) litter, (g) soil water and nutrients.

10.2 Assume a steady state (storages constant); the diagram is labeled to show the amounts of energy flowing in the system.

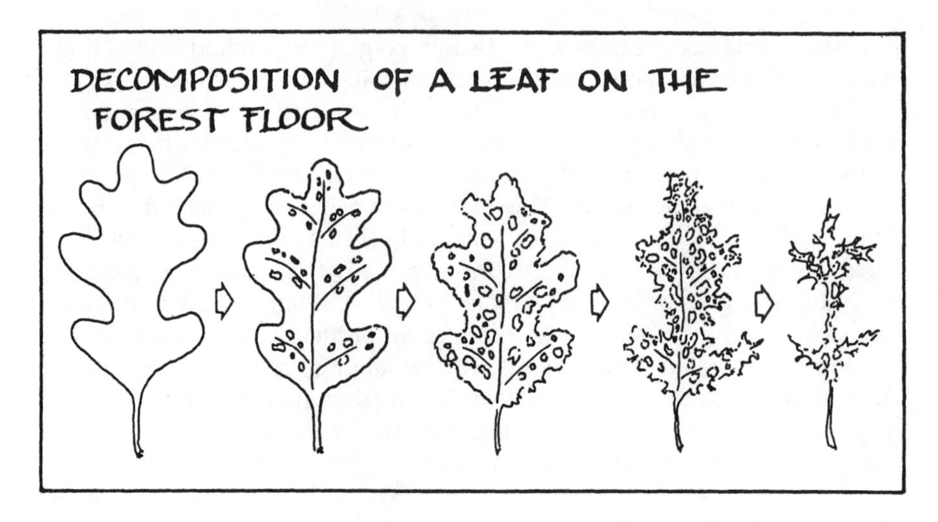

Figure 4.6 Diagram for Question 6. (From Wharton, C.H., *Forested Wetlands of Florida — Their Management and Use*, Center for Wetlands, University of Florida, Gainesville, 1976.)

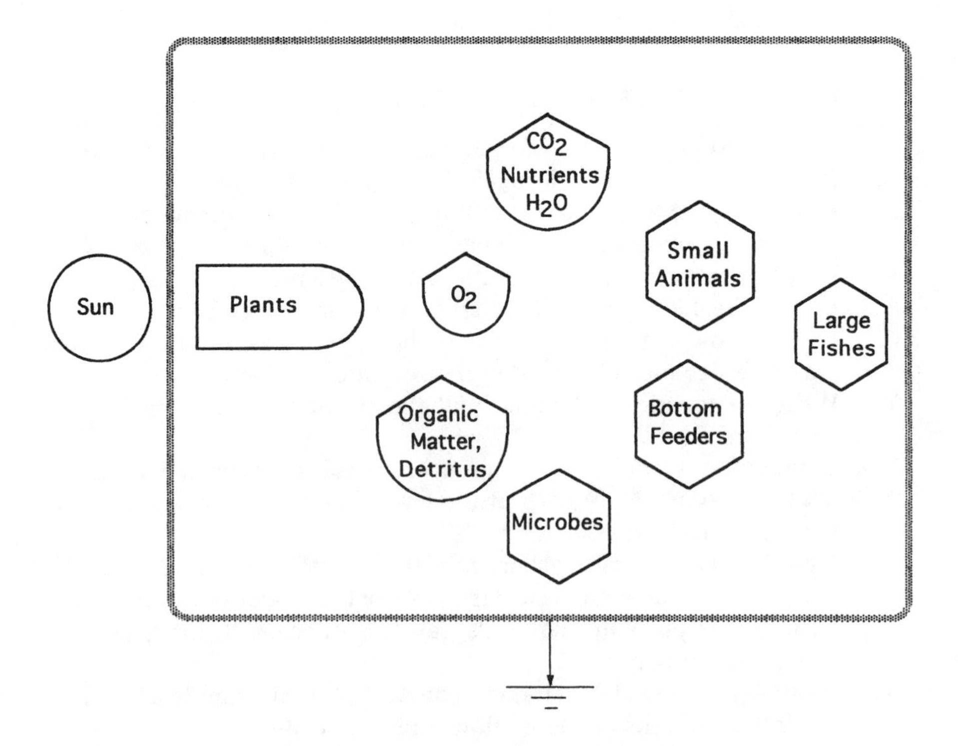

Figure 4.7 Diagram of a pond ecosystem.

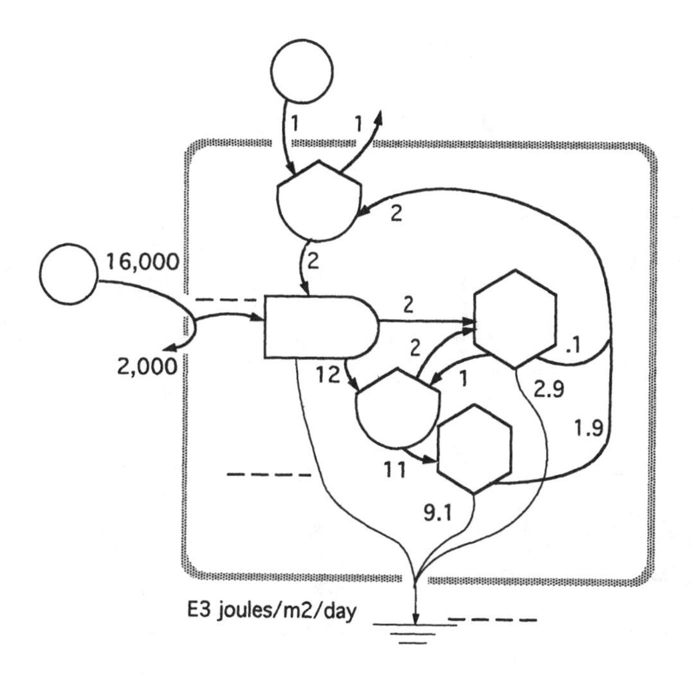

Figure 4.8 Diagram for Question 10.

(a) Put the correct numbers on the lines by the sun's inflow (light used), the heat sink from the producer, and the heat sink from the whole system.

(b) Calculate the percent of the light used by the system that goes out as waste heat from the pine production.

(c) Calculate how many joules of the light used by the system goes into tree biomass becoming litter (directly and indirectly), food for squirrels, and food for microbes.

10.3 Suppose only half as much energy was being dispersed as heat from the system. How much energy would be stored per square meter per day?

11. Prepare for a forest walk by bringing identification books of plants, animals, and animal tracks. If you can, bring a bag of plaster of Paris; water; a container to mix them; a container to carry finished, hardened footprints; and a pen to record information on the back of your casts. A camera and a notebook are also needed to help you keep a permanent record of your discoveries. Take an early morning walk in a forest or, even better, get up early on a camping trip. Find a path no one has trampled yet and watch carefully for footprints and piles of droppings. From your identification books, identify those you see. Mix the plaster of Paris and water to a thick soup consistency and pour it into the track. In a few minutes you will have a hard plaster cast of the track. Several examples of tracks are shown in Figure 4.9.

Figure 4.9 Tracks: (a) armadillo, (b) white-tailed deer, (c) dog, (d) opossum, (e) house cat, (f) raccoon, (g) great blue heron, (h) wild turkey, (i) mourning dove, and (j) Florida gallinule. (From Headstrom, R., *Identifying Animal Tracks*, Dover, New York, 1983. With permission.)

chapter five

Cycles of materials

Systems utilize materials for their structures and processes. Many different kinds of materials are required. To continue operating, a system must have an inflow of each kind of material to balance those necessary materials that are lost. Well-organized systems usually develop material cycles, so that much of what is needed in production is reused. In this way, the system is not so dependent on the rates of supply of materials. In this chapter, we will examine the material cycles in a forest ecosystem. Although we usually study the whole system first, in this case we will first look at each material separately and then see how they interact together. Since some energy accompanies everything, we can use the energy systems diagrams to represent the materials as we did in Chapters 3 and 4.

5.1 The water cycle in the forest

Ecosystems require water. Water comes into the forest as rain. Figure 5.1 is a picture of rainfall in a forest. Some of the rainwater flows downhill, leaving the area as runoff. Some percolates downward through the soil, becoming groundwater. When rain falls, some is caught on leaf and limb surfaces, and some of this water evaporates before it reaches the ground. Because the sun heats the leaves and the wind blows dry air through them, water is also evaporated from within the leaves which diffuses out through microscopic holes in the leaves. This process of water from the soil passing up the stems and out through the leaves as water vapor is called *transpiration.*

The amount of water that flows through trees as transpiration is much larger than the small amount of water used in photosynthesis. The sum of transpiration and evaporation is called *evapotranspiration.* The water stored among the soil particles as soil water is small compared to the amounts flowing through runoff and evapotranspiration. Figure 5.2 shows the flows and storages of water in a square meter of forest ecosystem.

Notice that the numbers in Figure 5.2 are in grams, a measure of weight, whereas the numbers in the summary Figure 5.6 are in joules, a measure of

Figure 5.1 Rainfall in a forest.

energy. Weight is easier to measure, but not useful when you are comparing different substances such as sun and rocks; however, the energy these components bring to the system can be compared.

5.2 *Principle of conservation of matter*

A principle of science is that matter is conserved. The matter going into a system is either to be found stored there or flowing out. If you compare the grams flowing in per day and the grams flowing out per day, the difference is what is being stored. In Figure 5.2, water inflow equals outflow. Showing materials only, Figures 5.2 through 5.5 have no heat sinks.

5.3 *The carbon cycle*

Drawn in Figure 5.3 is the carbon cycle. Carbon dioxide from the air is drawn into plant photosynthesis through the microscopic leaf holes and converted into organic matter that is used throughout the rest of the ecosystem by the consumer parts of the plants and the animals and microorganisms. These release carbon dioxide as part of their consumer metabolism. Much carbon dioxide diffuses up from the soil where there are many roots, microbes, and small animals. As the wind blows through the forest, some carbon dioxide is carried in and out.

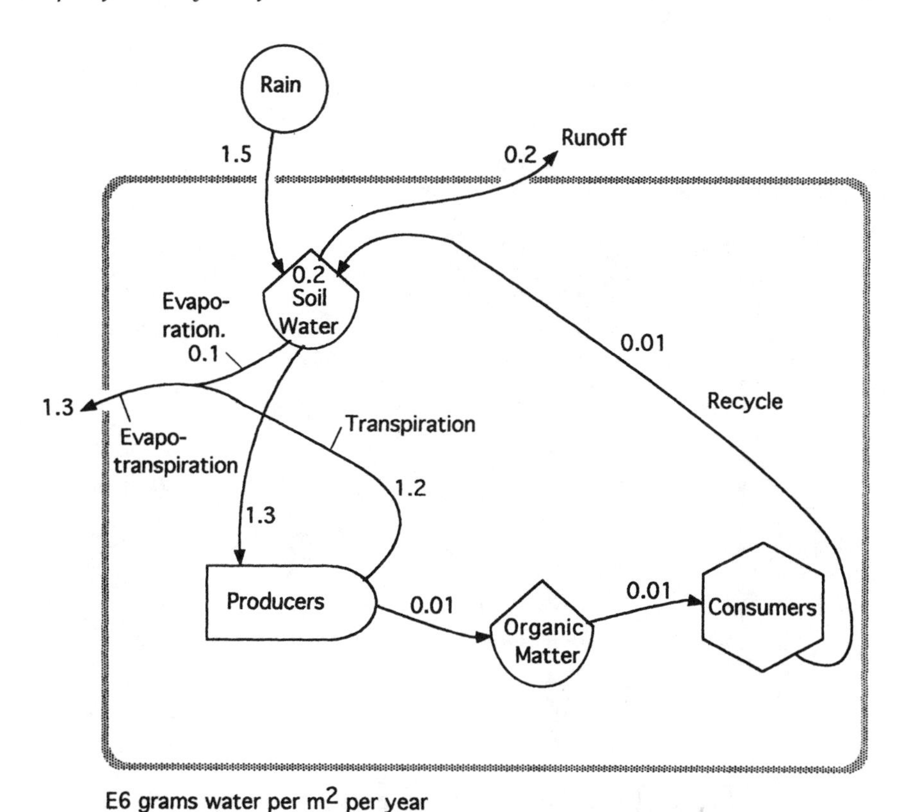

Figure 5.2 Water storages and flows in the forest ecosystem.

5.4 *The phosphorus cycle*

Chemicals (inorganic nutrients) are also required for the storages and processes of an ecosystem. One of the most important nutrients necessary to build healthy organisms is phosphorus, which is required for ATP, the energy processor in cells. Phosphorus is generally more scarce than other nutrients, such as nitrogen or potassium (potash). If the forest system did not recycle phosphorus, it would be scarce enough to limit the growth of forest plants.

Flows and storages that contain phosphorus nutrients are included in Figure 5.4 as the phosphorus cycle. The diagram shows outside sources of phosphorus to be rain and rocks. The phosphorus is present as inorganic phosphates, which the plants use to produce organic compounds necessary for life. The phosphorus in these compounds is part of the biomass which is returned to inorganic form by the consumers when they use the biomass as food. The inorganic phosphorus released becomes part of the storage of

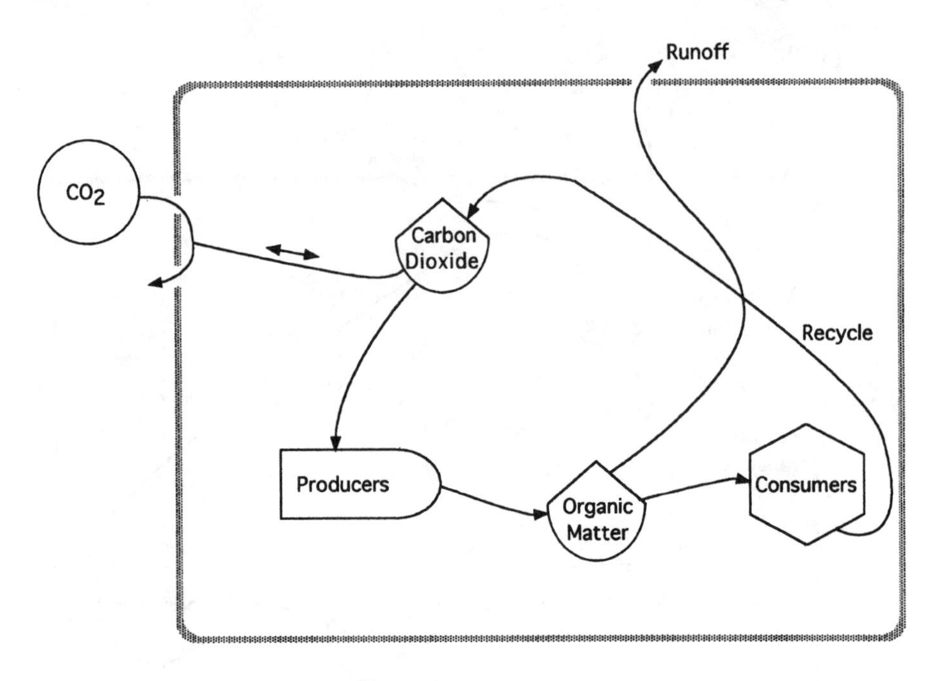

Figure 5.3 The carbon cycle.

nutrients in the soil. Thus, the phosphorus goes around in a circle, as Figure 5.4 shows. Some flows out of the system with the waters that run off the land surface or seep down into the groundwaters. Phosphorus has no gaseous phase in its cycle.

5.5 *Phosphorus in Florida*

The phosphorus cycle is particularly interesting in Florida. Topsoil is often low in phosphate, whereas the underlying rock is rich in phosphate. The limestone rock under much of Florida's topsoil is high in phosphate and is mined and used as fertilizer here and for export (phosphate mining is discussed in Chapter 25).

Because the sandy soils of peninsular Florida do not have much clay, they do not bind and store phosphorus well. The acidic water leaches out nutrients as the rainwaters percolate through. The phosphorus is washed downward, leaving surface soil deficient in this important nutrient. Florida plants have adapted to the phosphorus shortage problem by accumulating phosphorus from the rain over long periods of time. The accumulation of phosphorus is accomplished by the development of tiny rootlets. This mat of rootlets is such a good filter that it captures phosphorus from many sources including rainwater dripping from trees, animal wastes, and decomposing

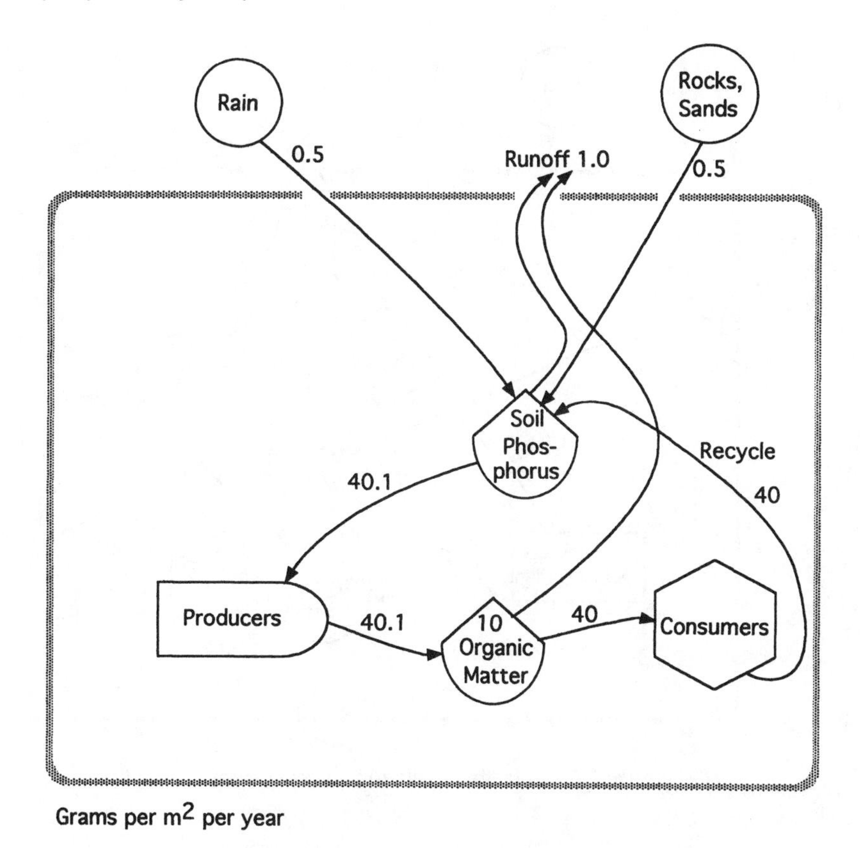

Figure 5.4 Phosphorus cycle in the forest ecosystem.

detritus. The phosphorus collected by these rootlets is used by the forest to create more biomass. Dead leaves and branches fall to the ground, returning the phosphorus to the detritus, where it is recycled to the rootlets. However, when the forests are cut and the wood is carried off, a considerable fraction of the phosphorus is removed from the cycle.

5.6 *The nitrogen cycle*

The chemical element nitrogen is essential for all life and is in all living organisms. It is one of the elements in proteins and genetic structures. Seventy-eight percent of the air is nitrogen gas, but most organisms cannot use it in this form. Gaseous nitrogen can be converted into usable forms (nitrates, nitrites, and ammonia) by special processes which require energy. For example, industrial processes use fuels to convert nitrogen gas into nitrogen fertilizer for farms. The energy of lightning converts nitrogen to nitrates in

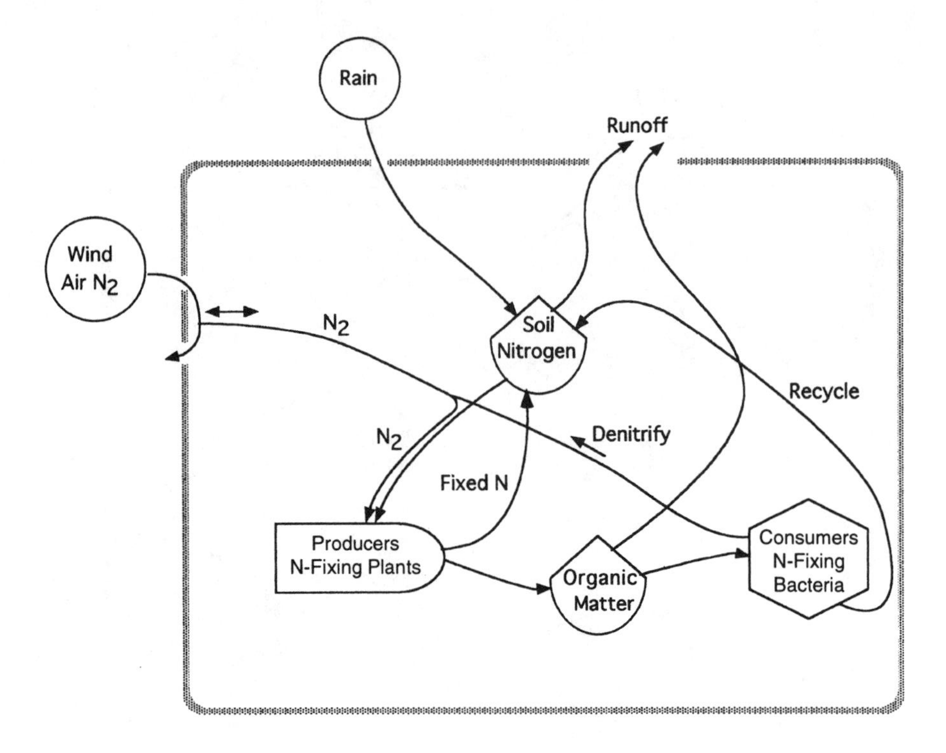

Figure 5.5 Nitrogen cycle for an ecosystem.

rain. The plants, algae, and bacteria that can do this are called *nitrogen-fixers*. Some plants and trees have nodules that contain bacteria that fix nitrogen into nitrates using sugars transported from the leaves as the energy source. Blue-green algae (cyanobacteria) can fix nitrogen using sunlight.

Figure 5.5 shows the cycle of nitrogen in ecosystems. Starting with the nitrogen-fixing organisms, nitrogen passes into plants and then up the food chain to the animals. In plants and animals, nitrogen is in the form of organic compounds such as proteins. Nitrogen returns to the soils and waters as waste products of animals and from the decomposition of plants and animals. The various nitrogen waste substances, such as urea in urine, are converted by bacteria into ammonia, nitrites, and nitrates. These are used by plants again to close the cycle. Some microbes return the nitrogen back to the atmosphere as nitrogen gas. This is called *denitrification*.

Figure 5.6 is an ecosystem diagram that has each of the separate mineral cycles included. By putting the material cycles together you can see how each is being driven by other processes. Photosynthetic production, for example, is drawing carbon, water, nitrogen, and phosphorus into organic matter, later to be released together during the consumption. The materials are coupled to the process being driven by the energy of sunlight, wind, and rain.

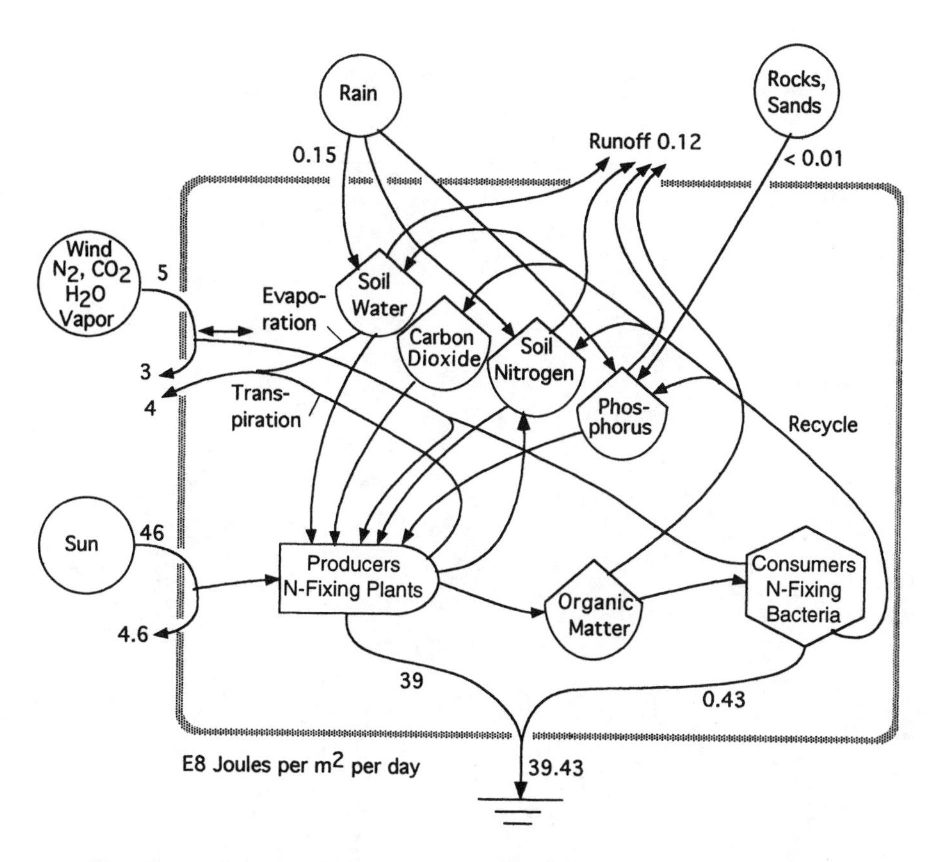

Figure 5.6 Summary diagram of nutrient cycles in a forest ecosystem.

Questions and activities for chapter five

1. Define: (a) nitrogen fixing, (b) nutrients, (c) microbes, (d) cycle.
2. Trace each cycle in Figure 5.6.
3. Differentiate among evaporation, transpiration, and evapotranspiration.
4. Explain why phosphorus is necessary in the forest ecosystem.
5. Explain conservation of matter using a forest as an example.
6. Describe the global water cycle. How does it differ from the Florida water cycle?
7. Explain nitrogen fixing and denitrification.
8. In Figure 5.4, balance all the inflows and outflows by putting numbers on the two parts of Runoff.
9. In Figure 5.6, what percent of energy inflows is dispersed out the heat sink?

chapter six

Production and limiting factors

6.1 Production

Production is the process by which two or more ingredients are combined to form a new product. For example, soil nutrients, water, carbon dioxide, and sunlight are combined to form organic matter during photosynthesis. Typically, industrial production involves the use of energy, labor, capital, and raw materials to form industrial products. The production process is illustrated in Figure 6.1. Notice the pointed interaction symbol with its inflows of ingredients and outflows of products. Whenever this symbol is used, it indicates that a production process is occurring.

During a production process, each inflowing ingredient carries energy of a different type and quality. During production, these energies are transformed to a new form. Some energy is degraded and flows out through the heat sink. These energy transformations during production processes are an example of work.

6.2 Gross and net production

Where there is a production process followed by a consumption process, as in photosynthesis and respiration in plants, we must distinguish between production and production minus the accompanying respiration. In Figure 6.2, *gross production* is the actual rate at which organic matter is made. Gross production is the flow at the point of the interaction symbol (5 grams per day, in this case). *Net production* is the production that is actually observed when production and some respiration are going on at the same time. In Figure 6.2, the gross production rate of biomass is 5 grams per day and the respiration rate is 3 grams per day. The net production is equal to gross production minus the accompanying respiration; therefore, net production is 2 grams per day.

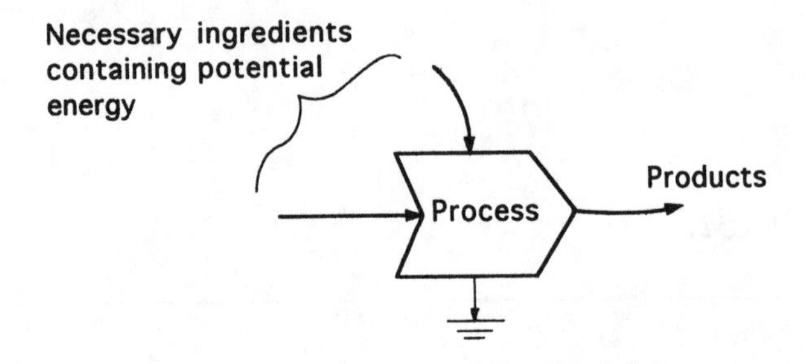

Figure 6.1 Production process with two interacting ingredients.

In more complicated systems, such as the forest, where there are several stages of production and consumption, there is more than one kind of net production — for example, net wood production, net litter production, etc. Net production also depends on the length of time it is measured. For example, plant cover in Florida produces much organic matter during daylight. This matter is stored in various plant tissues, such as in fruits, roots, and new leaves. During the night, however, most of it is consumed just to keep the plant tissues, the animals, and the microbes alive. During a year, there is more net production in the summer than the winter. By the time a whole year is passed, the net products of photosynthesis are likely to be the growth rings of trees, added organic matter in the soil, or peat of the swamps.

6.3 Limiting factors

Most production processes go faster when the ingredients are available in larger quantities; however, a reaction can only proceed at a rate that is

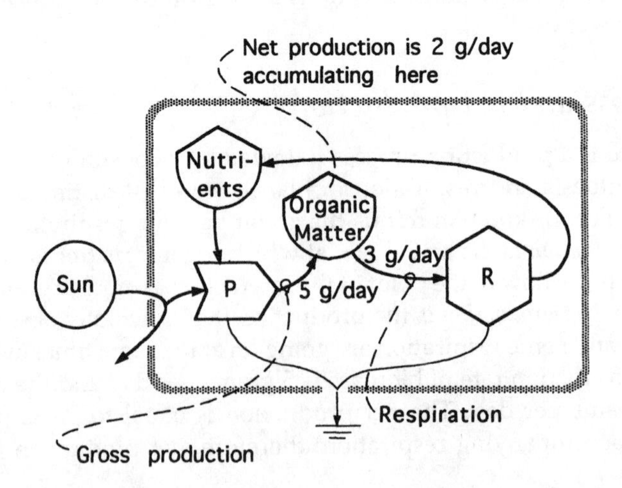

Figure 6.2 Gross and net production. P = production; R = respiration; g = grams.

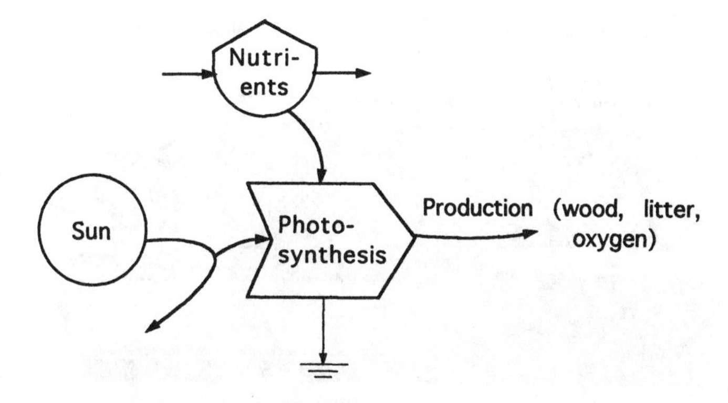

Figure 6.3 The sun is often the limiting factor in the process of photosynthesis.

determined by the ingredient that is least available. This ingredient is then called a *limiting factor*. For example, since light is necessary for photosynthesis, the process slows down and stops at night; sunlight is the limiting factor which controls this process.

In Figure 6.3, production is dependent on sunlight and nutrients. If nutrients are increased, photosynthesis will increase until there is not enough sunlight to stimulate it. Sunlight, then, is the limiting factor.

In Figure 6.4, production is graphed for various values of nutrients. As nutrients are increased, the rate of production increases; however, as light becomes limiting, the production rate slows its increase. This graph, typical of limiting factors, shows a relationship that is called the *law of diminishing returns* in economics. For example, production of furniture by a factory is able to be increased when there is an increase in the available labor, but the production ultimately is limited by the wood supply.

An example of limiting factors in Florida is the shortage of plant nutrients in the sandy soils — especially in those soils where the cover materials

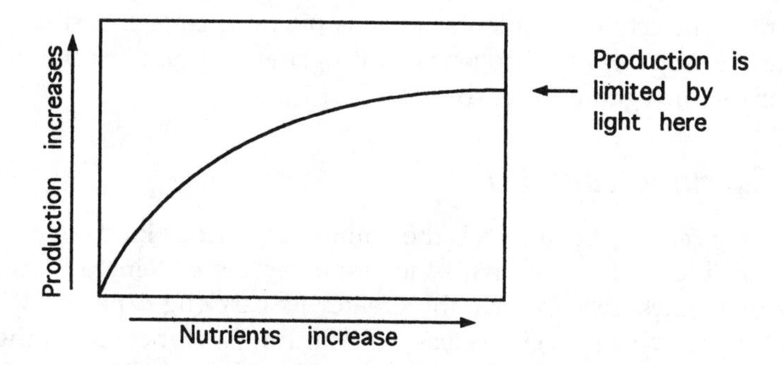

Figure 6.4 Graph of production rate of the process in Figure 6.3 as nutrients are increased and light becomes limiting.

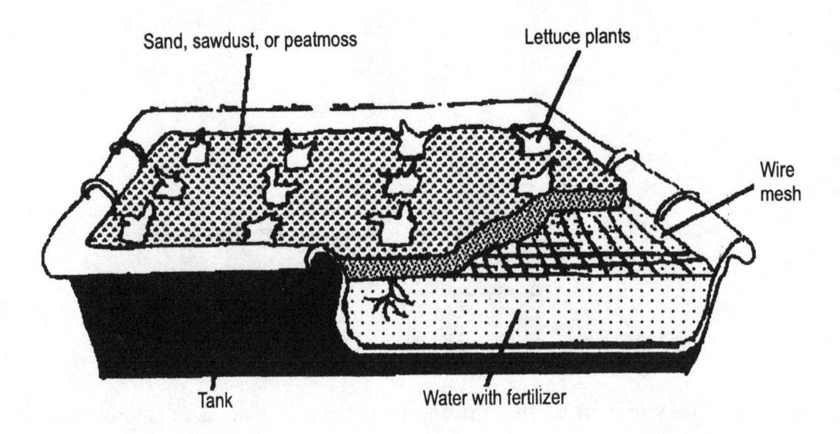

Figure 6.5 The hydroponic garden at Epcot Center in Orlando.

have been removed. This shortage limits the growth of new plants. Another example is Epcot Center at Disney World, where lettuce and other vegetables are grown hydroponically in water without soil (Figure 6.5). There, fertilizer is added until light becomes the limiting factor (Figure 6.4).

6.4 Soil

Soil is an important part of an ecosystem and is also a storehouse of the materials used by the ecosystem for living and nonliving processes. In a forest, various substances in the soil may be limiting to production. In many Florida soils, nitrogen is the limiting factor to plant productivity. Nitrogen may be particularly limiting when phosphorus is in excess, as in many Florida waters and in soils around phosphate mining.

When nitrogen is limiting, then nitrogen-fixing plants become common. Examples of nitrogen-fixing plants in Florida are the floating fern, *Azolla*, wax myrtle bushes in the wetlands, and many leguminous plants in long-leaf pine forests. Insectivorous plants, such as pitcher plants and sundews, are also found in bogs where nitrogen is limiting; they capture insects and digest them for the nitrogen in their tissues.

6.5 Carrying capacity

The *carrying capacity* of an area is the amount of various kinds of organisms which can live in that area with the resources there. Generally, the more energy that flows into an area, the greater its carrying capacity. With less energy, the carrying capacity is less. Any limiting factor can limit the carrying capacity. For example, if the amount of sunlight falling on a forest is decreased because of dust in the air, the carrying capacity of the forest will

Figure 6.6 Sparse Everglades vegetation in a drained area compared to increased production and carrying capacity with steady water flow.

be decreased. Resources such as nutrients also contribute to and can limit carrying capacity. Figure 6.6 shows the difference in the carrying capacity of the Everglades with and without steady flows of water.

In state and city planning, the carrying capacity for humans is often discussed. It is easier to determine the carrying capacity of a forest for squirrels because we know their basic requirements for food, water, and space. But, when the carrying capacity of an area for humans is considered, we have to ask the question, carrying capacity at what standard of living?

Questions and activities for chapter six

1. Define: (a) gross production, (b) net production, (c) limiting factors, (d) law of diminishing returns, (e) litter, (f) carrying capacity.
2. Give an example of a limiting factor in your life. What can you do to make it less limiting?
3. Which producer would have the most net production, a corn plant or a mature live oak tree? Explain.
4. Draw a graph showing production (photosynthesis) and consumption (respiration) in a pond as a function of time over the period of 1 day.
5. Choose a land area near you. Discuss what you think is limiting the production and carrying capacity there.

6. Ask your city planner about the "right" number of people per square mile for your city. What criteria were used to calculate this carrying capacity?
7. Give an example of the law of diminishing returns in economics.
8. Describe at least two factors that nitrogen-fixing plants and insectivorous plants have in common.
9. Run the program on limiting factors (Figure A6 in Appendix A). Load BASIC, then type in the program and "RUN". Explain what you observe.

chapter seven

Energy webs and transformations

The systems of Florida's environment and economy are a network of energy transformations. For example, energy is transformed when cars use gasoline to run, when people use food to power their activities, or when clouds use heat energy to make rain. In these examples, energy is transformed from one kind of energy to another and can be represented by Figure 7.1. The box represents the transformation process. Energy from the left is transformed into another kind that is shown leaving the box to the right. Some of the available energy in the inflow is degraded and goes out through the heat sink at the bottom of the box. Usually there is some control input which is shown on top of the box.

7.1 Structure of a food web

The pine forest food web is shown in Figure 7.2. There are many transformations and branched connections, as each group of consumers eats several other groups. Figure 7.3 is a picture of the organisms in their ecosystem.

7.2 Energy chain

To investigate energy flows within the food web, it is often convenient to combine units of the web into a simpler chain. Figure 7.4 simplifies the web in Figure 7.2.

In a simple food chain, a plant producer is eaten by a plant consumer (herbivore), which in turn may be eaten by a carnivore, which is eaten by another carnivore. In this example (Figure 7.4), insects eat parts of the tree, an armadillo eats the insects, and a bobcat eats the armadillo. In each step of the food chain, some of the food becomes part of the tissue of the next consumer.

Because it takes many joules of energy at the left end of the chain to make a few joules at the right, we regard the energy at the right as being of higher

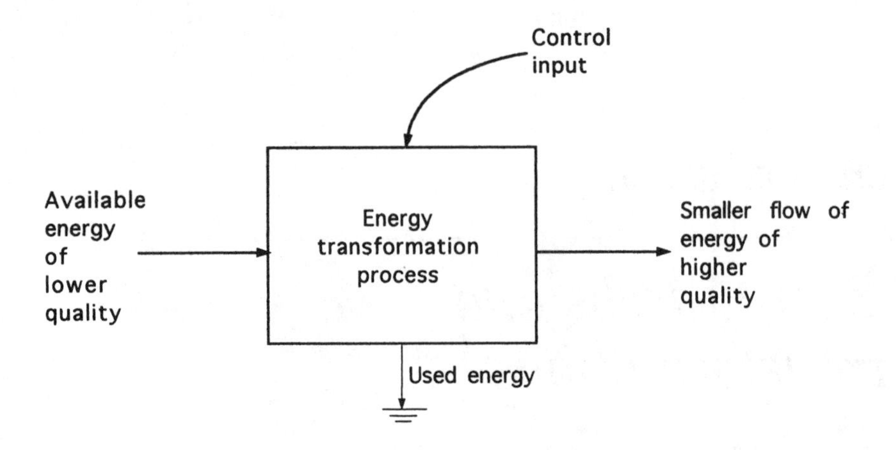

Figure 7.1 Energy transformation.

quality. A gram of bobcat took more energy to produce than a gram of tree; therefore, the bobcat is higher quality energy. The energy quality is lowest at the left and rises with each step along the chain.

The forest model in Figure 7.4 is further simplified in Figure 7.5 to a straight chain of energy transformations without branches. It has numerical values for energy flow on the pathways between the transformation steps. The energy relationships between the parts of the chain can easily be seen. About 1,000,000 (1 million) joules of sunlight are shown contributing to photosynthesis. Part of this is direct sunlight, and part is the sun's energy that has fallen on the ocean to send rain to the forest. About 1% of this energy is transformed by the forest producers into tree biomass. In other words, about 10,000 joules of new trees and other plants are produced per year. 999,000 joules go down the heat sink as necessary energy used during the production process.

Efficiency is defined as an output divided by an input (output per unit input). In Figure 7.5, the efficiency of the use of sunlight to produce wood is, therefore:

$$\text{Efficiency} = \frac{10{,}000}{1{,}000{,}000} = 0.01 = 1 \text{ percent } (1\%)$$

Fractions are converted into percent by multiplying by 100.

The range of efficiencies for photosynthesis in different plant species is between 0.01 and 2%. These efficiencies are low because sunlight is very dilute, and many successive steps and extensive chlorophyll-containing cellular machinery are necessary to concentrate sunlight into higher quality energy. Plants have evolved the photosynthetic process over several billion years, so it may be the most efficient way of using the sun's energy. This idea

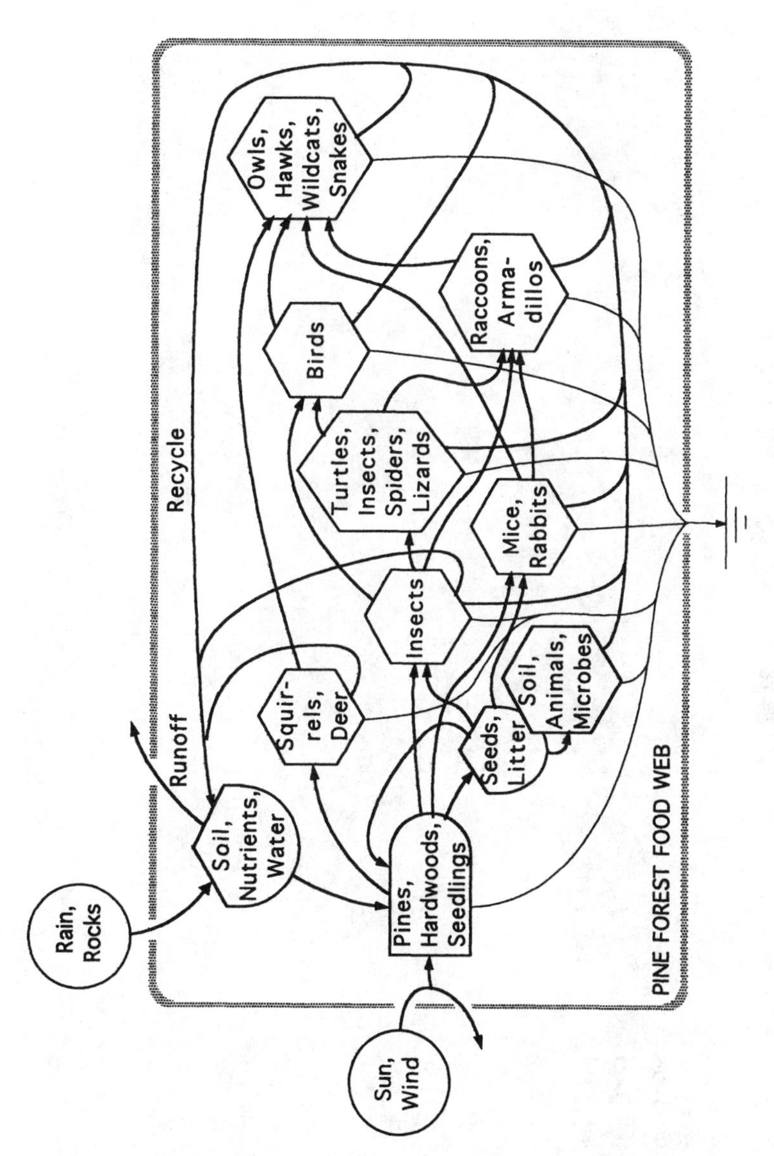

Figure 7.2 Food web of a forest.

Figure 7.3. Food web of a deciduous forest (illustration by E.A. McMahan

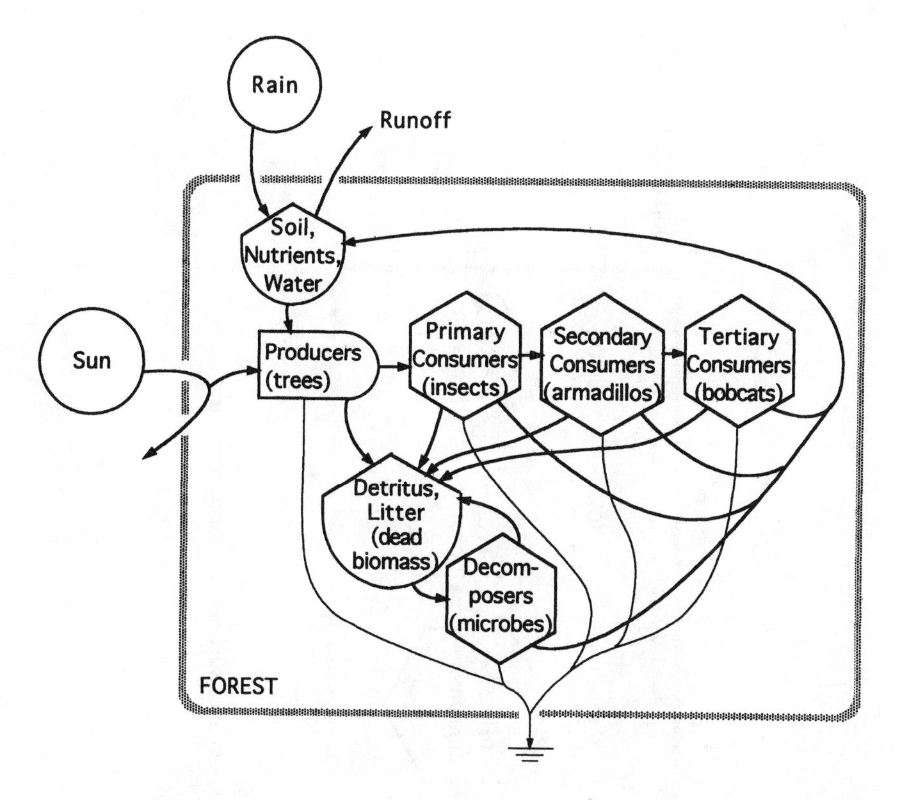

Figure 7.4 Energy systems diagram of a forest food chain.

of the efficiency of sunlight use is important when sunlight is considered as an energy source to replace fossil fuels for human systems.

At each succeeding level of the forest food chain, about 10% of the energy available to that level is converted to new biomass. This ratio also applies to many producers, which consume 90% of their own production during respiration. As summarized in Figure 7.5, the 1,000,000 joules of sunlight that support the forest directly and indirectly in one year become:

10,000 joules of energy in gross production

of which:

1000 joules become new net producer biomass

which is consumed to become:

100 joules of new primary consumer biomass

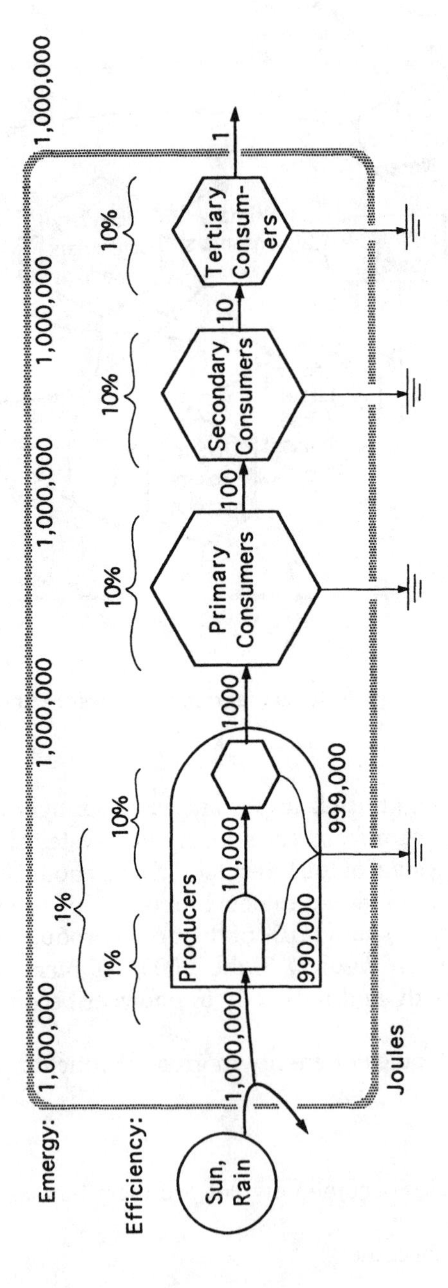

Figure 7.5 A forest food chain with levels of successive energy transformation.

which is consumed to become:

10 joules of new secondary consumer biomass

which is consumed to become:

1 joule of new tertiary consumer biomass.

This may be summarized by saying that to produce 1 joule of tertiary consumer, such as a bobcat, it takes 1,000,000 joules of sunlight.

In the forest example (Figure 7.5), it took 1 E6 joules of sunlight to produce 1 joule of bobcat activity. In a subsistence farming system, it takes the same quantity of sunlight to produce 1 joule of farmer's labor. In other words, the bobcat and the farmer work at similar levels of energy quality.

The carrying capacity of an area for certain organisms depends on where they are on the food chain. Generally, an area can support more producers (at the left end of the food chain) and fewer high-quality consumers (at the right). For example, a ranch will grow more grass than cattle.

7.3 Energy relationships in modern society

Energy transformations in the environment generate products used in the human economy, where they are transformed further in support of human consumers. Figure 7.6 shows the longer, industrial energy chain which began with green plants making organic matter. This was transformed into coal and oil in geologic processes, then to electricity and fuels such as gasoline, and finally to support people.

Figure 7.6 shows that about 20 million solar joules are required for one joule of human work. A joule is very small; a human puts out about 30,000 joules per hour of physical work.

7.4 Emergy

The word *emergy* (spelled with an "m") is defined as being one kind of available energy used up directly or indirectly in transformations and work to make a product. It is a measure of value based on what went into a product. The unit for emergy is the emjoule. For example, in Figure 7.5, 1 million joules of solar energy available to do work entered the chain of processes on the left, mostly being used up to make consumers. The emergy of the consumers that resulted to the right is 1 million solar emjoules. A joule refers to energy that is still present; an emjoule of emergy refers to available energy used up and no longer there. In Figure 7.6, the 20 joules of food, clothing, and housing that went to humans have an emergy value of 20 million solar emjoules.

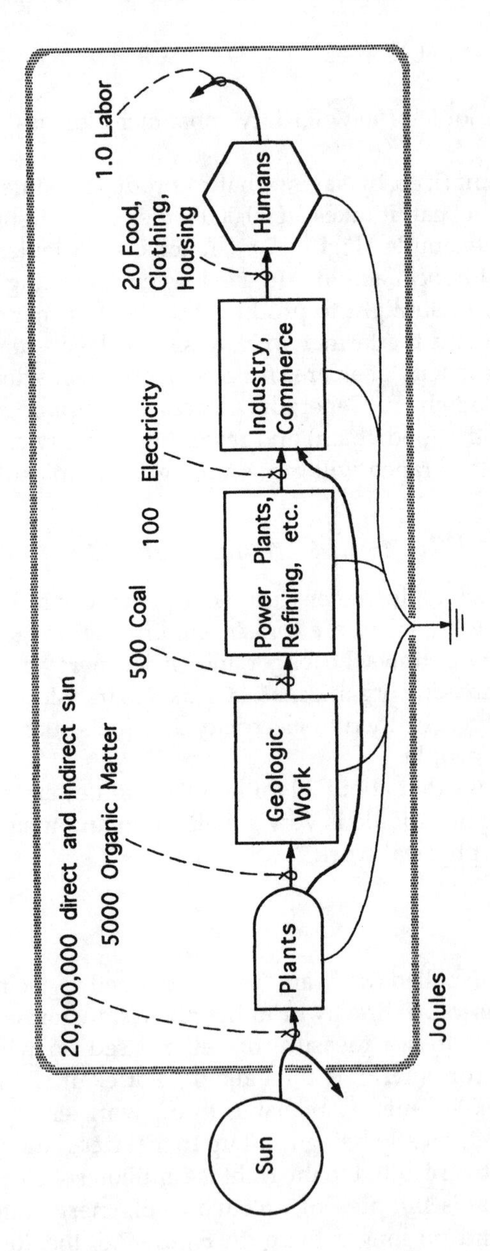

Figure 7.6 Energy chain for fuel-based urban people.

Table 7.1 Solar Transformities

	Solar emjoules per joule (sej/J)
Sunlight	1
Plant production[a]	4000
Wood[a]	21,500
Coal[a]	40,000
Oil[a]	53,000
Electricity[a]	200,000

[a]Includes indirect solar emergy of rain.

It takes a lot of low-quality energy (i.e., solar) to make a smaller amount of higher quality energy (Figures 7.5 and 7.6). To compare different kinds of energy, therefore, they are converted to a common basis, such as the solar emergy required to make each.

7.5 *Solar transformity*

The solar energy required to make 1 joule of some other type of energy is the *solar transformity* of that type of energy. The units are solar emjoules per joule (abbreviated sej/J):

$$\text{Solar transformity of energy of Type A} = \frac{\text{Solar emjoules required}}{\text{1 joule of energy of Type A}}$$

In Figure 7.5, 1 million (1,000,000) solar emjoules generated 100 joules of primary consumers; therefore, the solar transformity of the primary consumers is

$$\frac{\text{1,000,000 solar emjoules}}{\text{100 joules of primary consumers}} = \text{10,000 sej/J}$$

The primary consumers' energy is 10,000 times more valuable than sunlight. The farther to the right you go in the food chains, the higher the transformity. See the list of transformities in Table 7.1.

7.6 *Energy hierarchy*

A *hierarchy* is a pattern of organization in which many units at one level converge to fewer units at the next. For example, in an army many privates

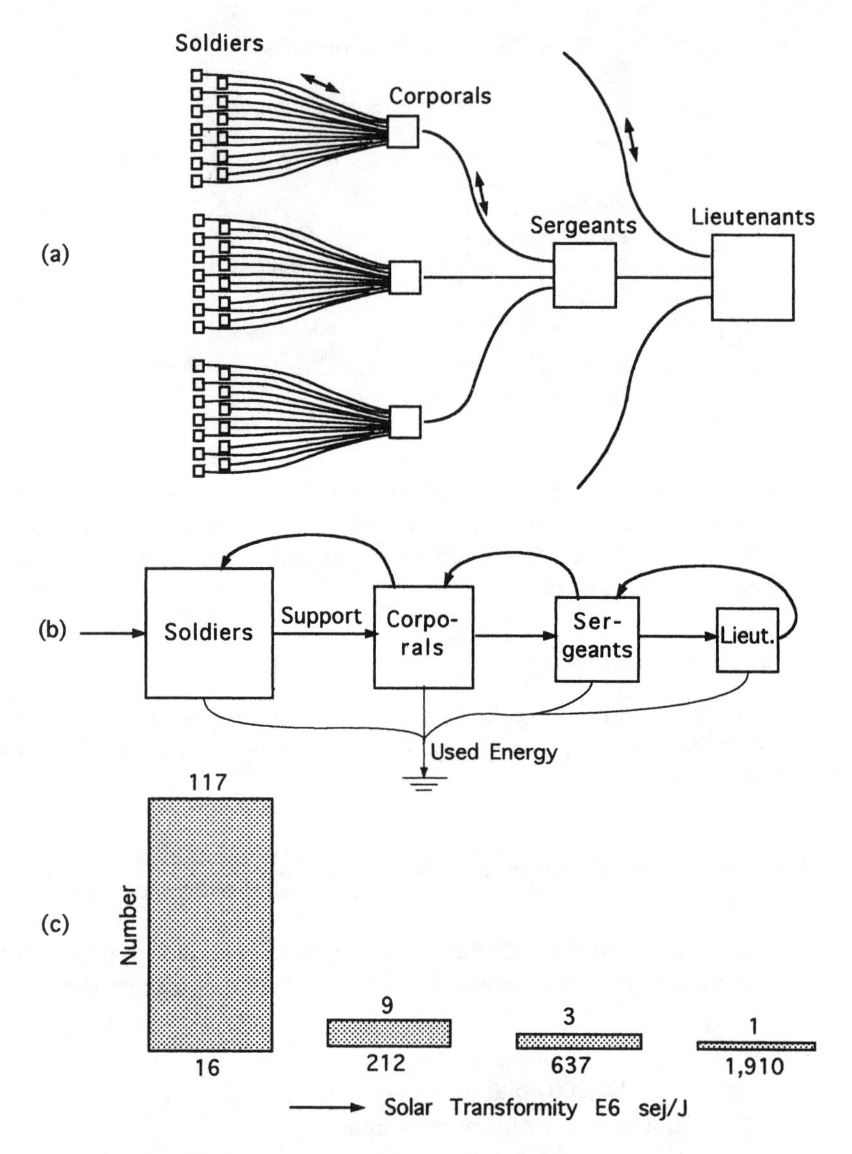

Figure 7.7 Hierarchy in the army. (a) Sketch showing the way in which people are organized in a military hierarchy; (b) aggregated energy systems diagram; (c) bar graph of numbers of people in each rank, also showing solar transformity of each rank.

report to fewer corporals, and these in turn report to fewer sergeants, etc. (Figure 7.7). Control and influence feed back in the opposite direction (lieutenants to sergeants to corporals to soldiers).

Since many joules of energy at one level are required through energy transformations to make fewer joules of energy at the next level, the energy

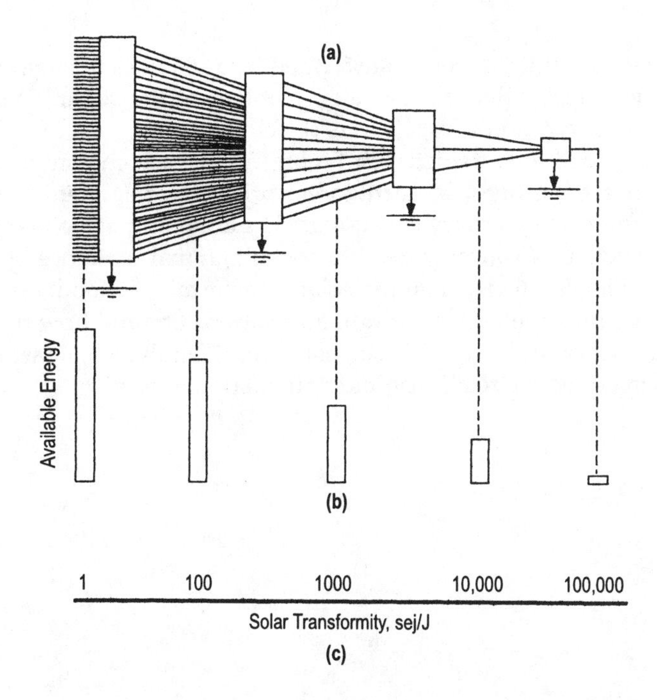

Figure 7.8 Diagram showing how a series of energy transformations is a hierarchy. (a) Lines of energy flow converging and diminishing; (b) bar graph of energy flow between each transforming unit; (c) solar transformities of these flows.

webs and chains are energy hierarchies. Figure 7.8a shows an energy hierarchy. Abundant energy on the left is required to generate lesser quantities at higher levels of the hierarchy on the right. Figure 7.8b is an aggregated energy systems diagram of the work and feedback controls.

Because the solar transformity of a product measures the solar energy that has been used to make that product, solar transformity indicates the level in the energy hierarchy. Notice the increasing solar transformity from left to right along the energy hierarchy in Figure 7.7c and in Figure 7.8c.

Questions and activities for chapter seven

1. Define: (a) efficiency, (b) web, (c) quantitative, (d) evolve, (e) energy quality, (f) emergy, (g) emjoules, (h) solar transformity, (i) hierarchy.
2. In a forest food chain, where is the highest quality of energy? Why?
3. Discuss the importance of the control input in Figure 7.1.
4. Explain why humans are placed at the end of the energy chain in Figure 7.6.
5. In terms of energy quality, how do the humans in Figure 7.6 compare to those living on a subsistence farm 300 years ago?

6. 1000 joules of light were transformed into 10 joules of sugar by algae in the water. What is the emergy content of the sugar? What is its transformity? Don't forget units (J, sej, sej/J).
7. For the food web depicted in Figure 7.3, list which components are producers, herbivores, carnivores, primary consumers, secondary consumers, and tertiary consumers. The animals may be in more then one category. (You may need to consult animal reference books.)
8. From Figure 7.6 calculate the solar transformity of food.
9. Discuss hierarchies in your human relationships and how they work. Examples include your school, classroom, family, job, apartment living, male-female relationships, state, nation, world.

chapter eight

Self-organization and succession

Self-organization is the process by which the various parts of a system become connected so as to work together. Those interactions that contribute to the better use of available resources succeed because they receive beneficial returns for their actions. For example, bees pollinate the flowers which produce nectar, so both the bees and those flowers survive. Grasses that can be eaten by grazing animals are reinforced by the wastes of those animals which provide nutrients for regrowth.

When humans colonize a new area, they self-organize. Different people try different things, set up various businesses, and experiment with new ideas for jobs. Those ideas that receive positive returns prevail, and those that are not suitable at that place and time are forgotten. The free market is one of the powerful mechanisms of self-organization. The services people perform that pay well are continued.

8.1 Self-organization in ecosystems: succession

The self-organizational process occurs all over the surface of the Earth, keeping it covered with ecosystems that are continually adapting to changing conditions. When a new ecosystem develops on bare ground or in a new pond, species move in and then are replaced by others. Usually the ecosystem is simple at first but becomes more complex as more organisms are added. The stages in this development are called *succession.*

Each stage prepares the area for the next stage. During succession there are usually growth of total biomass, an increase of nutrients, and an increase in the diversity of species, until the stage when further growth stops, called *climax.* Then, if outside factors stay about the same, the ecosystem will not change much, a condition known as *steady state.* Climax may continue until upset by catastrophic actions (for example, hurricanes, cutting by humans, insect epidemics). Figure 8.1 illustrates forest succession.

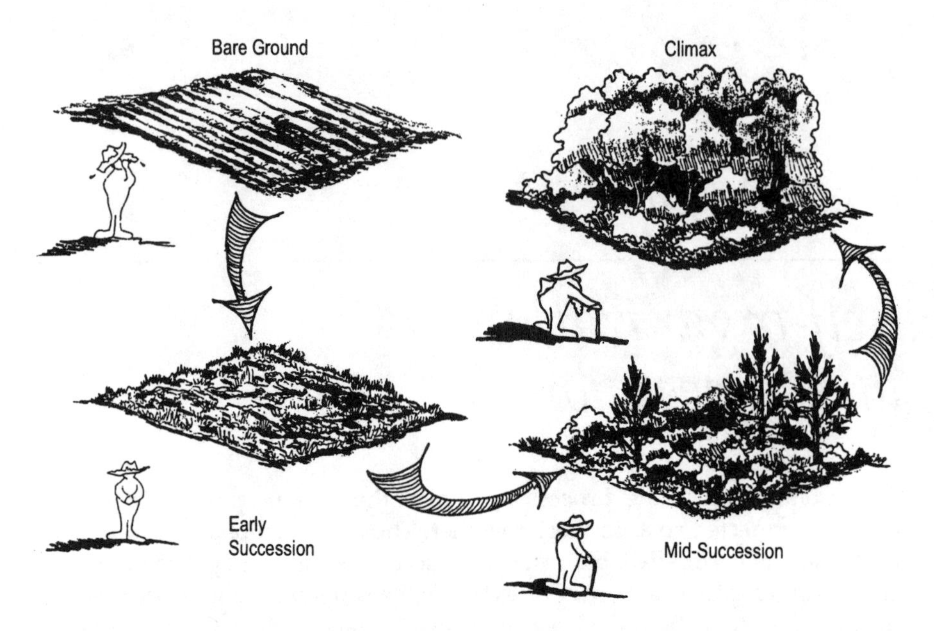

Figure 8.1 Forest succession.

8.1.1 Early succession

Gross production of plants is the organic matter they make. As succession starts in a new area, gross production is small because there is not yet a good cover of plants, a good mineral cycle is not started, the weedy colonizers are not efficient, and the consumers have not developed yet. Since plants adapted for this stage make much more organic matter than is used, the system grows.

In the initial period of this self-organizational process of succession, there is competition by alternative species and processes to use the available resources of energy, land, and materials. The ones best adapted to these initially undeveloped conditions grow very rapidly, overgrowing others and covering the area. For example, *weeds* (successional plants) first appear in a newly bare land area, covering the ground quickly. These *pioneer species* (colonizers) are specialized to produce rapidly, usually with inferior, temporary, and inefficient structures. By specializing in high growth rates, these species dominate briefly, while there are still new resources to be captured.

At this early stage, competition tends to reduce *diversity* (numbers of different kinds), because some species are driven out. The competition does tend to maximize production because energy resources are brought into use quickly by the species that overgrow the others.

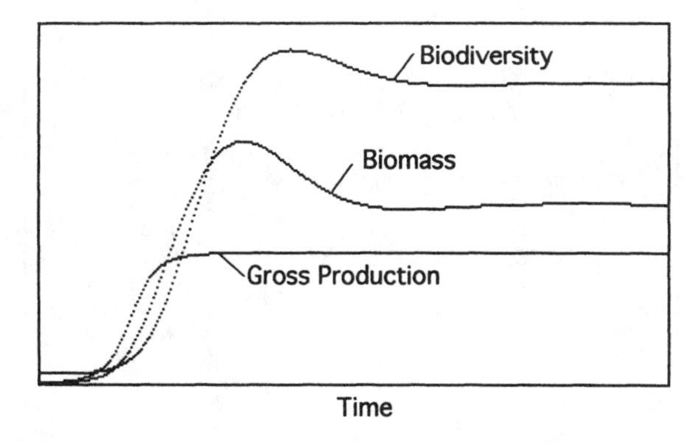

Figure 8.2 Succession showing increases in gross production, biomass, and diversity.

8.1.2 Later states of succession

As development continues beyond the colonization stage, species that are specialized to be more efficient replace the weedy growth specialists. Material cycles develop, and species prevail that contribute to each other rather than compete. In this way, a division of labor develops, and diversity increases.

At these later stages, nearly all of what is produced is being consumed by the many organisms that are contributing to the complex work of the system. Now production is maximized not only by having all the available sources in use, but also by having these used as efficiently as possible. Figure 8.2 is a graph showing succession with increases in gross production, biomass, and diversity.

At the climax state of succession, each organism has its *niche* (special place in the ecosystem) and there is very little competition. Figure 8.3 represents a system for which the food of the tree is divided among the birds that do not waste their energy in competition.

For the development of human societies, the succession process is similar. Emphasis on growth and competition is followed by more emphasis on cooperation and specializing to be efficient in the various sectors of the economy. As the economy of Florida slows in its growth, emphasis on being more efficient becomes more and more important.

8.1.3 Diversity

As an ecosystem goes through succession toward climax, the diversity of organisms increases. *Diversity* is the number of different units; *biodiversity* refers to the diversity of living organisms. An ecosystem develops to use all the different available energies by using many different plants and animals.

Figure 8.3 Woodpeckers with separate niches. (From Wharton, C.H. et al., *Forested Wetlands of Florida — Their Management and Use*, Center for Wetlands, University of Florida, Gainesville, 1976. Illustrated by Joy Bartholomew.)

Greater diversity increases the stability of the system. For example, in a forest with many kinds of trees, an insect which preys on one will not greatly disturb the whole forest system. Whereas, in early succession with fewer different species, the ecosystem is more susceptible to overgrowth by one species. Because agricultural systems are usually monocultures, more care and expensive insecticides are used to protect them from insect epidemics than are necessary for mature diverse forests.

8.1.4 Reorganization and oscillation

Climax ecosystems are not permanent, because they are disturbed by external influences such as storms, hurricanes, clear-cutting, and floods or by internal processes such as insect epidemics, diseases, and fire. For example, after early Florida grasslands were built up, they were consumed by roving herds of bison or by fire. Climax rivers have been channeled; climax saltwater systems have been overfished; climax swamps have been drained.

After a climax is disturbed by an outside or inside factor, succession starts up again, if environmental conditions are unchanged. For example, if a storm knocks down a big live oak tree, grasses will start succession again in that open space. However, if groundwater levels are changed by ditching, a different successional system may develop.

Ecological succession often follows the sequence in Figure 8.4, starting with a simple colonizing species, then more and bigger species. At the most fully developed state there is a large and varied biomass of living organisms as well as much dead organic matter in logs, soils, and sediments. As you go right in Figure 8.4, the structure and diversity of the ecosystem is shown coming down again due to the catastrophic effects of larger scale phenomena. Then it starts up again in a new succession. This repeated up and down is called *oscillation.*

Table 8.1 compares some characteristics of early succession with the fully developed climax and decline. A detailed example of forest succession is provided in Chapter 19.

8.2 Maximum power principle

During self-organization, systems are guided by the *maximum power principle,* stated as follows:

> System designs organize so as to bring in energy as fast as possible and use it most efficiently.

The maximum power principle indicates why certain patterns of system organization develop and others do not. The principle explains why successful systems have similar patterns. A successful design is one that has survived the test of time. The system design that prevails has each part of the network reinforcing others so as to draw the most power, using it to meet all other needs. To maximize your learning power as a student, you repeat whatever study pattern you have that helps you learn information which, in turn, can be used to learn even more.

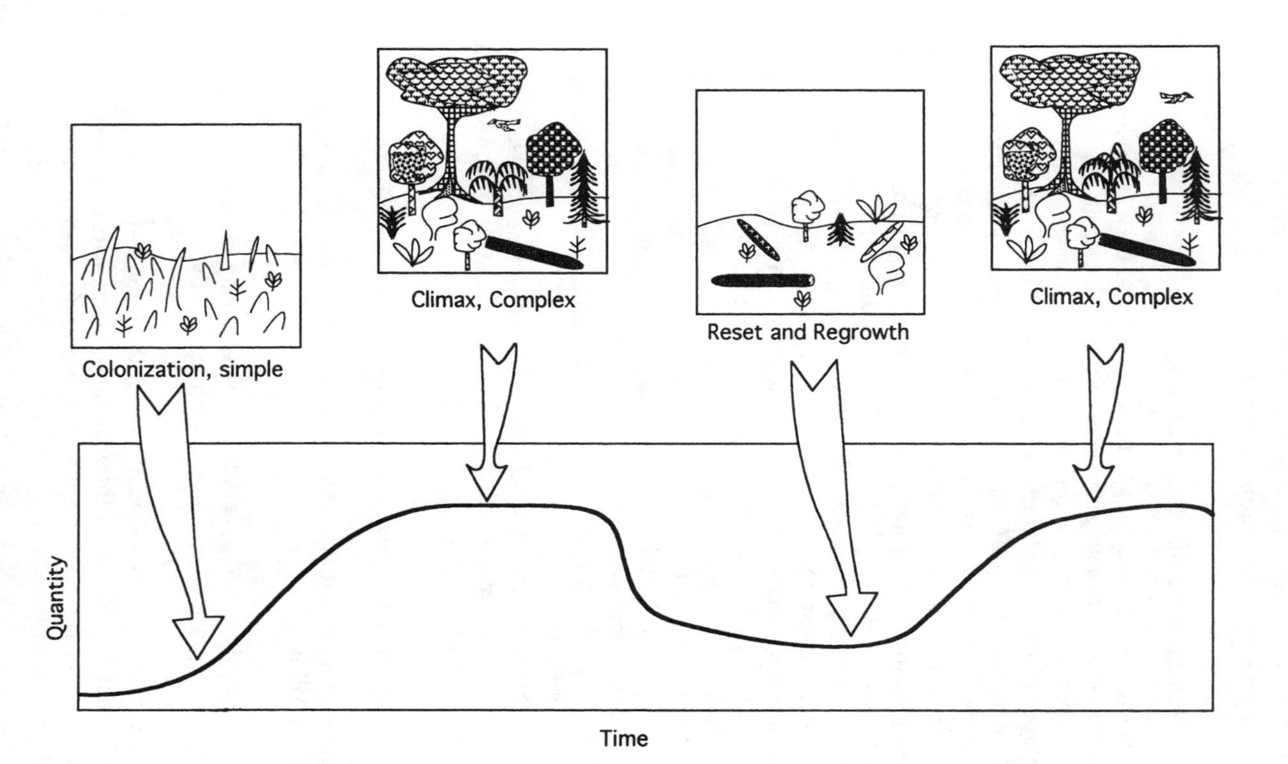

Figure 8.4 Repeating pattern of succession, destruction, and regrowth, showing change in complexity.

Table 8.1 Characteristics of Stages in the Cycle of Ecosystems

Characteristic	Succession	Climax	Decline
Biomass quantity	Low	High	Medium
Biomass growth	Net increase	Little change	Net decrease
Gross production	Increasing	Maximum	Decreasing
Consumer metabolism	Low	Maximum	Decreasing
Nutrients	Accumulating	Recycling	Leaking
Species diversity	Low	High	Resting[a]

[a] Species diversity may remain in the seeds, ready for regrowth.

The systems that maximize power are also organized to contribute to the larger system of which they are a part. For example, in the forest system, a tree uses the energy of the sun by growing leaves which increase in size and number to catch more of the energy of the sun. This process of the tree supports the forest system by building soil, making a stable microclimate, recycling nutrients, and providing food for animals. Thus, the tree maximizes both its own power and that of the larger system of which it is a part.

To maximize power in an economic activity, local resources are used to make a profit which is used to produce more. For example, consider a farm on which crops are planted at the best time in relation to rain and sun. The best fertilizers are applied to make the crops grow, and crops are grown that people will buy. This farm will produce enough financial return for the farmer to live well, maintain the soil, and repeat the process year after year. He may even be able to expand his system by buying up less efficient farms. The successful farmer's way of farming will survive and will then be copied by other farmers. It helps both him and the whole economy to be successful.

Another example of the maximum power principle is the reaction of the system after water levels were lowered by ditches in many south Florida areas, causing productivity of native trees to decrease. These drained areas were invaded by exotic trees, such as the *Melaleuca* from Australia and the Brazilian pepper from South America, which are better adapted to grow in the new conditions. These newcomers grew rapidly where drainage had left the area dry and displaced the native trees. Since the new system produces the greater production, it survives.

8.2.2 Good uses and feedback

A good use is one that contributes back to the system. As shown in Figure 8.5a, uses that do not feed back to reinforce some other part of the system are a drain on the production, and both that production and use tend to be eliminated. A system where everything is reused or recycled maximizes power (Figure 8.5b).

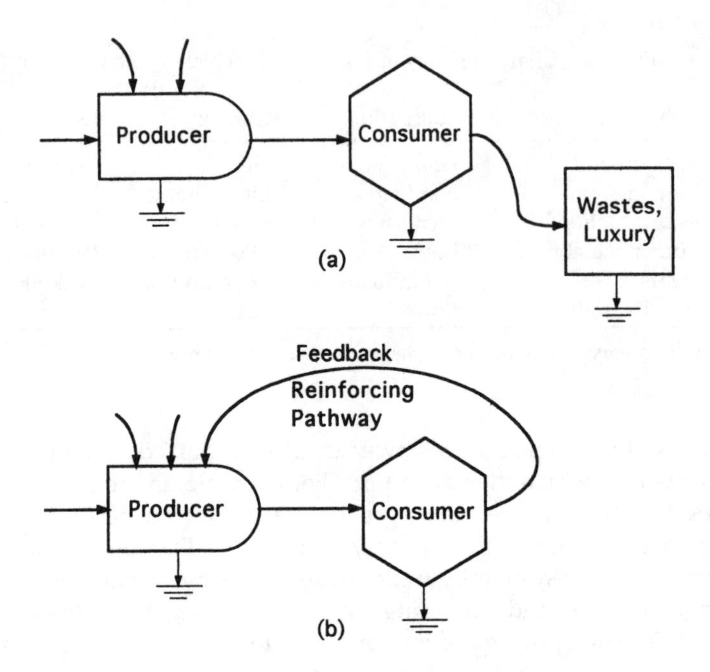

Figure 8.5 Comparison of consumers who reinforce production to those who waste it.

Examples of bad uses are those that end in dead storages such as waste dumps. Poor uses include luxuries that do not contribute to the rest of the system (Figure 8.5a). During times of excessive energy in this century, resources were used for luxuries. Better use of these resources is required to make the system more efficient and productive and for people to be better educated and in good health.

8.2.3 Feedback control

One population of organisms often controls another, as shown in Figure 8.5b. The controlling species, such as an herbivore, may help spread the plants it consumes. Whereas the food moves from left to right, the controlling action goes back from right to left. For example, squirrels plant acorns by storing more than they need, and birds spread seeds by eating fruits and then releasing the seeds in their wastes.

Control of population numbers is another example of feedback service. When one species of plant becomes numerous, the population of insects which feeds on it also increases. By feeding heavily on the plants, the insects regulate the numbers of plants of that species, allowing an increase of other species. As a result, the forest maintains greater diversity and better overall production.

8.2.4 Converging and dispersing of material in cycles

Another kind of feedback which maximizes power is the converging and dispersing of materials (mineral cycles introduced in Chapter 5). Raw materials used by a system are often collected from broad areas and brought inward to the center of the system's activity. Later, the waste materials may be dispersed back out to the larger area again. For example, the roots of a big tree gather the nutrients over a large area and bring them together to the trunk of the tree and to the rest of the plant. Later, when the leaves fall and decompose, these nutrients are dispersed back to the ground over a broad area. Chapter 34 describes the need for converging and dispersing of materials in a town.

Questions and activities for chapter eight

1. Define: (a) succession, (b) self-organization, (c) climax, (d) efficiency, (e) maximum power principle, (f) feedback, (g) niche, (h) population control, (i) diversity.
2. Give an example of the increase in diversity in succession of a forest.
3. Draw a system that maximizes power. Explain how your system makes use of the maximum power principle.
4. Why is feedback reinforcement necessary for the survival of an ecosystem? Give an example.
5. List ten luxuries in your life. Which of these feed back to make you learn more or work better? Which are a drain on the system around you?
6. Right now there are thousands of homeless people in Florida whose presence makes each of us feel uncomfortable, helpless, guilty, angry, or all of these. Each of them feels cold, hungry, sick, helpless, guilty, angry, or all of these. From the overall system view (Figure 8.5), they are not contributing to the larger system. Discuss and come up with possible plans of action.
7. Set up microcosms in the classroom to illustrate ecosystem principles. Start early in the term to give them time to grow and change. If each group sets up two microcosms, they can be treated differently to test the effects of a treatment. Toxic wastes, sewage, water, solid wastes (beer cans, paper trash, etc.), saltwater intrusion, fertilizer pollution, and acid rain are among Florida's problems. It is important to have only one experimental treatment, so you can be fairly sure that any change has been caused by that treatment. Of course, there may be other unanticipated changes, as you cannot make both containers exactly alike.

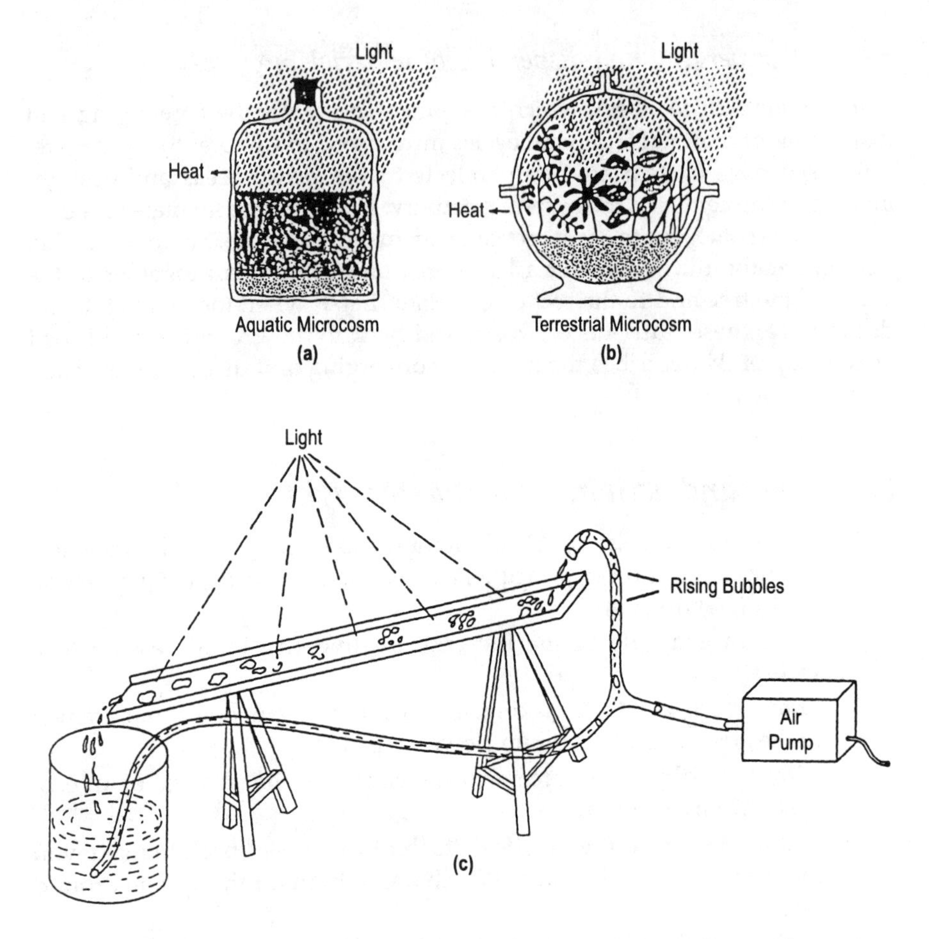

Figure 8.6 Examples of microcosms (a, b) and a diagram of a simple stream microcosm (c).

Directions for making a pond-type microcosm (Figure 8.6a)

From a body of natural water, dig up some bottom sediments with the animals and plants that live in them and place them in a plastic (not metal) bucket. Organisms can be taken from the water column with a dip net, plankton net, or seine. Enough water should be collected to fill the microcosm containers. Glass aquariums covered with a glass plate or plastic film held with a large rubber band are good (Figure 4.2). Transfer the water and organisms carefully. Evaporation should be replaced with distilled water or rain water, not chlorinated tap water.

Three types of pond microcosms can be started with the same natural material:

1. One kind has more animals than can be supported by plants, respiration exceeds photosynthesis (R > P), oxygen comes in from the air continuously, and food for the excess animals must be supplied.
2. The second type is constructed with bright light, plants, and plant nutrients (fertilizer); photosynthesis exceeds respiration (P > R).
3. A third type, a balanced system where photosynthesis equals respiration (P = R), may result if the container is sealed (Figure 8.6a). It can be sealed with plastic wrap which is not permeable to oxygen and carbon dioxide, not polyethylene film.

Directions for making a terrestrial microcosm (Figure 8.6b)

1. Choose open or closed type. Select containers that hold water, are transparent, and have openings big enough for a hand to go in and out. The covers can be tight enough to stop most exchange while letting a little air come and go as the air expands due to heat or shrinks as it cools down.
2. Arrange light conditions. Indirect daylight is best, because the direct sunlight may make the ecosystem too hot and stress animals, and indoor light has too little energy for fast growth. Microcosms can be made without light to illustrate what goes on in dark places such as caves or down in the soil. Such microcosms run on decomposing organic matter.
3. Decide on an ecosystem type and put in some soils from that environment. Interesting microcosms have been started from cattle dung, desert soil, a block of lawn grass turf, a sand bed with beach plants and animals, a patch of farm soil, peat, gravel, vermiculite, garbage, or shredded plastic (to simulate a dump). Arrange the soil materials on a slant, so one side can be drier.
4. Add a little water and possibly some fertilizer. Do not add so much that the soil is saturated (unless you want your microcosms to simulate a wetland). If you are creating an open system, add a little water every week (artificial rain). In a closed microcosm, the evaporated and transpired water will condense on the top and sides and drip and run down like rain.
5. Seed with many kinds of plant and animal life from the same environment as the soil. Avoid large individuals. The soil probably added many kinds of microorganisms. Do not be disappointed when only a few species survive.
6. Study the ecosystems: count and measure things; estimate growth rates by recording weights and dates on which organisms appear; test pH, oxygen, chemicals; identify organisms; make graphs of the differences between the two microcosms; and write a paper with your methods, data, and conclusions.

Directions for making a stream-type microcosm (Figure 8.6c)

A simple stream microcosm is diagrammed in Figure 8.6c. Support a stream bed within an aquarium above the level of water. Use a simple bubble pump to lift water from the aquarium to the head of the artificial stream. The return is screened to prevent clogging. The supports and stream bed should be constructed of some inert material such as brick, fiberglass, ceramic roofing tiles, etc. The stream bed should be filled with small rocks, gravel, leaves, and small sticks from a natural stream to provide seeding and a place for attachment of organisms. The aquarium water simulates lakes above and below the run of the stream. Glass and tygon tubing are suitable for use; metals are not. Excess organic matter at the start will tend to make respiration exceed photosynthesis (R > P).

chapter nine

Simulation of quantitative models

The biosphere is made of systems that change as time passes. Both environmental systems and human systems can be described by the way they change. The way a system changes depends on the system's organization and on the kind of energy source that is available. For example, some ecosystems increase in size and complexity while others stop growing. Some small towns can grow into cities, whereas other towns seem to remain the same size for decades. They appear to have reached a steady state. Other towns decrease in size and complexity, factories close, and people move away.

The energy language diagrams are a useful way of visualizing the ways in which systems function. The diagrams become much more powerful, however, when actual numbers are used and simulated by computer. A computer simulation draws a graph to show how properties vary over time.

The energy language diagrams that we have already used are really mathematical statements. We have been using symbols to represent quantities and relationships, which become mathematical statements by putting numbers on the symbols. If we do this, we have a language that is very close to the language understood by computers.

If you have access to a computer, you can do the simulation exercises in this section using a computer and the programs described in Appendix A. If you have limited use, you can arrange a demonstration of the computer simulations. If you do not have access to a computer, you can do a simulation by hand calculation.

9.1 TANK, a sample model

Think of a kitchen sink (Figure 9.1a); it has a faucet and an open drain. Let's start with the sink empty and turn on the faucet. Water will flow into the sink and then out the drain. The quantity that drains out increases with the quantity in the sink; as the water in the sink increases, the quantity that drains out increases. As the water flows in and out, the total quantity of water in the basin will increase and then level off when the amount of water inflow

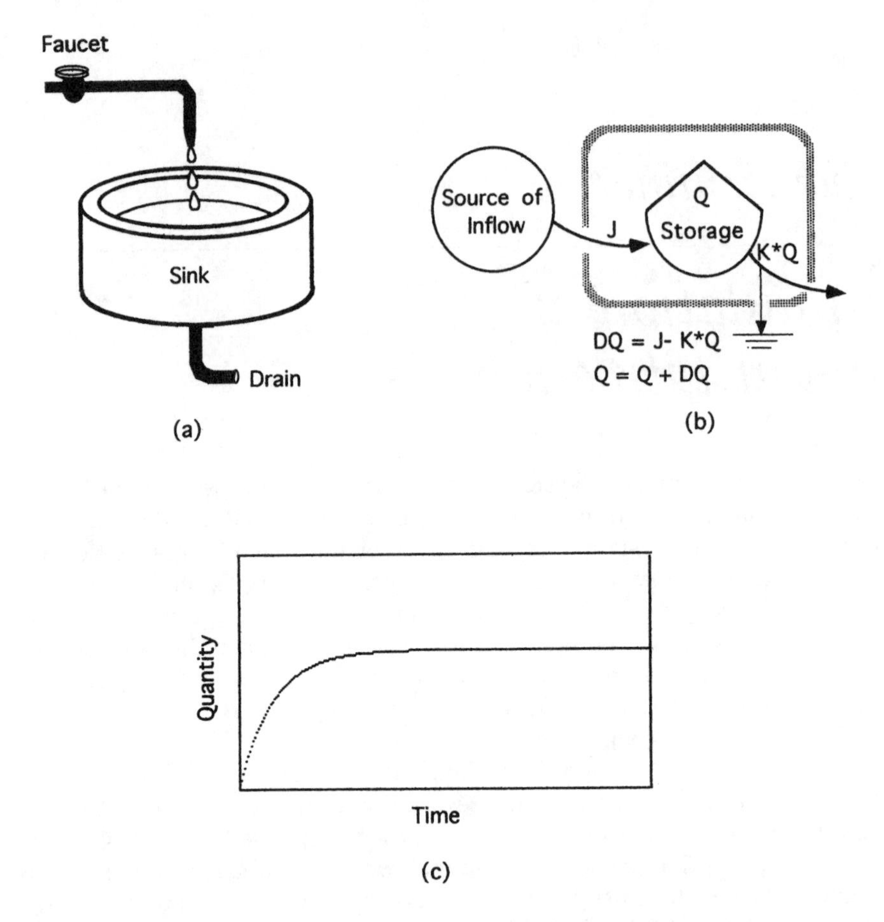

Figure 9.1 Simulation of a storage. The computer program TANK is listed in Table 9.3 and in Appendix A2. (a) Sketch of a sink with inflow and drain; (b) systems diagram of the model and change equations; (c) simulation graph.

equals the amount of water outflow. Water continues to flow in and out, but the level stays the same.

The graph in Figure 9.1c shows these changes. The change in time is graphed on the horizontal axis and the change in quantity of water on the vertical axis. The graph shows the change in quantity of water as it increases quickly, then more slowly, and finally reaches a level steady state.

Now look at the diagram in Figure 9.1b. J is the inflow which comes from outside the system. In our sink example, J is the water from the faucet. The tank is labeled Q, representing the quantity of water in the sink. The outflow through the drain is labeled K*Q since it is a proportion (K) of the quantity in the sink (Q). Although the drain size stays the same, the more water there is in the tank the heavier it is and the faster it will flow out.

Table 9.1 Hourly Calculations for the TANK Model[a]

Time (hours)	Hourly Change[b] (DQ = J – K*Q) (liters/hour)	Q at the Hour (Q = Q + DQ) (liters)
0 (start)		1.00
1	3.95	4.95
2	3.75	8.70
3	3.56	12.26
4 (etc.)		

[a] See Figure 9.1.
[b] J = 4; K = 0.05.

A heat sink symbol is drawn in every system diagram to show that there is energy dissipated as depreciation in every process and storage, although calculation of heat dispersed is not included in most of the simulations.

Below the diagram are the equations for the change in the quantity of water in the sink (Q) over a unit of time. The first equation (DQ = J – K*Q) says "the change in Q (DQ) is J minus K times Q." The second equation (Q = Q + DQ) says "Q at this time is what Q was plus the change in Q (DQ)." In terms of the water example, the equations say, "The quantity of water in the sink now is what it was an hour ago plus the change caused by the inflow and the outflow."

9.2 Simulate by hand

To show how successive changes are calculated and graphed by the computer, you can calculate a few numbers by hand (Table 9.1). The TANK model is in Figure 9.1b with its equations: DQ = J – K*Q and Q = Q + DQ.

1. Start with inflow J = 4, quantity Q = 1, and outflow constant K = .05. Plot successive Qs on a graph with time (Figure 9.1c).
2. Plot the first point: time = 0, Q = 1.
3. Now calculate and plot Q at time 1:

$$DQ = J - K * Q = 4 - (.05)(1) = 4 - .05 = 3.95$$

$$Q = Q + DQ = 1 + 3.95 = 4.95$$

4. Calculate Q at time 2 and plot it:

$$DQ = J - K*Q = 4 - (.05)(4.95) = 4 - .25 = 3.75$$

$$Q = Q + DQ = 4.95 + 3.75 = 8.7$$

Table 9.2 Some Instructions in BASIC Language to which Computers Respond

Command	What It Does
LIST	Lists the program in the working memory
RUN	Runs the program, working through the instructions in numerical order
GOTO	Goes to a designated instruction number and performs that action next
IF	Provides a condition for doing something, such as going to another designated line (e.g., IF T is less than 20, GOTO ...)
PRINT	Shows on the screen the numerical value of the quantities that you list after the PRINT command
PSET	Shows on the screen a graph of the changes in values of the quantities you list after the PSET command (the plot command varies for each kind of computer)
=	Sets a quantity equal to what is specified
+	Adds the next quantity
–	Subtracts the next quantity
*	Multiplies the next quantity
/	Divides by the next quantity
<	Indicates less than
>	Indicates more than

If you plot 100 points, your graph will look like Figure 9.1c. You can see how tedious this is. The computer can calculate and graph hundreds of these points in seconds.

9.3 *How you make the computer simulate your model*

When the relationships in the model shown in the energy diagram are put into the computer as a program which is run, the resulting graph (Figure 9.1c) shows what happens with time (called *simulation*). The program is written in BASIC language, which is available to load on most computers. A computer program contains the directions which tell the computer what to do. It is like a cookbook recipe with the quantities listed and then the directions given. In BASIC, each instruction statement is given a number. You usually start using 10, 20, 30, etc. so that you can add statements in between your originals. When all the statements are written and you press RUN from the Menu, the computer will start at the first number and follow your directions in sequence. In QUICKBASIC the line numbers are optional.

The words and symbols you need to use in instructing a computer are given in Table 9.2. They are part of the BASIC language. The TANK program statements and their explanations are given in Table 9.3.

Load BASIC or QBASIC into your computer. To get your program on the screen, press LIST. To save your program on a disk, press SAVE AS and type

Table 9.3 Program in BASIC for Simulation of the Model in Figure 9.3[a]

Program		Explanations
2	REM TANK	REM means remark; does not affect program
5	CLS	Clears the screen
10	SCREEN 1,0	Sets medium resolution graphics
20	COLOR 0,0	Sets color
30	LINE (0,0) – (320,180),1,B	Plots a border around the graph
40	J = 4	Enters the value of inflow J
50	Q = 1	Enters the beginning quantity for Q
60	K = .0.05	Enters the coefficient K
70	PSET (T,180–Q),2	Tells computer to plot a point for T and Q; (180 – Q) arranges zero at its normal position
80	DQ = J – K*Q	Equation for the change of Q over the time interval
90	Q = Q + DQ	Equation for Q after the change
100	T = T + 1	Equation to increase time by one unit
60	IF T < 320 GOTO 70	Runs the program from statement 70, plotting points until T = 320

[a] For a PC; see Appendix A for instructions to adapt the program for Macintosh.

the program name. To print the program list, press PRINT. Each kind of computer has a different way to print the graph (see Appendix A).

9.4 Spreadsheet calculation

Another way to simulate models is to let the computer calculate tables (like Table 9.1) using a spreadsheet program such as Excel or Lotus. These programs can print out the table or neat labeled graphs.

9.5 "What if" experiments

This model and those in the next chapter represent many different kinds of systems. Simulation models can be used for controlled experiments, in which you change something in the program and run it again to see the effect.

Suppose that the tank were full at the start instead of empty. What would happen then? As Figure 9.2b shows, if you start with the tank full, the level will decrease until the same steady state is reached. What would happen if the water inflow is turned off? As Figure 9.2c shows, the level in the tank decreases quickly at first, and then more slowly, because, as the amount of water decreases, its pressure on the drain becomes less.

An example in nature is a stream flowing steadily into a pond that also has a stream flowing out of it. When the stream first starts flowing, the pond fills up to a level where the flow into the pond equals the flow out (Figure 9.2a). Figure 9.2b illustrates the situation of the pond after a large rainfall. The

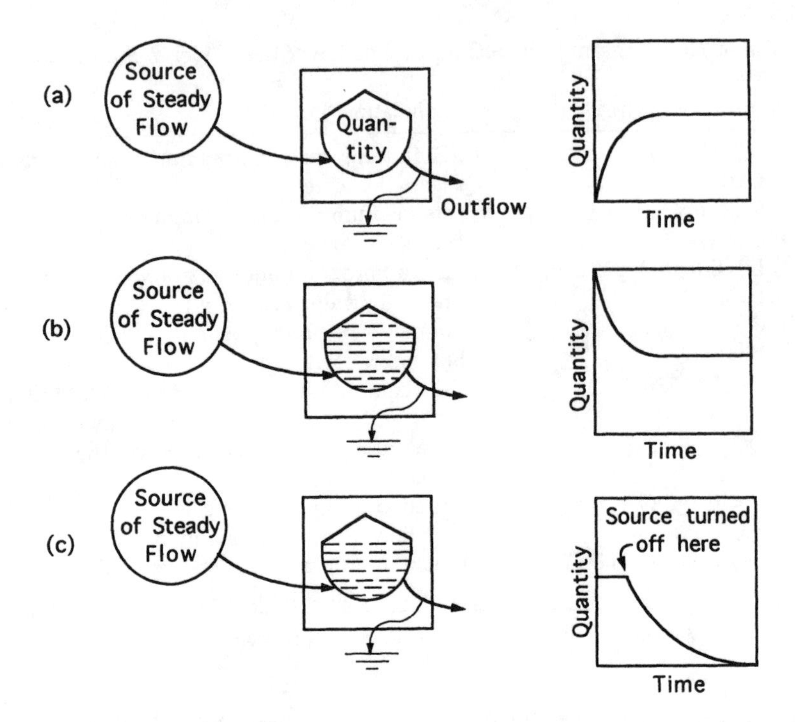

Figure 9.2 Simulation graphs for the storage model (TANK program) with different starting conditions: (a) starting with empty tank; (b) starting with more stored than at steady state; (c) tank at steady state initially, after which the inflow is turned off.

quantity of water stored in the pond is high (because of the rain) but soon comes down to the same level as before. If the inflow stream is suddenly diverted, the water in the pond will drain away until there is none left, as is shown in Figure 9.2c. The balance of water that develops in any pond, lake, or groundwater reservoir is an example of the storage tank model.

Another example is the buildup of leaf litter on the floor of a forest. Litter builds up a layer which continues to grow until the rate of loss from decomposition equals the rate of gain from leaf fall (Figure 9.2a). If a sudden gale dumps a load of leaves on the ground, the change in litter quantity would look like Figure 9.2b. In some forests, the leaves stop falling in the winter; the pile of leaves in the forest would then decrease as shown in Figure 9.2c.

An economic example might be your bank account. If you put in a steady income and spend it at a steady rate, the graph of your money in the bank would look like Figure 9.2a. Figure 9.2b would represent a large sum to start your account, and Figure 9.2c shows what would happen if your income stopped.

Questions and activities for chapter nine

1. Define: (a) simulation, (b) quantitative, (c) program, (d) spreadsheet, (e) BASIC.
2. Try the experiments in Figure 9.2 using a real sink with a faucet and drain.
3. Think up and explain a TANK model system in your favorite subject.
4. Investigate where to find BASIC and how to print graphs on the computer you have available. Start with Appendix A. You may have to read the computer manual or even call the company.
5. As you probably know, there is much more to computer programming even in simple BASIC, but you already have enough to simulate the models in Chapter 10. Try them, using the programs in Appendix A.

chapter ten

Growth models

The models in this chapter show how systems grow, become steady, and then decline, depending on their energy sources and their use of the sources. The computer programs for these models are in Appendix A. If you can run them on the computer, you can ask what would happen with various experiments, make the changes, and see if your predictions were right.

In these models, the label on the storage symbol is Q for "quantity". This general term for contents of a storage may refer to population numbers, biomass, or energy stored.

10.1 Model 1: exponential growth

The first model is shown in Figure 10.1. It represents a population growing on a *constant-pressure source.* A constant-pressure source can supply as much energy as is needed. As an example, think of a population of rabbits growing on a hopper of food which is replenished regardless of how fast it is eaten. Follow the flows in the diagram to see that, as the rabbit population increases, it feeds back to bring in more energy (by eating more) to make even more rabbits. Suppose a system starts with a male and female rabbit which produce four bunnies, which then pair and produce eight offspring. At the same rate of increase, the next generation will produce 16, the next 32, the next 64, and so on. As the number of rabbits increases, they use more of the energy source and the number increases even faster.

You can see that there is an acceleration of growth of the rabbit population as long as the same concentration of food supply is maintained. The curve of a population under these conditions is said to be *exponential growth.* Exponential growth is at a constant percent increase per time.

In practice, a constant-pressure energy source cannot be maintained indefinitely, so that perpetual exponential growth is impossible. However, during the early stages of population growth, when the demand for food is small compared to the amount available, energy may be available at constant pressure and growth may be exponential. Eventually, though, food would become limiting and the situation would have to be represented by a different model.

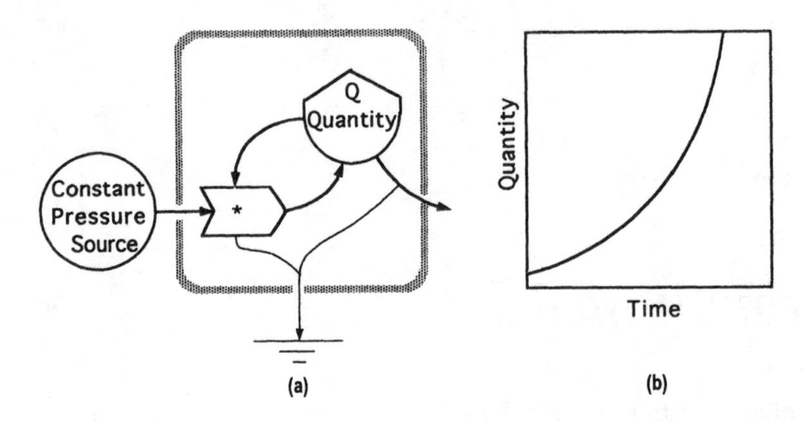

Figure 10.1 Exponential growth of a system with an energy source that maintains a constant pressure (Model 1). (a) Systems model; (b) growth curve. The computer program EXPO is in Appendix A, Figure A3.

10.2 Model 2: logistic growth

Populations initially growing quickly on a constant-pressure source may become so crowded that losses develop due to interactions among members of the population. A steady-state, level population results. This kind of growth is called *logistic growth.* Logistic growth is the balance between production in proportion to population and losses in proportion to the effects of the individuals interacting (crowding).

The growth process may be understood with the help of the diagram of the model in Figure 10.2. An example is the growth of yeast in a fermenting

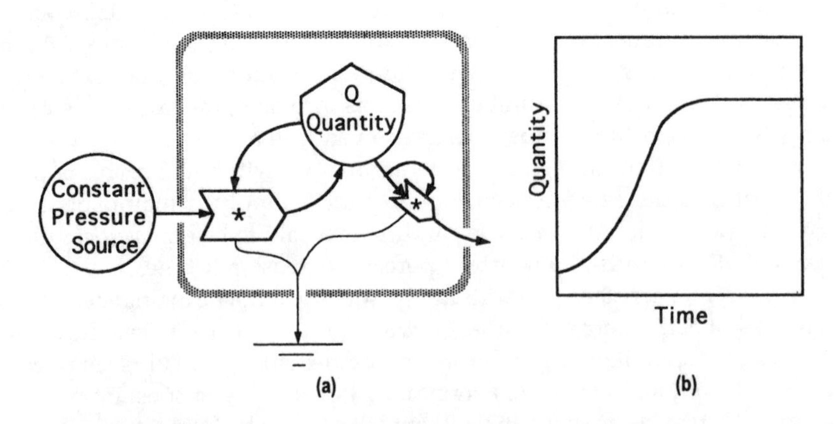

Figure 10.2 Logistic growth: growth of a system with a constant-pressure energy source and a self-interaction on the outflow drain (Model 2). (a) Systems model; (b) growth curve. The computer program LOGISTIC is in Appendix A, Figure A4.

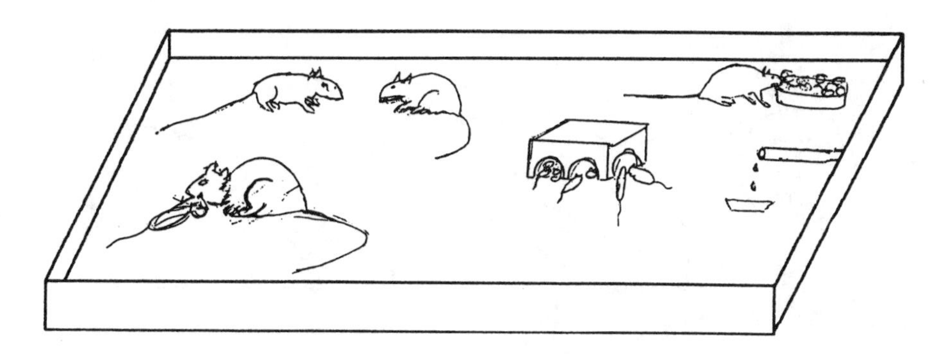

Figure 10.3 A mouse experiment in which the mice became so crowded the population stopped growing.

brew. First, the growth of the population is almost exponential. Food availability is constant, and as the population grows it feeds back to eat more and more. However, as the yeast cells become crowded, their by-products begin to interfere with their own growth. A steady state results between production and loss of cells.

In Figure 10.2, notice that the production part of the model is the same as that in Figure 10.1. The energy supply is a constant-pressure source, and the population is drawing in energy and feeding back to draw in more. Population growth is therefore exponential at first; however, as Figure 10.2 shows, the population, by interacting with itself, creates an accelerating energy drain which will eventually draw off enough energy to stop population growth. Thus, the graph shows exponential growth which slows and eventually levels off to a steady state. This system has a constant-pressure source and a *self-interactive drain*. An example is a mouse experiment (Figure 10.3) in which mice were set up in a large cage with food and water always available. They reproduced and their population increased until some mice started to move their young, some started fights, and others became sterile as their cage became too crowded. The population stopped growing and leveled.

Another example of Model 2 (Figure 10.2) is the growth of a human population and its services in a city. Growth may increase exponentially until the crowding of houses, streets, stores, and cars starts to increase the negative factors of dirt, noise, crime, and pollution, and the cost of dealing with these becomes progressively greater. The more the population builds up, the greater the drain until the growth of the city levels off.

10.3 Model 3: growth on a constant-flow source

Ecosystems use many sources whose flows are steady and controlled by outside systems. Examples of *externally limited sources* are sun, rain, wind,

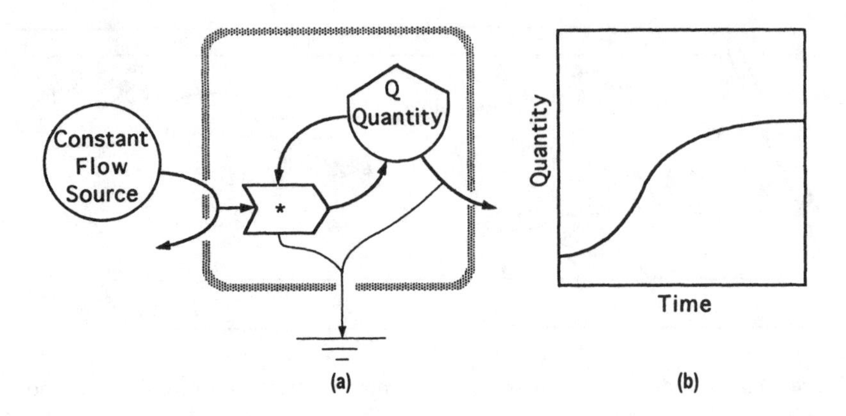

Figure 10.4 Growth of a system with a constant-flow energy source (Model 3). (a) Systems model; (b) growth curve. The computer program RENEW is in Appendix A, Figure A5.

and flowing streams. Populations in the system cannot increase the outside inflows. Their growth is limited to that which can be sustained on the inflowing energy. Trees using sunlight are an example. There is nothing that the trees can do to increase or decrease the incoming sunlight. This kind of source is also called a *renewable source.* In Model 3, the inflow is constant.

Figure 10.4 shows the way an externally limited constant-flow source is drawn in symbol diagrams. A pathway from the source is shown passing into the system, with some of it continuing on out of the system again. The energy use is shown as a line from the side of the inflow path. You can think of it as a pipe connected to the side of a stream to draw out water.

Now consider growth that occurs on such a source when the inflow is constant, and the pumping is in proportion to the population using the stream. Model 3 simulates growth on a *constant-flow source* (Figure 10.4). The model is like the exponential growth model except there is a constant-flow source instead of a constant-pressure source. As the population grows, more and more of the inflow is diverted until almost all of it is being used as fast as it is flowing in. After that, no more increase in growth is possible, and the population is at steady state. This can be called a *dynamic equilibrium.*

An important example in nature is succession, as in the growth of a forest. When the forest is young, light energy is not limiting. Growth of the small trees is rapid and much of the surplus light passes by unused. As the forest grows, however, the trees use more and more energy, and less and less escapes unused. When almost all of the light in the forest is used, growth slows and levels. The forest becomes a balance between growing and decomposing. Succession is discussed in Chapters 8 and 15.

Another example of growth on a constant-flow source is the building of towns along a river. Towns use water for drinking, farming, fishing, and

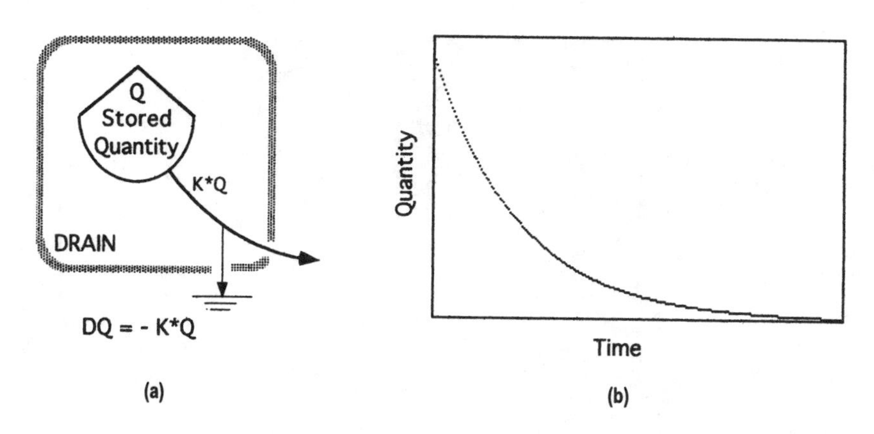

Figure 10.5 Model of a storage with an outflow depending on storage. (a) Systems model; (b) simulation graph. The computer program DRAIN is in Appendix A, Figure A6.

treating sewage disposal. New uses can be added until almost all the water is being used as fast as it is inflowing.

The graph of growth on a constant-flow source is "S" shaped (Figure 10.4). It has the same shape as the logistic growth (Figure 10.2) but for a different reason. The logistic model in Figure 10.2 is not limited by its source (constant-pressure sources do not limit growth) but is limited by crowding. The constant-flow source model is limited by the rate of supply of its source.

10.4 Model 4: storage loss

The model of growth in a storage tank simulated by the TANK program was considered in detail in Chapter 9. A variation of this model in Figure 10.5 considers only the losses from a storage without inflow. The graph generated by the program DRAIN decreases rapidly at first and then at a slower and slower rate. This curve is observed in many fields and is sometimes called *exponential decay*.

An example is water in a temporary pond which is percolating into the ground. Another example is tide-pool water diffusing through beach sand. An example from nuclear physics is the decay of radioactivity of elements such as uranium. Other examples include the decline in weights of animals during starvation and the decay of a pile of leaves in the forest.

10.5 Model 5: growth on a nonrenewable source

Some systems depend on resources drawn from a *nonrenewable* source — for example, a population of beetles growing on the energy available from a decaying log (Figure 10.6). N represents the log and Q the beetles. When the

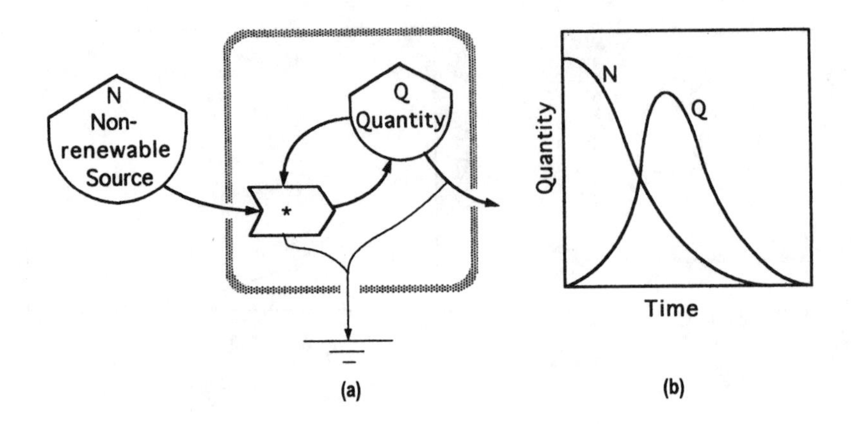

Figure 10.6 Growth of a system with a nonrenewable energy source (Model 5). (a) System model; (b) growth curve. The computer program NONRENEW is in Appendix A, Figure A7.

beetle population is small, there is ample energy, and growth is exponential. Later, as the log begins to disappear, growth of the beetle population slows. As the log continues to get smaller, the number of beetles decreases until there is no more log and no more beetles. On the graph, the line, Q, represents the population numbers. The line, N, represents the energy remaining in the log at any given time.

Another example is a mining town with a single nonrenewable economic resource such as a coal deposit. It will become a ghost town when the coal is all mined, as in Figure 10.7.

10.6 Model 6: growth on two sources

Our sixth model has two sources, one renewable and one nonrenewable (Figure 10.8). Both sources interact with the tank quantity which grows and provides feedbacks to the processes. Growth thus uses both sources. As the nonrenewable source runs out, growth declines until it reaches a steady state using just the renewable source. This model is formed by the combination of the models of a nonrenewable energy source (Figure 10.6) and a renewable (constant-flow) source (Figure 10.4).

An example of Model 6 is a population of fish living in a pond to which a batch of fish food is added. The two energy sources are the renewable solar energy coming into the pond from the sun and the nonrenewable energy in the added fish food. The fish population will grow exponentially at first until the fish food gets scarce. Then the population will decline to a level that can be supported by the renewable food chain based on the pond's use of sunlight for photosynthesis.

Another example of a system that grows and levels this way is the economic system created by human societies. Our economic system has been

Figure 10.7 A town that mined coal became a ghost town when the coal deposit was used up.

growing on nonrenewable fossil fuels as well as the renewable sources of sun, rain, and wind. As the nonrenewable fuels are used up, our economic system may have to decrease in quantity and settle into a steady state, living only on the agriculture, forestry, and hydroelectric power supported by

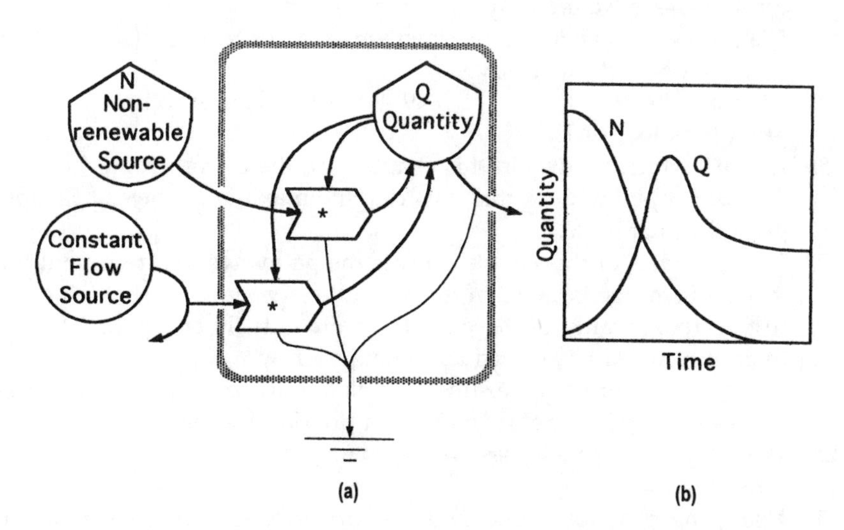

Figure 10.8 Growth of a system with renewable and nonrenewable sources (Model 6). (a) System model; (b) growth curves. The computer program 2SOURCE is in Appendix A, Figure A8.

renewable energies. However, if new sources are found, a different model would be needed.

Florida's economy, like that of the world, may fit the overview of the two-source model (renewable and nonrenewable sources). Later chapters will show the way Florida's economy is largely dependent on nonrenewable fuels. However, there are other models of the future which allow a different point of view.

Questions and activities for chapter ten

1. Define: (a) model, (b) exponential growth, (c) constant-pressure source, (d) acceleration, (e) logistic, (f) self-interactive drain, (g) constant-flow renewable source, (h) dynamic equilibrium, (i) nonrenewable resource, (j) renewable resource.
2. Draw your own model of a living system. Be sure to label all the parts of your model.
3. Describe three examples of living systems that show exponential growth in their early stages.
4. A living system undergoing exponential growth suddenly levels off to a steady state. What are two possible causes?
5. In Florida there are many introduced exotic (non-native) plants and animals. The populations of these organisms are rapidly expanding, similar to the exponential growth curve. (a) Discuss your ideas about why these organisms are so successful in Florida. (b) In the long run, will you expect these organisms to continue exponential growth to reach a steady state? Why?
6. Diagram your own model of growth in a storage tank. Explain whether your model ends at steady state.
7. Diagram the "gold rush" of 1849. What would a graph of your diagram look like? Why?
8. Explain which models in this chapter use their products to increase production. How does this illustrate the maximum power principle given in Chapter 8?
9. In the models of this chapter, locate the pathways that represent the losses of storage because of the second energy law.
10. If the storage tanks of these models were initially empty (quantity = 0), in which models would the quantity grow?
11. Using agriculture as an example, explain how renewable and nonrenewable energy sources interact and provide feedback.
12. Which of the models best represents growth and succession in a forest?
13. Using the programs described in Appendix A, run the simulation programs for the models in this chapter.

chapter eleven

Oscillating systems

In Chapter 10, we considered systems that undergo a period of natural growth, after which they level off to a steady state. Succession of a forest from a bare field to a steady-state climax is simulated by a model of one of these systems (Figure 10.5); however, most systems do not develop steady states for long.

Instead of leveling off, systems considered in this chapter develop repeated *oscillations*. Quantities are always changing up and down. For example, oscillations are observed in lakes in suspended algal populations (called *phytoplankton*). When the phytoplankton cells get abundant, the small herbivore animals dispersed in the water (called *zooplankton*) increase and eat the phytoplankton cells until they are scarce. After this, the zooplankton population has to decrease until the phytoplankton regrow, then the zooplankton population can become abundant again. Thus, the producers and consumers go up and down, each out of phase with the other.

Similar oscillations are observed in herbivore-carnivore and host-parasite relationships. Examples are spruce budworms and spruce trees in Canada, and pine beetles and loblolly pines in Florida. In economics, oscillations occur when merchants build up their stocks of goods and consumers buy these stocks.

The ecosystems of Florida have fairly complicated webs. Prey-predator kinds of oscillations have not been demonstrated as often in Florida as they have been in the Arctic, where the ecosystems have fewer species. In south Florida, the growth of populations of small fishes in wetlands is sometimes followed by a sharp increase in feeding and reproduction of wading birds such as wood storks (Figure 11.1). Another example is the buildup of insects in fields and along roads, followed by a surge of feeding by cattle egrets. Much money is spent by agricultural managers and state agencies to prevent plant insect oscillations from occurring in orange groves, where there is the potential for insects or disease parasites to multiply rapidly.

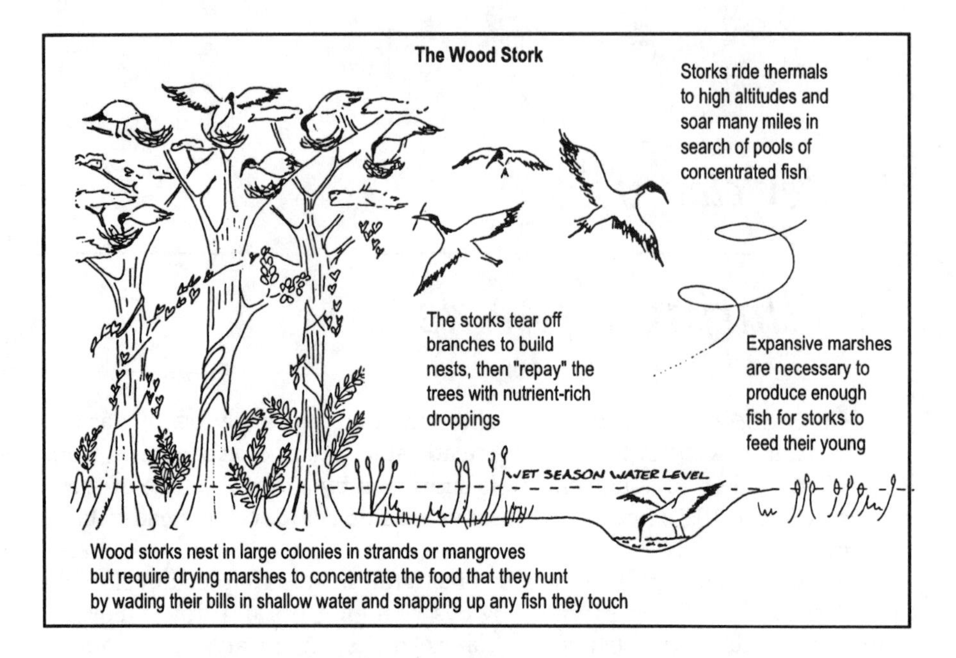

Figure 11.1 As water dries up in the Everglades, fish are crowded into small ponds where wood storks can feed voraciously. (From Wharton, C.H., *Forested Wetlands of Florida — Their Management and Use*, Center for Wetlands, University of Florida, Gainesville, 1976. Illustrated by Joy Bartholomew.)

11.1 An oscillating model

Long used in ecology textbooks to represent oscillating systems is the prey-predator model in Figure 11.2. Pioneer mathematical ecologists suggested this model might account for the observed oscillations of Arctic animals such as the snowshoe hare and its predator, the lynx. The regular oscillations of their populations were recorded by pelt counts made in Canada by the Hudson Bay Company from 1845 to 1935 (Figure 11.3).

As shown in the systems diagram Figure 11.2a, there is a storage of plants available to the prey population. As the prey population starts to grow exponentially, the population of predators grows so fast that the numbers of the prey population are reduced. With less to eat, the predator population declines again. The model produces the alternating pulse of the two populations graphed in Figure 11.2b.

11.2 Two-population plot

Instead of graphing both populations with time as in Figure 11.2b, you can plot a graph with the quantity of one population on the horizontal axis and

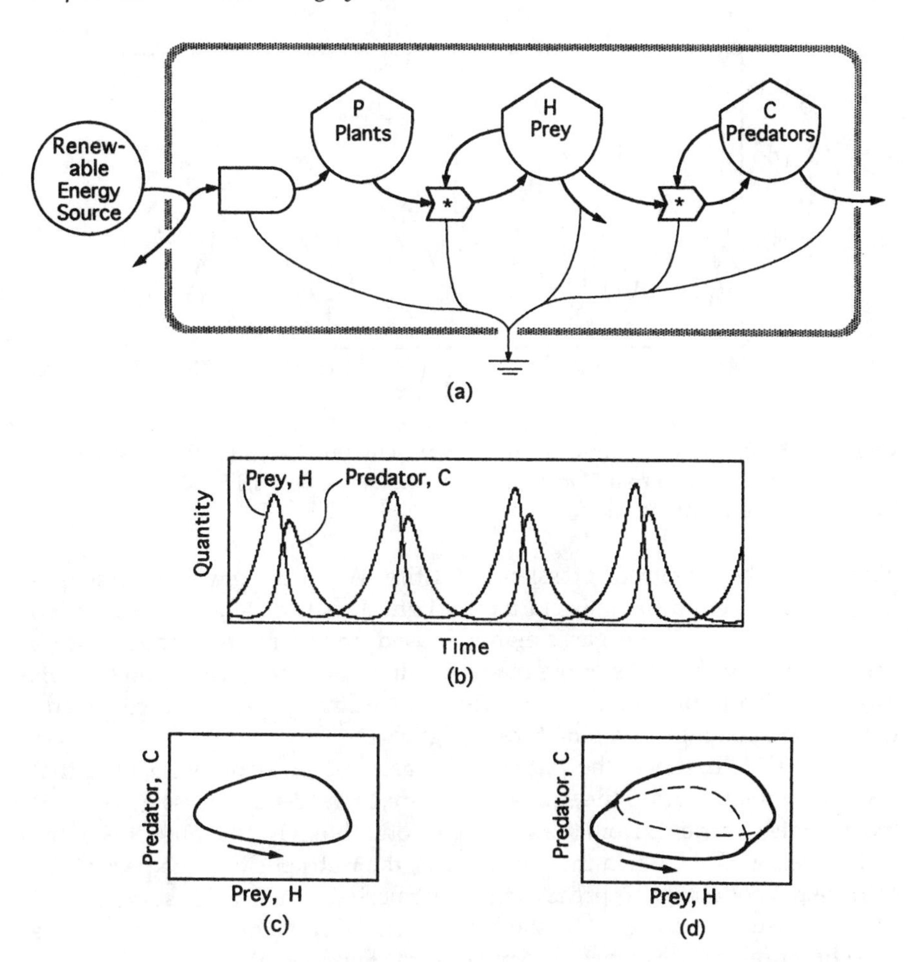

Figure 11.2 Prey-predator oscillation model showing prey (herbivores, H) being consumed by predators (carnivores, C). P is the plant food for the prey. (a) Model; (b) results of simulation; (c) same simulation as (b) on a graph with each axis representing one of the populations; (d) shifting loop when oscillations are chaotic.

the quantity of the other plotted on the vertical axis. The result is Figure 11.2c, where a circular line is traced as the oscillation proceeds. This way of graphing the population shows that the oscillation is repeating. The simulation program for this prey-predator model is given in Figure A9 in Appendix A.

11.3 Switching model

Another kind of oscillating model represents the consumer action with a switching pathway that turns on when the quantity of products reaches a threshold. This model is shown in Figure 11.4a with a switch symbol. An

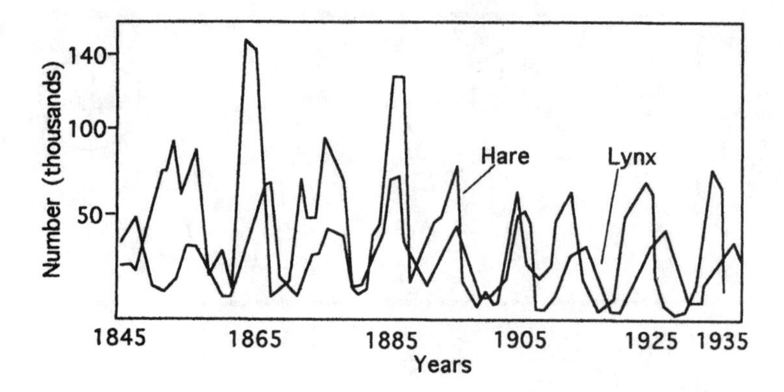

Figure 11.3 Graph of the number of animal pelts purchased by the Hudson Bay Company in Canada. (From MacLulich, D.A., Biological Series, No. 43, University of Toronto Studies, Canada, 1937.)

example is the system of grassland and fire. When the grass biomass gets large enough and a flame is started by lightning or matches, consumption occurs by the fire. The organic matter is used up and many of the nutrients are returned to the soil storage ready to stimulate more plant growth. This model is also appropriate for the repeated burning of ground cover and hardwood seedlings in a pine forest (Figure 4.5).

Figure 11.4b shows the pattern of grass and nutrients with time that results from simulating the model in Figure 11.4a. As the grass grows and incorporates nutrients from the soil as part of its biomass, the nutrients in the soil decrease. Then, when the grass burns, the nutrients become part of the soil again. Production is spread over a long period, whereas consumption by fire is an intense pulse for a short time. The simulation program for the grassfire model is described in Appendix A, Figure A10.

11.4 *Pulse model*

Figure 11.5 is an important model that represents the pulsing of many kinds of systems alternating among the predominance of production, consumption, and recycle. For example, in the PULSE model, grassland plant producers build up a storage of dead grass and organic matter that holds the nutrients. Then, at a certain stage of growth, there is a rapid pulse of consumption by the grass consumers. Nutrients are released to stimulate another cycle of growth. Notice there are two consumer pathways. One of these operates at lower energy and one at higher energy which turns on a frenzy of consumption. The simulation program for this pulsing model is given in Appendix A, Figure A11.

Examples are the plagues of locusts that occur in many countries and sometimes in a few counties of Florida. One theory regarding the Mayan Indian civilizations of Mexico is that they built a civilization in a pulse by fast

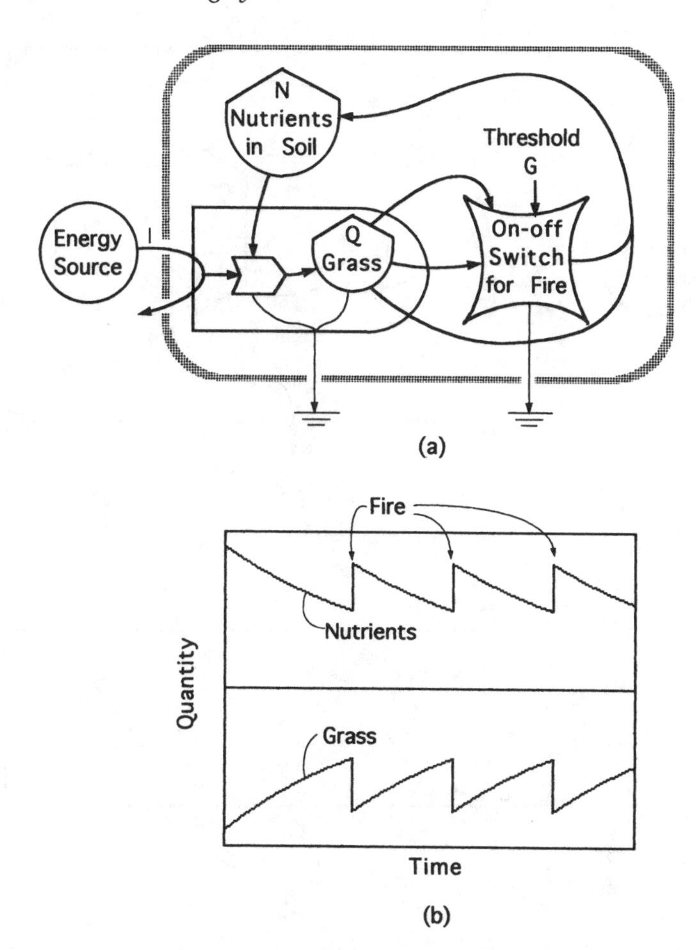

Figure 11.4 Switching model FIRE. (a) Systems diagram; (b) simulation result. I = energy sources (sun, wind, rain); Q = quantity of grass per m^2; G = threshold amount of grass to turn fire on; N = nutrients in soil.

consumption of their natural resources, especially soil. Then they went back into a more dispersed way of life while the soils were rebuilt by the natural ecosystems. After that, there was a later pulse generating more civilization, temples, and culture. A similar model explains the oscillating chemical reactions that generate flashing colors in the chemist's laboratory solutions.

The model can be used to think about our own society. The slow processes of the Earth build up storages of resources such as coal, oil, and phosphate. Our civilization has been in a rapid pulse based on frenzied consumption of these resources, including the mining of phosphate in Florida. Later we may have to change to a lower population and consumption rate while the Earth rebuilds storages for a later pulse of civilization. Models like

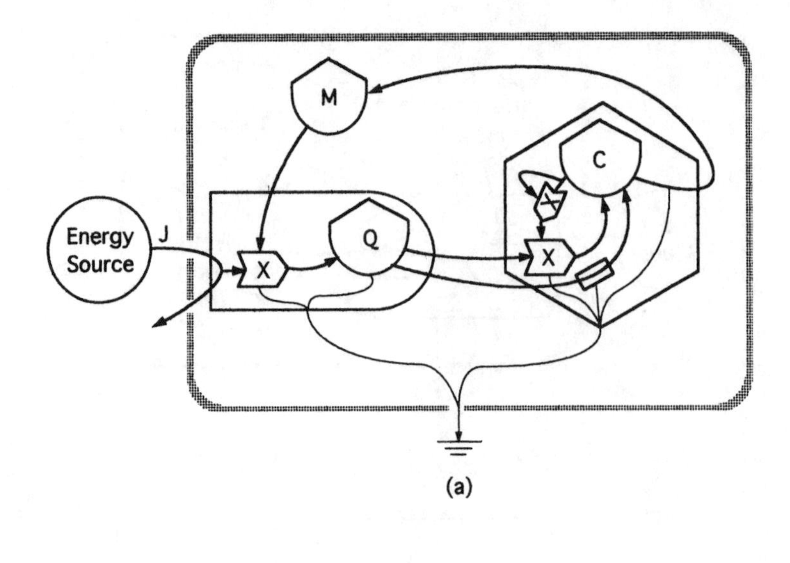

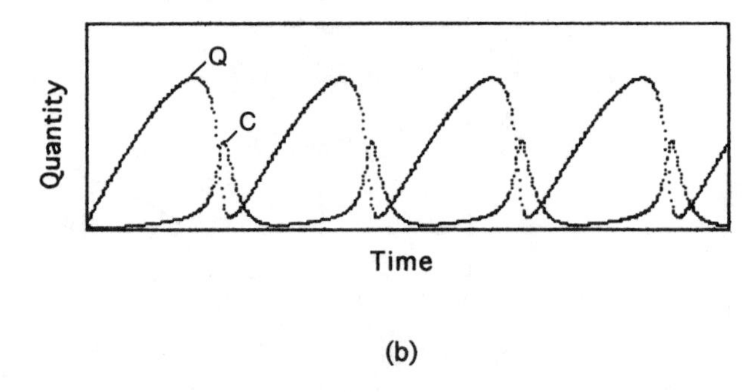

Figure 11.5 Model PULSE. (a) Systems diagram; (b) simulation result. J = energy flow; Q = producers; C = consumers; M = materials.

this are just ideas for considering possible ways that our environment and economy work. In Figure 10.8 and Chapters 36 and 37 are models for considering other ideas about the future.

11.5 Chaotic oscillations

According to the dictionary, the ordinary meaning of the word "chaos" is confusion, an undetermined state; however, the word chaos is now being used in science for oscillations that seem irregular but nonetheless follow the equations of a model. For example, consider the prey-predator model again (Figure 11.2). Especially if there is a high energy input, the storage from production can go very high, which causes the consumer stock to go very high. This pulls down the producer so much that the next growth cycle is small and the consumer growth is small. With few consumers, the next cycle of production can go high again, etc.

A chaotic oscillation is not really in a confused state, but the values jump around so much that it looks irregular. With a simple regular oscillation, the pathway on a two-population graph is a simple loop (Figure 11.2c). With a chaotic oscillation, the loop on the two-population graph shifts its position back and forth within a larger area, as in Figure 11.2d. The data on animal pelts (Figure 11.3) is somewhat chaotic.

Questions and activities for chapter eleven

1. Define: (a) predator, (b) prey, (c) oscillation, (d) phvtoplankton, (e) zooplankton, (f) epidemic, (g) chaos.
2. Diagram a prey-predator system. Identify the prey and the predator.
3. What would happen to your prey-predator system if the energy source increased?
4. What would happen to your prey-predator system if the number of prey in the system were decreased?
5. Diagram an oscillating system that contains a pathway for recycling of materials.
6. Under what conditions is fire most likely to occur in a typical oscillating system? Be sure to mention threshold value and recycling of nutrients in your answer.
7. Compare the concept of chaos described in this chapter with your idea of chaos and with the dictionary definition.
8. Run the computer programs for the models in this chapter (prey-predator, fire, pulse). See Appendix A.

part two

Environmental systems

In Part II are examples of Florida ecosystems and some important issues that affect their management and use. Before ecosystems are discussed, however, to set the stage, Chapters 12 to 14 are devoted to Florida systems of weather, water, geology, and oceanography. Understanding the influences such as weather that organize both terrestrial and marine ecosystems and the large-scale processes that shape the lands and nearshore waters will lead to a better understanding of why different ecosystems are where they are and how they process energy and materials. Chapters 15 to 19 consider the main kinds of ecosystems, and Chapter 20 provides a summary by showing the locations of the various ecosystems in Florida.

chapter twelve

Weather and climate

The Earth is covered with air, a gaseous mixture of about 21% oxygen gas (O_2), 78% nitrogen gas (N_2), and a few other gases in small quantities, including argon, carbon dioxide, and up to 1% water vapor. The weather systems of the world are composed of masses of circulating air that carry heat and water vapor and use them to make wind.

Weather starts with the energy of the sunlight, which heats the Earth and drives the process of photosynthesis. Photosynthetic production by the plants makes all the organic matter of the Earth on which life depends (Chapter 6). Differences in the amount of the sun's energy received makes the tropics warm and the polar regions cold. These differences in temperature operate the global weather system of winds and rains.

In principle, the atmosphere and ocean are a heat engine like those used in industry, using temperature differences as their fuel. In a steam engine, we heat a fluid which circulates and causes wheels to move. Since global air circulation is driven by the sun's heating, the atmospheric weather system is a kind of heat engine. Air circulation develops when warm fluids expand and rise while cold fluids contract and sink. For example, along the sea coast heated air rising over the land draws a sea breeze in from the cooler waters which at night reverses.

On a larger scale, where there is enough energy (in temperature differences), storms develop. Storms are spinning air masses, where air currents converging at the bottom are carried upward and out the top, causing strong winds, heavy clouds, rains, snow, and other weather phenomena. The sun's heating evaporates water from the ocean which storms release as rain. There are several kinds of storms. Thunderstorms are 0.1 to 30 miles (16 to 48 km) across. Tropical storms and hurricanes may be 30 to 500 miles (48 to 800 km) across. Winter storms may be 1000 miles (1600 km) wide. The weather in Florida is part of the general circulation over the whole Earth which brings several kinds of storms to Florida.

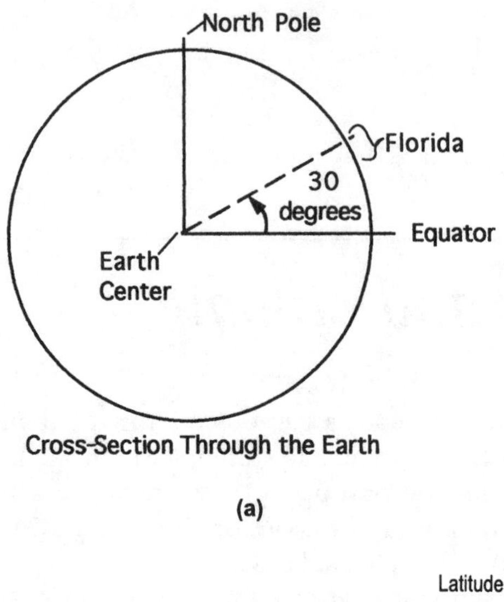

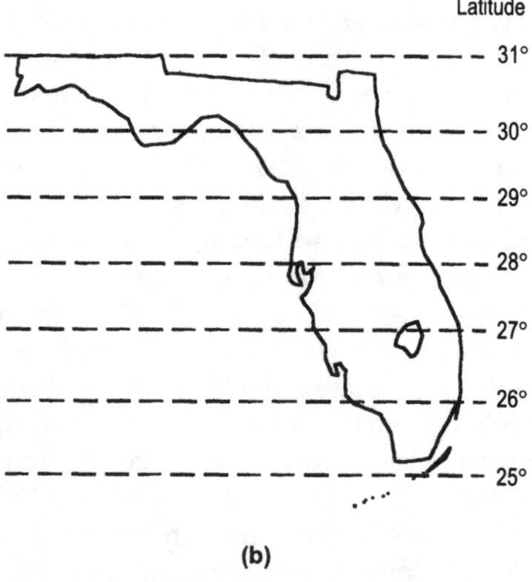

Figure 12.1 Latitude in Florida. (a) Cross-section of the Earth showing 30 degrees north latitude for north Florida. Latitudes are given in degrees according to their angle in an Earth cross-section between the line from the center of the Earth to the equator and the line to the latitude; (b) north latitude lines crossing Florida.

To understand how global weather affects Florida weather, notice where Florida is on the globe (Figure 12.1a). The position between the equator and the poles is indicated by the *latitude*. Latitude lines are parallel to the equator, about 70 miles apart, and go from 0 degrees at the equator to 90 degrees at the poles. Latitude lines crossing Florida range from 25 degrees in the Keys to 31 degrees at the northern border (Figure 12.1b).

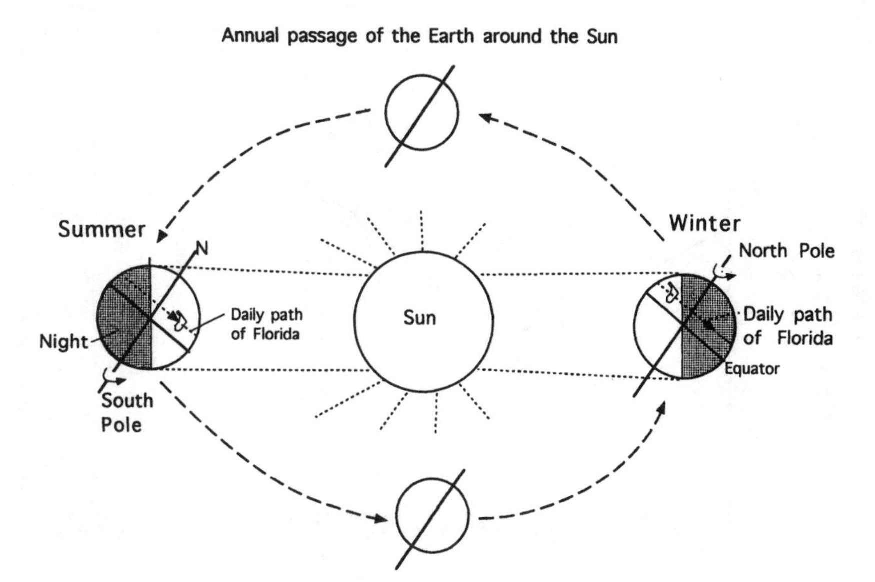

Figure 12.2 The effect of Earth's motions on the sunlight reaching the Earth's surface.

12.1 Energy from sunlight

The quantity of sunlight that falls on an area varies each day because the Earth rotates. Florida faces the sun in daytime and is in the Earth's shadow at night. The sunlight reaching the Earth varies with season as the Earth goes around the sun because of the angle of the Earth's axis of rotation (Figure 12.2). The Earth keeps its same axis of rotation as it completes one circle per year. In summer, the tilt of the Earth faces the sun so that in Florida the days are long and the sun is overhead, sending more energy per area to the ground. The north pole is continuously lighted (left view of Earth in Figure 12.2).

Then, in winter, with the tilt of the Earth's axis facing away from the sun, Florida has short days and the sun is at a low angle. The energy of the sun received per area per day is less. The north pole is in continuous darkness. In summer, when more solar energy comes to the northern hemisphere, heavy clouds in Florida reflect more of the summer sunlight back out to space. As a result, at the ground, Florida has a rather even quantity of sunlight through the year (Figure 12.3).

In sunlight are streams of energy units called *photons* (also called energy waves) of different length. Some of the shortwave blue light in sunlight is scattered about by the air, making the sky blue. About half of the solar energy is in longer wavelength infrared sunlight, invisible to the eye and not used in photosynthesis. But this radiation heats the land and waters when it is absorbed. The light with shorter wavelengths, which is visible to the human

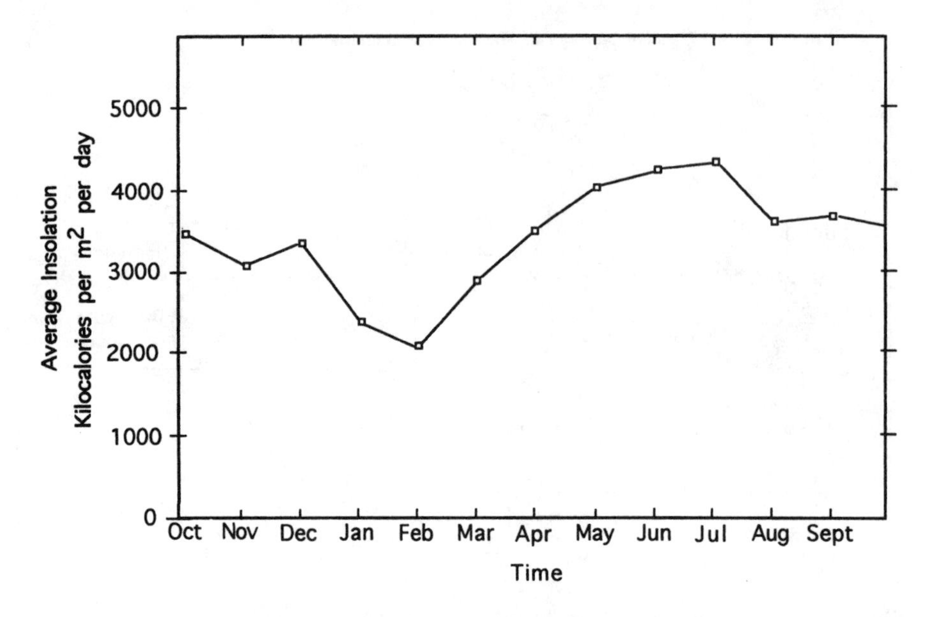

Figure 12.3 Annual record of sunlight energy received in Gainesville, Florida, 1974 to 1975. (From the Agronomy Department, University of Florida, in cooperation with the National Oceanographic and Atmospheric Administration.)

eye, is used by green plants in photosynthesis. In a rainbow, we see the light of different wavelengths separated; the longer ones are red and the shorter ones are violet. Most of the visible photons, as well as the invisible ones, are eventually transformed into heat as they are absorbed at the Earth's surface.

Heating causes molecules to move faster. Since hot air molecules move faster than cold ones, hot air molecules push each other apart and take up more space than cold air molecules. A volume of hot air weighs less than the same volume of cold air. A thermometer measures an air temperature increase because hot air heats the fluid molecules in the thermometer, causing them to vibrate faster and take more space, pushing the fluid up its tube and indicating a higher temperature on the thermometer scale

Some of the heat energy (molecular vibrations) in everything is continuously turning into invisible photons and radiating away in all directions. At the top of the atmosphere, some heat in the air goes out to space in this radiation. Thus, the sunlight comes to Earth in solar photons of short and medium wavelength, heats the Earth, and then goes back out to space as longer wave photons.

12.2 Global atmospheric circulation

Because sunlight heats tropical areas more than polar areas, the tropical air expands. As a result, tropical air rises. Some of the air spreads out and piles

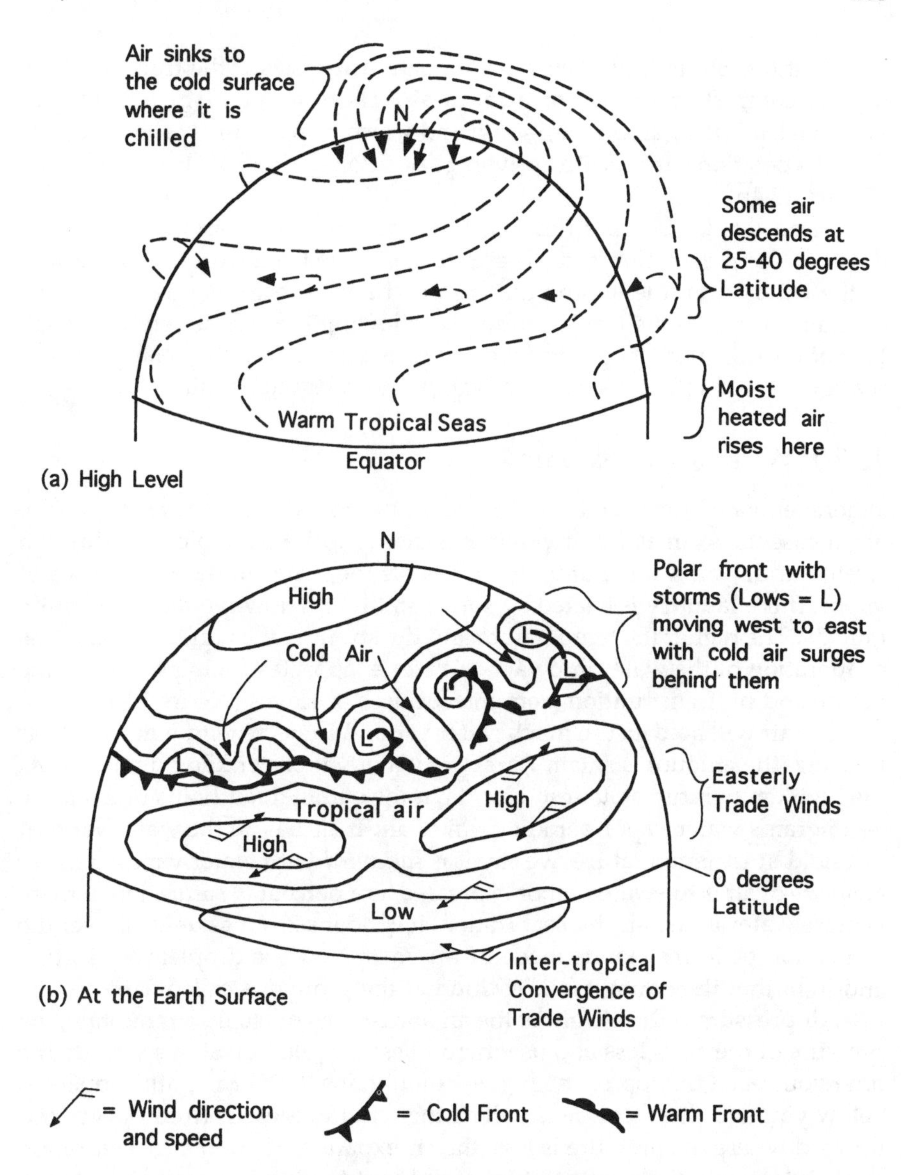

Figure 12.4 General circulation of air due to tropical heating from the sun. (a) High-level flow of air rising from the tropics and circulating to the poles; (b) weather systems on the Earth's surface showing the polar front separating cold and warm air masses. Lines are isobars (lines of equal pressure). "High" is a center of high pressure.

up, causing some of it to descend into the zone between 25 and 40 degrees latitude because of the weight above it (Figure 12.4a). The rest of the tropical air circulates towards the poles in the upper air. At the poles, heat energy goes out to space by the process of radiation, and the polar air cools and sinks. At the ground, the dense polar air returns toward the tropics, where it is in contrast to the warm tropical air. The semi-permanent boundary between cold polar air moving south and warm tropical air is called the *polar front*. Its position varies, with an average position at about 45 degrees latitude (Figure 12.4b).

Because of the sharp difference in temperature of the air masses, storms develop along the polar front. These storms move from west to east because of the winds from the west in the upper air above them (Figure 12.4a). The cold air returns to the tropics in bursts following the storms (Figure 12.4b). Part of the energy that operates these storms comes from the heated tropical sea transferred to storms as water vapor and released as rain.

12.2.1 Water vapor and rain

Wherever water is exposed to the air, some of the water molecules go into the air in gaseous form as water vapor. For example, the wind blowing through fresh laundry on a line outdoors carries off the water in the form of water vapor. If the laundry is heated by sunlight, the water evaporates even more quickly. The higher the temperature and the stronger the wind, the faster the evaporation of the water into vapor. Water evaporates from lakes, from the ocean, and by transpiration from the surfaces of leaves (Figure 5.2).

The air will hold just so much water vapor. When the temperature is near freezing, the amount is small, 1 gram of water vapor per kilogram of air. At the high temperatures of air in Florida in the summer, air holds much more — 25 grams water vapor per kilogram of air. If air has all the water vapor it can hold at its temperature, we say it is *saturated* with water vapor. *Relative humidity* of air is the water vapor expressed as a percent of saturation. If there is more water in the air than saturation, it condenses out as dew all over the vegetation or forms clouds made of tiny droplets. The droplets are so tiny and light that they do not fall. A cloud at the ground is called *fog*.

Air pressure is the weight of the air above. As you go up a mountain, the pressure of the air is less because there is less air piled up above you. By the time you reach the top of the highest mountains, half of the atmosphere is below you and the pressure is half. If the wind carries air from the ground upward where the pressure is less, the air expands. There the concentration of air is less, and the concentration of the heat (molecular motion) in the air is less. In other words, the temperature is less. Maybe you have already experienced how driving up a mountain brings you into cooler air. If air that is near saturation is carried upward so that it expands and is cooler, some of the water vapor condenses out as clouds. Wherever there are up-currents of air, clouds tend to form.

Because they have electric charges on them, cloud droplets do not easily combine into larger droplets, since objects with the same charge repel each other. Some special processes are required to make cloud droplets combine into drops large and heavy enough to fall as rain. Such conditions often occur in storms. For example, air rising 3 or 4 miles (4.8 to 6.4 km) reaches freezing temperatures and droplets begin to freeze (snowflakes). Because the vapor release tendency of a solid is less than a liquid, the ice crystals grow, with a net exchange of vapor from water droplets to the crystals until they are large enough to fall. Before they get to the ground the crystals melt into raindrops. Strong turbulence in storms also converts tiny cloud droplets to larger rain drops. The pattern of rains is controlled by the pressure and winds in storms. The cycle of water is discussed in Chapter 13 (Figure 13.1).

12.2.2 Air pressure and winds

An area of low pressure exists at the ground when the column of air above is less dense or not as tall as that over the surrounding areas. If lines of equal pressure (called *isobars*) are drawn, they will make a ring around the low-pressure area at ground level (Figures 12.4b and 12.5a). The low-pressure center is called a *low*. The higher pressure of the surroundings starts to push air towards the lower pressure area. Air starts to flow from high to low pressure across the isobars. However, except at the equator, the landscape is rotating out from under the airflow to the left (in the northern hemisphere). Consequently, the air flow seems to be pulled to the right as if there were a force on it. From the point of view of a person on the rotating Earth there is a force to the right of the motion, and it is named the *Coriolis force* (see Chapter 14). At the ground there is also the force of friction against the flow. The result of the three forces (pressure force, Coriolis force, and friction) is air flow along the isobars but slightly across them, towards low pressure. Notice the wind arrows in Figure 12.4b.

Wind arrows point in the direction of the wind, and the velocity of the wind is indicated by the number of lines on the shaft. The custom in describing the direction of wind is to indicate the direction it is coming from — north winds are from the north. (The custom in describing water currents is the reverse, using terms referring to the direction in which it is going.)

The winds converging into a low cause air to rise, expand, and form clouds and rain. A storm begins to form. When water vapor condenses and falls out as rain, the energy in molecular motions that was previously used to make the water into gaseous form (water vapor) is released as heat. Rain formation heats the upper air where it forms, and that air expands upward, causing air to spread out at the top of the storm. Then there is less weight of air above the ground, and the air pressure at the ground becomes less. With lower pressure at the center, pressure differences increase and stronger winds develop. Storms are strong winds circulating around centers of low pressure. A line of storms centers (lows) is shown in Figure 12.4b.

The air that goes up in low-pressure areas spreads out on top of the surrounding air, giving those areas high pressure at the ground (highs) in Figure 12.4b. Here the air sinks. Wherever air sinks downward to lower levels where the pressure is higher, its molecular motions become more

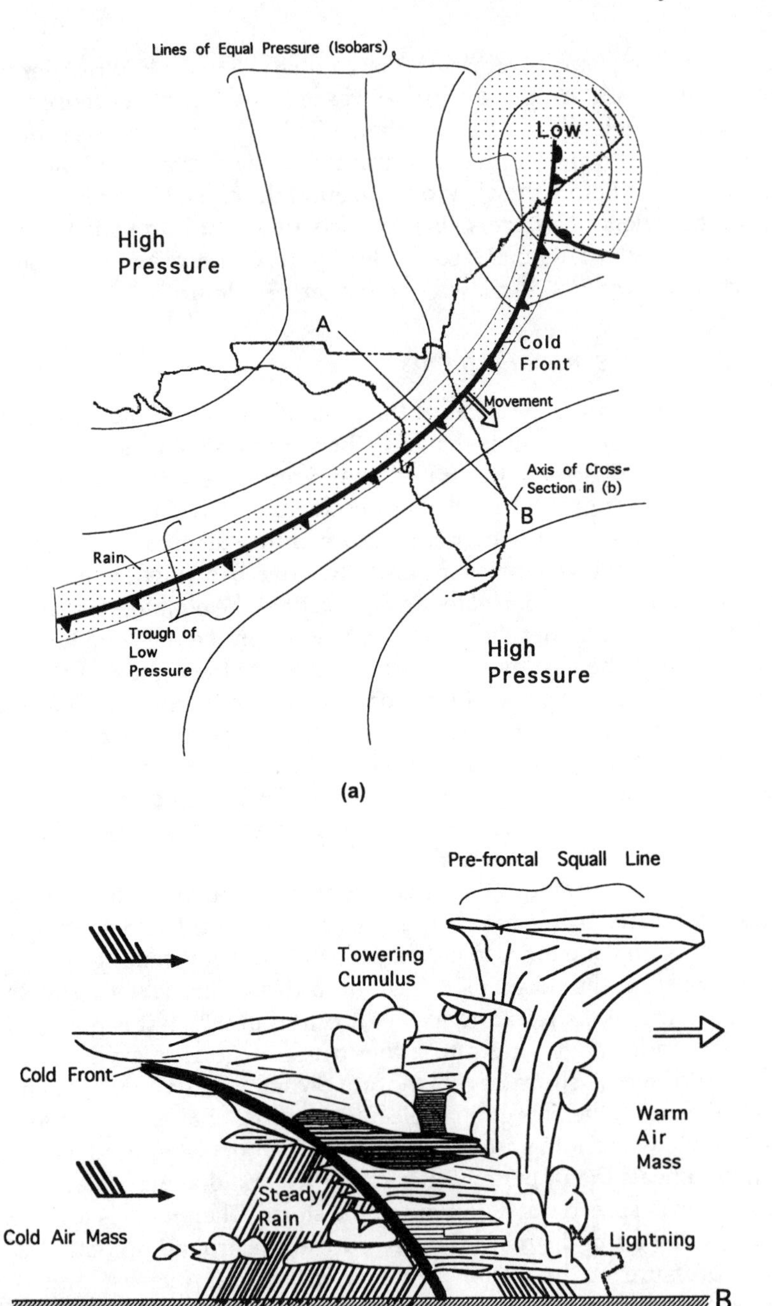

Figure 12.5 Storm low and cold front over north Florida in winter. (a) Map view; (b) side view of a cross-section through the frontal zone from A to B.

concentrated, its temperature rises, and clouds that are in the air evaporate into water vapor. The sky clears, and the air becomes less saturated, less humid. Air coming down tends to dry out vegetation, sometimes causing deserts.

In a global view, the air that rises at the equator and piles up in the upper air at 25 to 40 degrees latitude produces large high-pressure centers at the surface of the subtropical oceans (Figure 12.4b). Over land, the sinking, clear air at these latitudes has created the great subtropical deserts such as the Sahara desert in Africa and the desert in the center of Australia.

12.2.3 Fronts in storms

Most storms involve the interaction of converging streams of cold and warm air. The boundary between the cold air and the warm air is called a *front*. This boundary slopes like a slanting wall with warm air over the cold air (Figure 12.5). A *cold front* is a front where the cold air is pushing the warm air back. On a map, a cold front is marked in blue or with a series of points in the direction of movement (Figures 12.4b and 12.5a). As the cold air cuts under the warm air mass, there are often lines of thunderstorms (squall lines) and sometimes rain falling into the cold air (Figure 12.5b).

When the warm air pushes the cold air back, the boundary is called a *warm front*. It is marked on maps in red or with rounded points in the direction of movement (Figure 12.4b). The warm air tends to slide up over the cold air with rain, sleet, or snow falling into the cold air. Usually there are a cold front and a warm front, each with one end in the low-pressure center.

12.2.4 Florida's winter weather

The most common weather regime in winter in Florida has rains from passing cold fronts, which are a part of the polar front. Often low-pressure centers pass from west to east just north of Florida, with a trailing cold front that moves south across Florida (Figure 12.5a). As the cold air mass moves southward into Florida, its front edge tends to go under the warmer air that is already there, forcing it up and causing rain to fall from the warm air. A cold front moving south over Florida is like a slanting roof sliding southward. Figure 12.5a shows the ground position of the cold front marked by a line with points. In Figure 12.5b is a cross-sectional view along a north-south line showing the position of the cold air as it moves south and the warm-moist air above the front with its many clouds from which rain is falling.

To the east of one of these lows there are winds from the south bringing warm moist air northward. As these storm lows pass, the northerly winds on the back side of the storm (winds from the north) bring down the bursts of cold air.

Sometimes when the air ahead is dry, the front passes rapidly with little weather. At other times, it causes squall lines. Sometimes it moves very

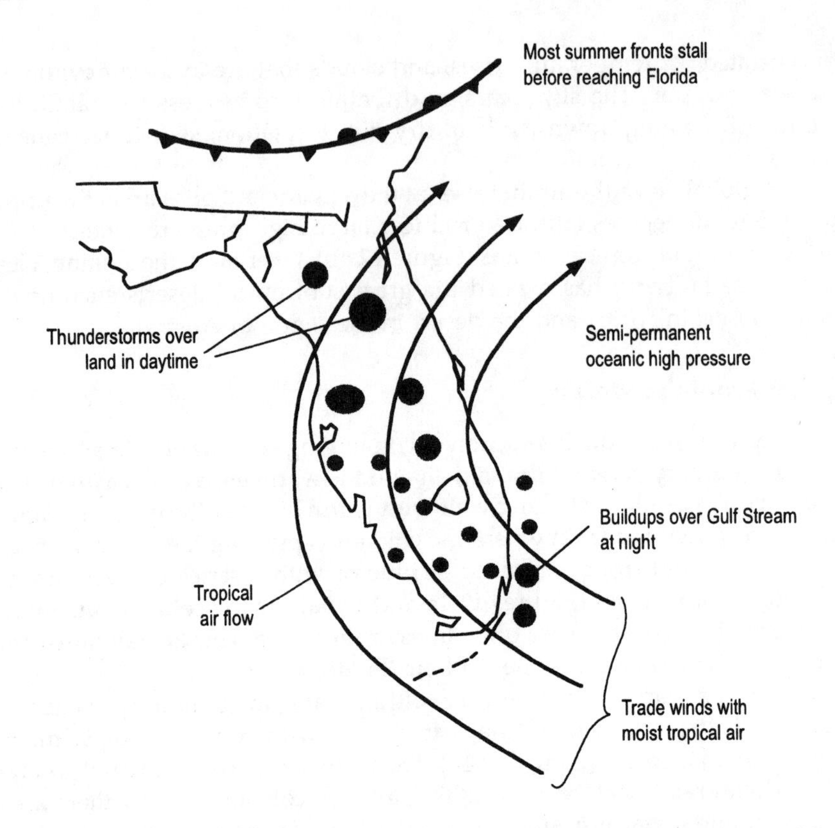

Figure 12.6 Peninsular Florida, with flow of moist tropical air in summer. Notice wind following the pressure lines (isobars).

slowly with warm tropical air sliding over the cold front with heavy, steady rains behind it. A cross-section from A to B in Figure 12.5a through a cold front shows the slope of the front over the cold air, some over-running clouds, and thunderstorms (Figure 12.5b).

12.2.5 Florida's summer weather

In the summer, the polar front zone is usually farther north, and we get the weather that goes with tropical air masses flowing in from tropical seas (Figure 12.6). This flow is part of the giant high pressure that is centered over the Atlantic Ocean at 35 degrees north latitude. Air flowing northwestward over Miami circulates back eastward to leave the state with a flow from the southwest at Jacksonville. The tropical air mass in this flow is full of water vapor and has the energy to form thunderstorms whenever a little more heating is added as it flows over sun-heated ground. Sometimes clusters of thunderstorms are formed. The highest number of thunderstorms in a year in the U.S. form around Tampa Bay. Within the air flow that comes in from the Caribbean Sea are zones of low pressure called *easterly waves* that have rain and thunderstorms that move across the state from east to west.

In May and rarely in summer, a cold front gets as far south as Florida, bringing dry clear air. It is initially cool, but after a day or two of solar energy and a clear sky, the temperature may get up to 100° Fahrenheit (38° Celsius).

12.3 Severe storms

Strong winds that can cause damage are usually found in severe storms.

12.3.1 Thunderstorms

When an air mass is warm at the bottom and cool at the top, it is unstable and may develop vertical convection (up and down circulation). Up currents are marked by towering cumulus clouds. If there is enough energy, the up currents go up 4 to 6 miles (6.4 to 9.6 km), spreading out at the top with a cloud that looks like the blacksmith's anvil of days past. When currents in a storm are large enough to develop lightning, the storm is called a thunderstorm.

You have probably experienced static electricity when you walked over a plastic rug. The friction with the plastic collects electric charges. Electric charges are either plus or minus. Plus attracts minus. If the charges are large enough, a spark (electric charge) jumps between two objects of different charge. After walking on the carpet, a spark may jump from your hand to the door knob. Lightning that forms in thunderstorms is like that. The strong rising currents of air collect and carry plus charges up, leaving the lower clouds with a minus charge. When the difference is great enough, the charges are equalized by a lightning strike. When the charge in the clouds is very different from the charge at the ground, a lightning charge jumps from the cloud to the ground, sometimes killing people. The echoes of the lightning noise bouncing off clouds is thunder.

Sound travels through air at about 1/5 of a mile per second (0.32 km per second), and light travels almost instantaneously (186,000 miles per second). The time between when you see lightning and hear its thunder indicates its distance. If there are 5 seconds between lightning and thunder, it is about 1 mile (1.6 km) away.

Most of the thunderstorms that form within summer tropical air flows in Florida are large and full of clouds but are of only moderate intensity. You can tell the intensity of a storm by the frequency of the lightning and the thunder it causes. In a weak storm, there are several minutes between lightning strikes.

12.3.2 Severe thunderstorms and tornadoes

Lines of stronger thunderstorms can form ahead of fronts when fast upper air winds blow cold air out ahead of the front. Note the faster upper air wind in

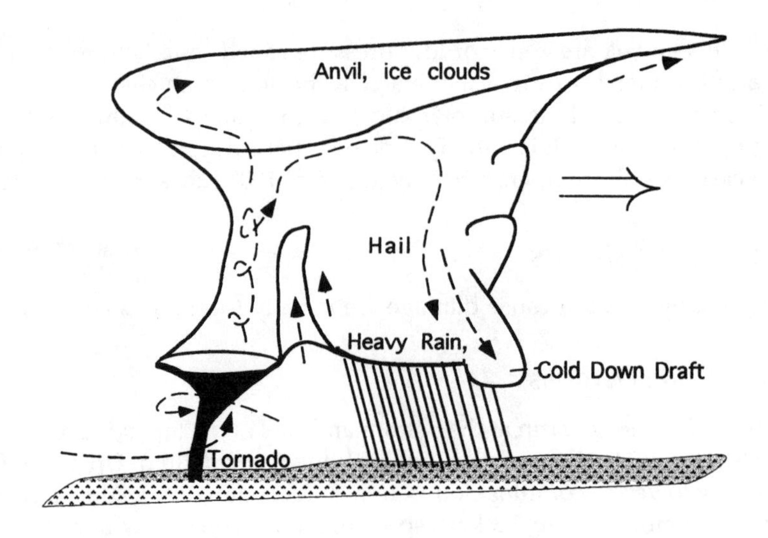

Figure 12.7 View of a strong thunderstorm with a tornado.

Figure 12.5. The cold air is heavier than the warm air below, causing a rapid turnover of air layers. Thunderstorms in dry air develop sharp down currents as well as up currents because the falling rain evaporates into and cools the dry air forming down drafts at the front of the storm. The strong up currents of severe storms carry rain into the freezing zone. Drops freeze and grow in size to form hail, sometimes golf ball size or larger.

The lightning activity of severe storms can be almost continuous. Tornadoes often move from southwest to northeast and may destroy houses, toss cars around, and leave a swath of overturned trees. Figure 12.7 shows a strong thunderstorm with a tornado. Very severe thunderstorms that form in air that already has a spin may form tornadoes. For example, tornadoes form on the edge of hurricanes.

12.3.3 *Tropical storms and hurricanes*

As the tropical sea heats up in summer, there is enough energy for tropical storms to form, sometimes in the easterly waves (low-pressure areas moving east to west). A tropical storm is a low pressure in a tropical air mass around which air circulates and into which air is converging. Some parts of tropical storms have almost continuous rain. As long as they are over the warm sea and upper level air is carrying upcurrents away, conditions may favor intensification. When the spiraling winds reach 75 miles per hour (120 km per hour) or more, they are called *hurricanes (typhoons* in the Pacific Ocean; Figure 12.8). Hurricanes have destructive winds that blow over trees, billboards, and housing. Hurricanes are given a rating from 1 to 5 according to their wind velocity and energy. Wind gusts of the strongest hurricanes

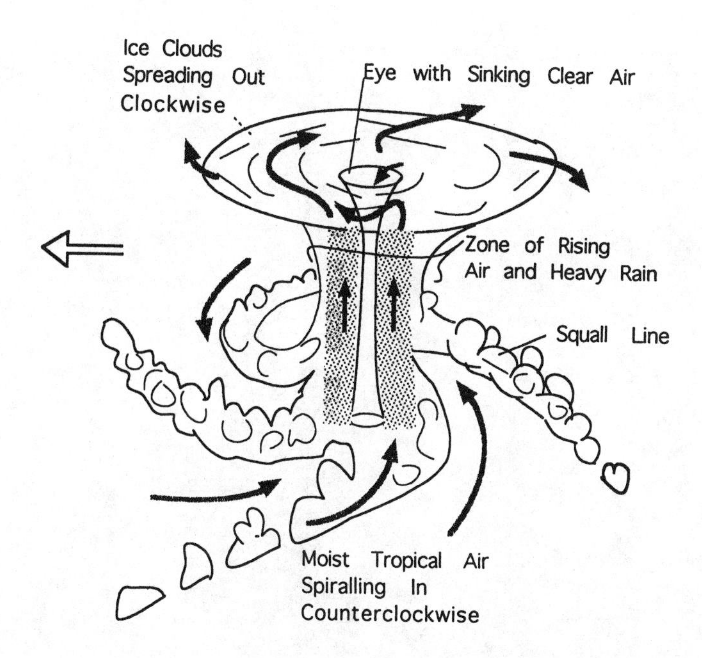

Figure 12.8 View of a tropical hurricane showing clouds and wind circulation.

may reach 230 miles per hour. Hurricane Andrew, rated a #4 type, in 1992 destroyed Homestead and Florida City (Figure 12.9).

Terrible devastation has been produced by past hurricanes in Florida when winds have piled up ocean waters 4 to 18 feet (1.2 to 5.4 m) above normal sea level, and waves to the right of the center sent these seas ashore as a surge with giant breaking waves. Hurricanes over Lake Okeechobee, in the days before it had a levee, sent waves over the shore that drowned many people.

Much of Florida's development along the coasts (and especially the Keys), which are exposed to hurricanes, has occurred since the last large hurricane hit that area. Many people in the government and the weather service are concerned that people may not be evacuated from low-level areas at the coast in time to prevent loss of life, due to highway bottlenecks. The water ahead of an approaching hurricane rises continuously a day or two before winds increase. Often bridges are blocked by the rising tides before people can leave the offshore islands.

Hurricanes tend to move along with the general flow of tropical air, but they may get in a situation where they are not in a regular air flow and stall. One hurricane at Cedar Keys stayed on top of the town for 7 days. Once a hurricane has been over land for 6 hours without the energy it gets from warm sea, the friction of the land causes the storm to lose its hurricane velocity winds, but the storm continues with heavy rains and lesser, gusty winds.

Figure 12.9 Homestead, Florida, after Hurricane Andrew.

12.4 Climate and biomes

As we have seen, local weather is made up of the passage of various storms, fronts, thunderstorms, hurricanes, etc. Between these storms there is usually a quiet air mass with high pressure, descending air, and few clouds. The climate of a place is found by summarizing averages (temperature, humidity, wind velocity, sunlight, frequency of storms, etc.). Many interesting maps of climate are found in the *Florida Atlas* (Fernald and Purdum, 1992) — for example, maps of annual average temperature show 65°F along the northern border of the state, increasing to 80°F degrees in south Florida, away from most cold front outbursts and warmed by the tropical ocean and Gulf Stream.

12.4.1 Microclimate

Sometimes what is important is not the characteristics of the climate as measured by a weather station in the open air of an airport, but rather what happens down within the vegetation or inside a city. The climate of small isolated places is called a *microclimate*. For example, on clear calm nights the outgoing radiation from ground to space cools a thin layer of air on the ground many degrees, sometimes freezing crops. On these cold nights, temperature is lower on the ground than in the air above. With this unusual condition, called an *inversion*, ground air does not mix with the air above. If toxic substances are released into the ground air while there is an inversion, it is trapped, accumulates, and can be dangerous for people.

Freezing temperatures from very strong cold fronts often kill orange trees and other crops in middle peninsular Florida, but not every winter. Locally, the presence of lakes and wetlands helps prevent local freezing, because the temperature of air over water cannot get below freezing until that water has all frozen, because water going from a liquid to a solid state (ice) releases heat. Growers spray water over orange trees and their soils to keep the microclimate conditions at freezing or above. Because of the widespread draining of wetlands, some of the local protection against freezing in Florida has been lost. In recent years, winter freezes have occurred farther south. The frost frequency is part of Florida's microclimate.

12.4.2 Climatic zones of the Earth

In a global view, we can summarize the climatic zones in terms of the pressure and winds responsible for them. Zones of high pressure and clear skies occur at the poles and the subtropics. Rainy belts are found at the equator and the mid-latitudes. The winds south of the subtropical highs are steady from the east and are called *trade winds*. The winds in the zone of polar front storms are variable but average from the west. The winds on the edge of the polar high are polar easterlies.

12.4.3 Land biomes

The main types of ecosystems are called *biomes*. Similar biomes develop in similar climates. The main land biomes for a generalized continent mapped in Figure 12.10 correspond to the zones of wind and weather in Figure 12.4 according to their north latitude and relation to the sea. In the southern hemisphere, climate and biomes are similar for corresponding south latitudes.

In the high-pressure zones at 25 to 40 degrees latitude, air descends (Figure 12.4a) and becomes compressed, heated, and clear of clouds. A dry climate, grasslands, and deserts are found; however, in areas such as Florida that are next to tropical seas, air flows bring summer rains, and a subtropical evergreen forest results.

In the low-pressure zones at the equator and in the storm belt at mid-latitudes (Figure 12.4b), air is rising, cooling, and condensing water to form clouds, storms, and rain or snow. Wet ecosystems are found in these low-pressure zones. The rainforest forms in the equatorial low-pressure zone where the trade winds converge from both hemispheres. Seasonal forest (tropical deciduous) and the grassy savannah (grassy with sparse trees) are between the rainforest and the desert zone.

In the zone where low-pressure storms pass, deciduous forests are found in the east and coniferous forests farther north. Where cool moist westerly

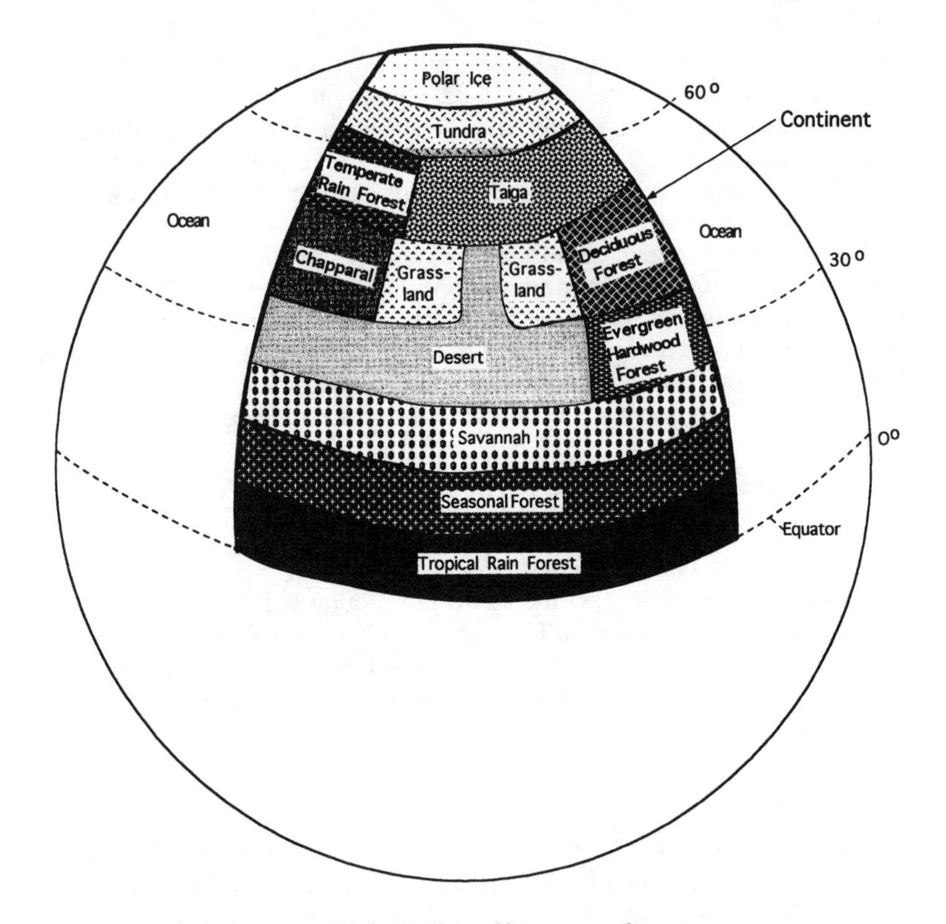

Figure 12.10 Relationship of biomes to climatic zones.

winds first come off the ocean on the west side of a continent, a temperate rainforest with conifers develops (45 to 55 degrees latitude). Farther inland, the trees are smaller (taiga). On the west coast of continents, where storms and rain come mostly in winter at 25 to 35 degrees latitude (for example, California), a Mediterranean type of climate occurs with vegetation marked as chaparral (a scrubby woodand) in Figure 12.10.

Florida occupies the zone of climate that supports deciduous forest in northern counties and subtropical evergreen forest elsewhere. Once every 3 to 10 years, the global weather pattern shifts to a regime called "El Niño", named for the arrival of hot ocean water on the coast of Peru at Christmas. In Florida, the north-south circulation is stronger and westerly winds are farther south, bringing more cold front rains that inhibit hurricanes. Land ecosystems of Florida and the effects of water and geologic substrate are considered in more detail in Chapters 18 to 20.

Questions and activities for chapter twelve

1. Define: (a) weather, (b) atmosphere, (c) microclimate, (d) water vapor, (e) tornado, (f) isobar, (g) hurricane.
2. Explain how Florida's weather is caused by the global weather.
3. Explain why air going up a mountain forms clouds and air coming down a mountain is dry.
4. Describe the buildup of a storm.
5. Describe how Florida's summer weather differs from winter.
6. Explain how clouds form.
7. Why does spraying water on plants and their soils during a freeze protect the plants?
8. What is the property of water that makes the land hotter than a pond or seawater in the daytime and cooler at night? Explain why orange trees near water are more likely to survive in a freeze.
9. With your back to the wind, in which direction is low pressure? Explain.
10. The winds of a hurricane over the sea drive currents in the waters below. Considering the Coriolis force, what happens to the water level under a hurricane in the middle of the Gulf of Mexico? In coastal areas around the Gulf? Why?

chapter thirteen

Geology and hydrology

When compared to the time scale of human life, land seems like a permanent base composed of earth with unchanging rocks. In fact, however, land is part of a cycle of land materials that rise from the sea; are eroded into sediments, wind-blown, or dissolved; and are washed to the sea shore again. Part of the work is done by the cycle of water. With water vapor from the sea, the restless atmosphere brings winds and rains to the land, followed by calmer weather before the next storm (Chapter 12). The ecosystems and the human economy use the rain water. Some evaporates and some drains back to the sea.

Although earth and water are sometimes shown as separate cycles, they are really coupled, a single process. For the global cycle, part of the energy comes from the sun's heating of the sea and land, and part comes from the heat of natural radioactivity deep in the Earth. Figure 13.1 shows the cycle of earth and water together.

In some places on Earth, the land comes up with violent thrusts and volcanic eruptions that build tall mountains of hard crystalline rock. The rivers there are heavily laden with sediments from rock decomposition and tumbling rocks returning to the sea. In dry areas, the rocks are worn down by the winds and blowing sands.

In Florida, however, the land cycle is very slow, the highest land being only about 300 feet (90 m). The land coming up from below is made of loose sands, shells, and soft rocks. Rain percolating through this layer dissolves out chemical substances. Much of the land materials returning to the sea are dissolved in sluggish surface streams and water flowing slowly below the surface, the groundwater. Only when stream currents are strong, as in floods, do they erode and carry suspended sands and other sediments to the sea, making the waters turbid until the sediments settle.

13.1 Soil formation and weathering

Uplands are covered with vegetation — some natural, some cultivated farms, and some yard plantings. The geological materials on top become a part of

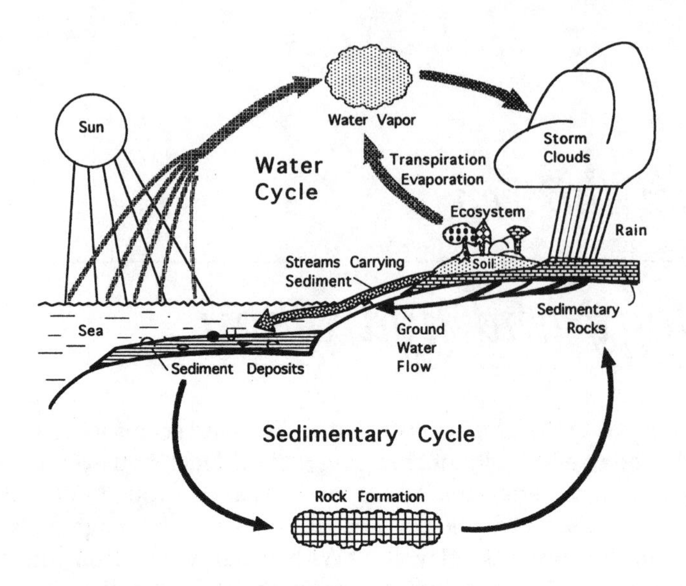

Figure 13.1 Cycles of earth and water.

these ecosystems. The life processes, including the roots, insects, earthworms, and microorganisms, interact with the geological materials to form soil with a loose texture. The respiration of the organisms releases carbon dioxide (Chapter 3), which makes the upper soil acidic and helps dissolve cementing substances in a process called weathering. More discussion of the soil part of ecosystems is found in Chapters 19 and 20.

13.2 Chemical processes

In the main process of solution of limestone, the most common Florida rock, rains percolating through upland soils and wetlands pick up carbon dioxide. Carbon dioxide (CO_2) plus water (H_2O) become carbonic acid (H_2CO_3), a weak acid that dissolves limestone, making bicarbonate ions (HCO_3^-) and calcium ions (Ca^{++}) in solution (an ion is an atom of an element that carries an electric charge). Limestone is calcium carbonate ($CaCO_3$). The reactions are

$$CO_2 + H_2O \rightarrow H_2CO_3$$

$$H_2CO_3 + CaCO_3 \rightarrow 2HCO_3^- + Ca^{++}$$

Waters that have percolated through the land into the groundwaters below are rich in calcium ions and bicarbonate ions. These kinds of waters are said to be *hard*. Because calcium reacts with soap, such water does not wash soap out of clothes easily. After solution, the waters are those we see coming out of the many hardwater springs of Florida (Silver Springs, Wakulla Springs, etc.). These spring waters are clear because the organic matter normally in surface waters has been filtered out and consumed before the water comes out in the spring.

Another result of years of solution is the network of caves and tunnels in Florida's limestone which may be as much as 20% of the land volume. Sometimes with all this space below, the land above may be too heavy, and the structure collapses, making a new *sink hole.* New sink holes particularly form during dry times, sometimes burying entire houses.

The original deposits of marine shells and other skeletons contained small amounts of phosphorus, 1% or less. In the process of solution, the phosphorus, which is not very soluble, combines with calcium carbonate to form calcium phosphate crystals (somewhat like your teeth). The carbonate ions wash through, leaving the calcium phosphate behind. After enough solution, the phosphate deposits of Florida developed which we now mine (Chapter 25). Calculations show it takes millions of years for such solution to form commercial phosphate deposits. Although the processes seem very slow on the scale of human lives, Florida geologically is a structure that is continuously being eroded and dissolved while being reformed with new depositions.

Fossils provide evidence of the sources of rock. Remnants of sea shells are usually visible. After limestone has dissolved, loose sharks' teeth can be found. Sharks swam in the warm waters over Florida when it was under water, and their hard teeth (calcium phosphate) are very resistant to destruction. Up to the end of the last ice age 9000 years ago, there were large land mammals in Florida that became extinct as humans moved in. Occasionally, the bones of mastodons, giant sloths, and other mammals are found in swamp deposits.

13.3 Geologic structure of Florida

The main structure of peninsular Florida is almost a mile thick, sitting on old crystalline rocks of former ages (Figure 13.2). It consists of layers of *sedimentary rock,* each one arching over the one below it. *Sediments* are the loose sands, clays, and shells that we find washing along in streams, along the ocean currents of beaches, and depositing on the bottoms of estuaries, lakes, and seas. When the sediments settle together, especially after they have been under the pressure of sediments deposited on top, they become cemented together as sedimentary rock. Each layer shown in Figure 13.2 contains sedimentary rocks deposited in the past when that layer was at the top of the land mass. The old layers are on the bottom, the more recently deposited ones are above them, and the most recent ones are at the surface. Where the sediments are mostly clay (tiny particles), the rock is called *shale;* where the rock is mostly cemented sand, it is called *sandstone;* where the rock is

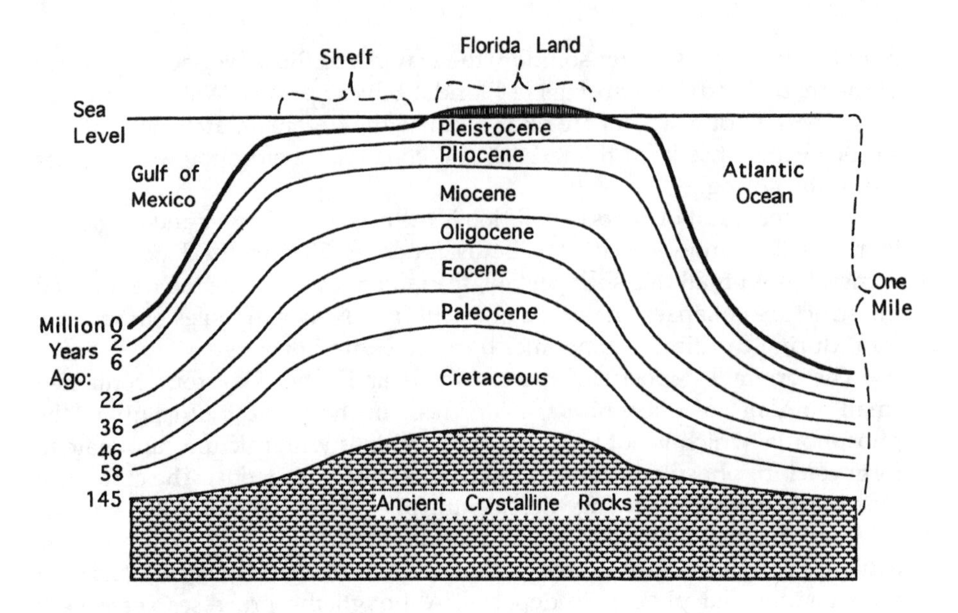

Figure 13.2 Cross-section view of the geologic structure of Florida showing the names and ages of the past periods.

cemented shells and other calcareous skeletons and reefs, it is called *limestone.* Clay sediments and quartz sands come down rivers from the erosion of hard continental rocks.

The materials for the massive structure of the Florida peninsula were added gradually. Because south Florida has been sticking out in a tropical sea a long way from hard continental rocks, the predominant sedimentary rock in its mass is limestone. Sea shells and other skeletons of marine animals formed limestone, taking their substance (calcium and carbonates) from the seawaters. In north Florida, currents along the shore brought sands from the north (see Chapter 14), including some containing titanium, which is valuable in industry. Strong winds from the north brought some clay particles.

Adding sediments a little at a time on top pushed the rocks that were once on top down to almost a mile below the surface after millions of years. Where pressures are great, rock is somewhat plastic (flexible). As some more weight is added to the mass, the whole mass sinks down a little, just as a boat floating in water would do if you added some weight to it. In other words, continental blocks (such as Florida) float on the plastic rocks underneath.

It is customary in geology to name periods of the past during which various events took place. In Figure 13.2 are the names given to the periods of time reflected in the various layers found in the mass of Florida. The ages of these layers are shown in millions of years.

Wherever weight is removed from Florida's geologic block of rocks, the block rises up a little. Over millions of years there have been more solution and erosion in the middle of Florida, where it has been sticking up, so the rocks have risen more there than out on the edges of Florida's land mass, which has been under water more of the time. At the coast, more sediments, seashells, and reefs have been deposited, making weight that presses down. This is why the land is hump-shaped and high in the middle of the state (Figure 13.2).

As Figure 13.1 shows, the sediments of earlier times come to the surface as sedimentary rock. The oldest rocks to outcrop (to reach the surface of the land) are Eocene rocks in the middle of the state at Ocala which are 90 million years old. They contain many old fossils of marine animals long since extinct. Here the younger rocks that were on top in Figure 13.2 have eroded away. The Eocene rocks contain some of the same fossils that are found in the rocks out of which pyramids were built in Egypt. There has been much speculation about what kind of long-extinct life they represent (Figure 13.3).

13.4 Hydrologic cycle in Florida

In Florida, rains are abundant in most months, with an occasional drier period or drier year. Figure 13.4 summarizes the rainfall from all the many kinds of storms. More rainfall falls in the panhandle, where more cyclonic lows pass, and along the Gulf Stream in southeast Florida, which receives moist tropical air from the Caribbean. Lower average rainfall in the central part of the peninsula favors drier ecosystems (Chapter 19).

Most of the rain falling on Florida vegetated areas is immediately transpired into the air, contributing to the productivity of the plants in the process (Chapter 6 and 19). A little runs off to form streams or lakes. In Florida, evaporation is rapid from any open waters or wet surfaces. Some sinks into the ground to become *groundwater*.

Groundwater is the water in the spaces between rocks and sand grains. The smaller the grains of sediment, the closer they pack, and the smaller the spaces that contain water. Clay sediments may have only 1% of their space for water; however, much of Florida's land is made of sands and limestones, which may have 10 to 35% space. Thus, Florida has more groundwater storage than many states. Most of Florida gets its drinking water from groundwater.

The top of the groundwater may be several feet below the ground surface. In low places, the groundwater level can be at or above the surface, making the area a wetland or a lake. These shallow groundwaters percolate down and recharge the deeper groundwaters. With more roofs, streets, and smooth lawns as a result of development, more water runs off just after heavy rains instead of recharging groundwater or making plants productive. New developments are required to include small basins to catch and hold some of this runoff to prevent flooding downstream.

Because of all the interconnecting spaces in the rocks and sands, Florida is like a big sponge. Rain is poured on top, soaks into the porous land, and under the pressure of more rain is pushed out the sides at lower elevation.

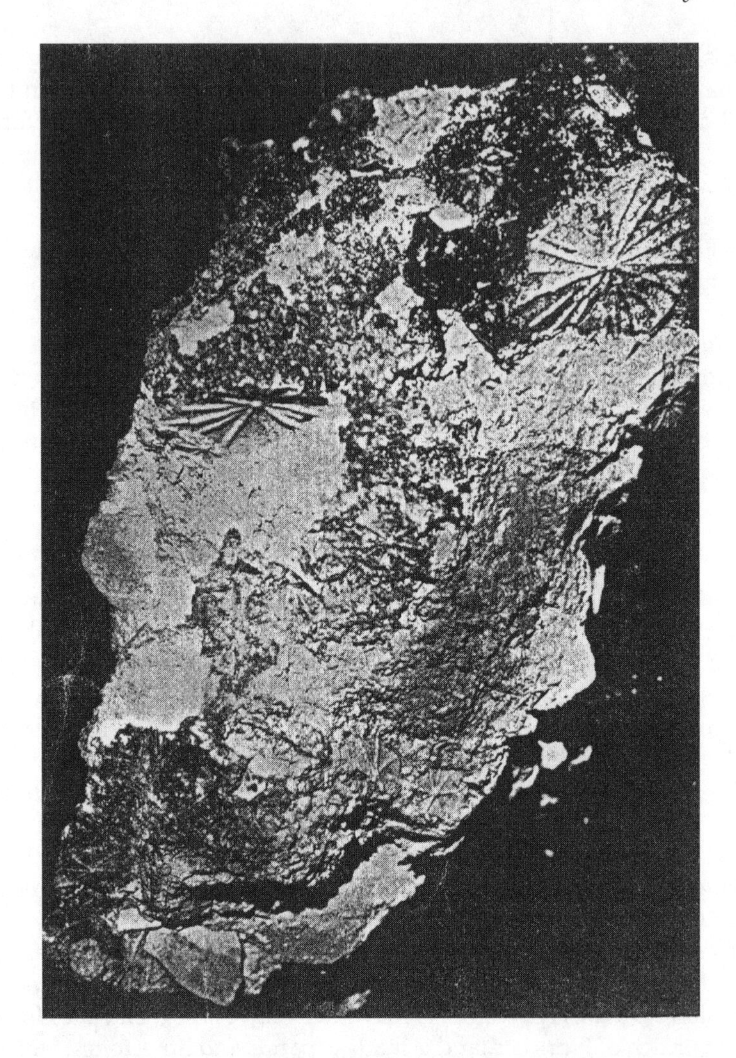

Figure 13.3 Fossils of giant Foraminifera (calcareous protozoa) in limestone from the Eocene Period near Ocala, Florida. (From Cooke, C.W., *Geology of Florida*, The Florida Geological Survey, Tallahassee, 1945.)

Some comes out into the flood plains of small streams, some as springs on land, and some in underwater springs under the sea.

Part of the rain falling on flat lands drains first into wetlands in depressions on otherwise high ground. Okefenokee Swamp in southern Georgia, Osceola swamps, Green Swamp, and the Big Cypress area swamp are examples. After being filtered by the wetland ecosystem, some of the water of these perched swamps drains downward into the groundwaters below. Even more groundwater recharge occurs through sandy uplands in the center of the state (Figure 13.5).

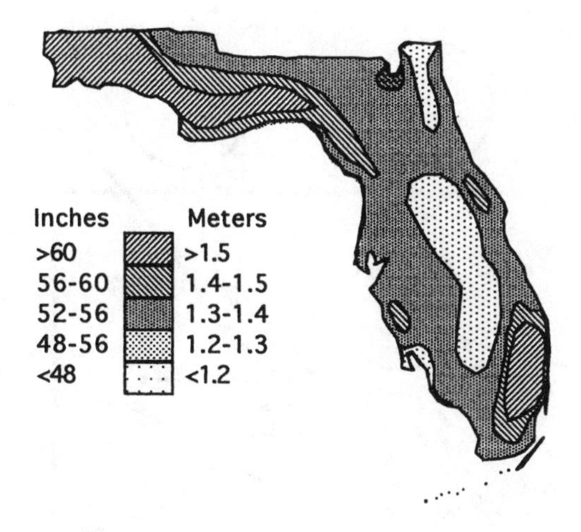

Figure 13.4 Map of average rainfall in Florida.

As long as freshwater is being added in the middle of Florida, the pressure is downward and outward. The freshwater keeps saltwater of the ocean from coming into the porous land mass. For every 1 foot of fresh groundwater above sea level, there are 40 feet of freshwater below sea level. If, however, the groundwater next to the shore is pumped out for use, there is no water pressure to keep the seawater out. Then there is *saltwater intrusion*

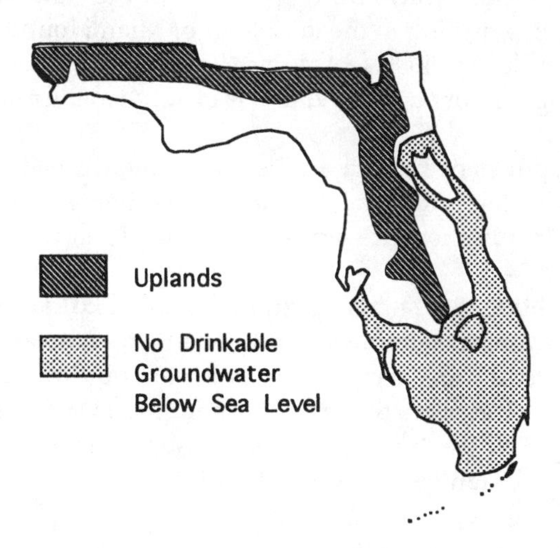

Figure 13.5 Map of Florida showing the land (shaded) less than 25 feet in elevation that was flooded in the ice ages and as a result has salty groundwaters. (Modified from Fernald, E. A. and E. D. Purdum, Eds., *Atlas of Florida,* University Press of Florida, Gainesville, 1992, p. 61. With permission.)

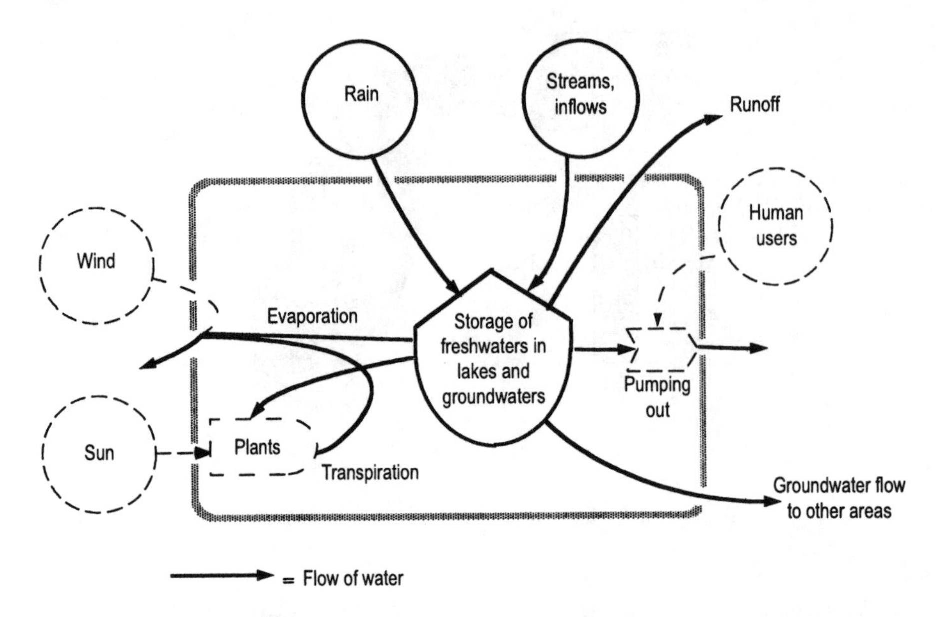

Figure 13.6 Systems diagram of processes contributing to the freshwater budget of an area.

when saltwater enters places where humans were getting their freshwater. Once the salt is in the ground, it takes a lot of flushing to get it out. Even when industry pumped groundwater from the center of the state, saltwater came up from below. Deep drilling in the land west of Miami found cold seawater in the porous rock layers 4000 feet down. To maintain good groundwater supplies, pumping cannot be allowed that is faster than the rate that the rains restore the water.

The water supplies of Florida are being threatened because wastes are being allowed to seep into the ground or are being pumped into the groundwaters. Already there are areas where the groundwater is becoming toxic (see Chapter 34).

In the water budget systems diagram (Figure 13.6) are the main processes that add (rain and inflows running off other areas) and subtract (evaporation, transpiration, runoff, groundwater outflow, and pumping) water from an area. The flows of water are solid lines. Dashed lines show the sources of energy that are driving the system: sun and wind driving evapotranspiration and human control pumps withdrawing water for use.

13.4.1 Variations in sea level

On a shorter scale of time, over intervals of thousands and hundreds of thousands of years, the sea has risen and fallen due to ice ages and other causes. During an ice age, part of the water of the ocean water stood over northern lands as glaciers, and the world's sea level was lower. Thus, the sea has closed over most of Florida at times and retreated to the edge of Florida's continental shelf at other times (Figures 13.2 and 14.1). The positions of shell deposits, reefs, beaches, mangroves, and river mouths have moved back and forth, leaving their particular kinds of deposits in different places.

In a time before the last ice age when the sea was 25 feet higher than it is now, much of south and coastal Florida was under the sea for thousands of years. During this time, the sediments and porous rocks underneath these low areas were filled with saltwater and only partly washed out in the thousands of years since. Consequently, most of the land under the shaded area in Figure 13.5 contains salty groundwater that is not suitable for most human uses. Springs in these areas are salty and contain marine animals.

13.4.2 Beaches

Much of Florida is bordered by sandy beaches, a main attraction for tourists and high-quality living. Florida beaches are among the best and longest in the world. Inland, there are many ancient beaches that mark the position of the shoreline when the sea stood higher.

Beaches exist where the forces that form the beach keep up with the forces that tend to disperse it. To have a beach, there must be a sand supply and a source of wave energy to concentrate, shape, and cleanse the beach. Much of the sand that forms Florida beaches comes down the rivers from farther north that bring sand from weathered rocks from the main mass of the southeastern U.S. In the sheltered northeast corner of the Gulf of Mexico the wave energies reaching shore are less, and the rivers drain swamps without much sand. There marshes can develop into the Gulf in place of beaches.

When waves strike beaches at an angle, they set up a beach current running in the direction that they are breaking. For much of the winter, beach currents run from north to south, carrying sand grains — sands made of quartz (silica) and feldspar minerals — from Georgia rivers down the peninsula. Then, in summer, the waves come from the southeast, and their breaking sends a beach current running northward carrying sands from broken-up shell depositions and coral reefs (calcareous sand). Over thousands and millions of years, Florida has accumulated sands from these sources.

Beaches along most of the Florida coast form barrier islands (Figure 13.7). The breaking waves and the resulting currents are on the open sea side. Marshes line the back side of the beach island on the estuary side. Rains falling on the sands percolate down and fill spaces in sand with freshwater that pushes back the saltwater, making a zone of fresh groundwater. Land plants with long roots catch wind-blown sand and sand dunes build up. Vegetated dunes provide protection against hurricane waves. When the vegetation is eliminated, the dunes move with the wind. Careless housing development has eliminated the beauty and protection of vegetated dunes in many areas.

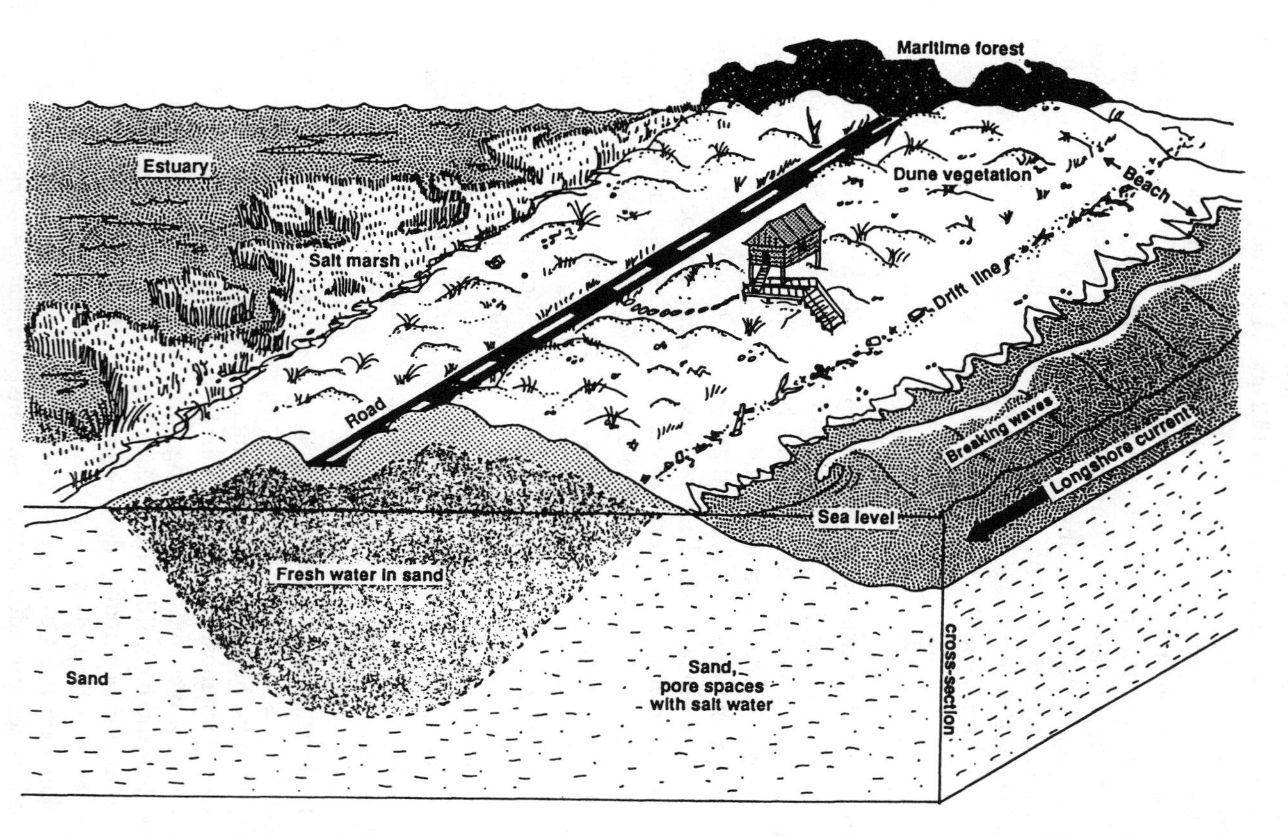

Figure 13.7 View of the beach, marshes, and groundwater on a barrier island beach.

In this century, there has been a general rise in the world sea level of about 1 foot (30 cm). The combination of changing level of the beach relative to sea level and the cutting off of sand supply has caused many Florida beaches to diminish, with the sea undercutting coastal hotels. In an effort to keep sand from moving, rock jetties were built which blocked the nourishment of new sand from the longshore currents, causing further erosion in front of hotels and houses. Some counties invest millions of dollars in "beach renourishment". Pumping sand back on the beaches only lasts several years, destroys the offshore bottom ecosystems, and makes waters turbid and less productive.

When groundwaters are drawn from beneath the shore, lands may sink. Extreme examples are those in western Taiwan and Venice, Italy. How important this is in Florida is not known. See more on beaches in Chapter 15.

Questions and activities for chapter thirteen

1. Find some native rock. To test if it is limestone, put several drops of vinegar (or other acid) on it. If it fizzes, the reaction is producing CO_2 and it is limestone. What is the chemical reaction? If it does not fizz, how can you decide whether it is shale or sandstone?
2. Go fossil hunting. A shallow stream or a recent road cut through limestone rocks is likely to have many fossils. You may find teeth of sharks and rays, clam shells, and snail shells. A good reference book is *Fossils* by Rhodes, Zim, and Shaffer, a small but accurate Golden Press Book.
3. Find out the source of your drinking water. Is it from the deep aquifer groundwater or from a surface water well? Ask what kind of treatment it is given and why.
4. Put out a container to collect rain water. Measure the depth of the water with a ruler and keep a record. How does your rainfall compare to that reported by your local television station or newspaper? If the container is left undisturbed without removing the water, does the level go up or down? What processes affect the water level? Figure out a way to measure the evaporation and try it.
5. Dig a hole in a local wetland. How deep down is the water level? Dig a hole in a nearby dry area and compare the water levels.
6. Collect rocks and sediments from your area. Identify them and estimate their ages. Use the *Florida Atlas* (Fernald and Purdum, 1992) to see what rocks are likely.
7. If you live near the beach, identify kinds of sand. Is the sand mostly calcareous or quartz? What is the source?
8. Visit a spring. What is the source of the water?
9. Discuss Florida's drinking water problems. What actions by human developments can interfere with the quality of groundwater supplies? Lake water supplies? How should the state ensure enough clean water for its population?

chapter fourteen

Oceanography

Seawaters surround the peninsula of Florida with surging tides, restless waves, and swirling currents. They moderate the climate, support fisheries, cleanse the coastal waters, help navigation, cool coastal power plants, and attract tourists. This chapter on Florida oceanography considers the environmental processes of seas around the state including physical, chemical, geological, and biological processes. Ecosystems with interesting organisms depend on the pattern of estuaries, continental shelf, continental slope, and deep sea.

Figure 14.1 shows Florida with the Gulf of Mexico on the west and the Atlantic Ocean on the east. Florida has 800 miles (1800 km) of coastline, much of it marked with beaches, small islands, inlets, brackish-water estuaries, and sometimes marshes and mangrove swamps. The continental shelf from 0 to 600 ft (180 m) deep is 60 miles (96 km) wide on the west coast but is narrow on the east. Beyond the shelf is the slope where the sea bottom drops rapidly to the ocean floor about 2.4 miles (4 km) deep.

14.1 Physical processes

The systems of seawater circulation are driven by the tidal forces from the moon and sun, the force of the winds on the sea surface, and the rivers running into the sea.

14.1.1 Tide

Due to the gravitational pull of the sun and the moon, the sea is higher on two opposite sides of the Earth and lower in between. As the Earth rotates through the high and low zones, the surface of the sea rises and falls about twice a day. As the sea rises, water raises the level of surf on the beach and rushes into the estuaries. Waters then drain out as the sea surface falls again. The range of the tide from high to low varies in different places because of the way water swirls against the land in ocean basins of various shapes. It is only about a foot (0.3 m) in some places in the Gulf of Mexico but is 3 feet (1 m) or more along the Atlantic coast of Florida.

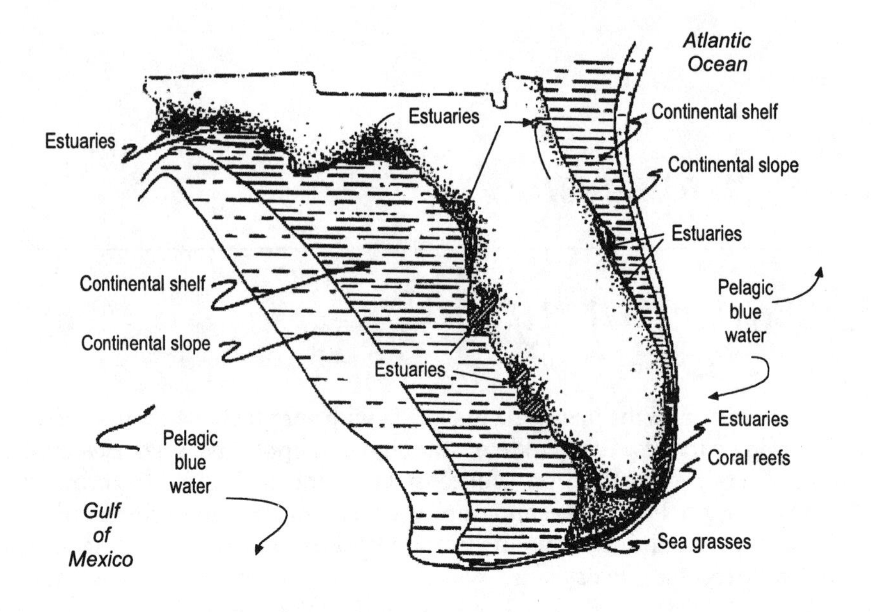

Figure 14.1 Undersea zones and oceanic ecosystems of Florida.

Figure 14.2 helps in understanding the timing of tides and their range. The moon goes around the Earth each month, and the Earth goes around the sun each year. Both of these circles are in roughly the same plane, but the sun, Earth, and moon are not exactly in line, except at rare times during eclipses. A lunar (moon) *eclipse* is when the Earth's shadow falls on the moon; a solar (sun) eclipse is when the moon's shadow falls on the Earth.

When the sun and moon are somewhat in a line, either on opposite sides of the Earth or on the same side of the Earth, their gravitational pulls on the sea are reinforced, and the tidal range is largest. The largest tidal range is called a *spring* tide; note two spring tides in Figure 14.2. (The use of the name "spring" for the tidal range has nothing to do with the springtime of the year; rather, it means to spring apart, to have a wider range.) A person looking at the moon when it is in line but on the sun's side (Figure 14.2a) cannot see the sunlit side of the moon and sees mostly a dark circle, a *new moon*. A person on Earth looking at the moon when it is behind the Earth but roughly in line with the sun sees a *full moon* (Figure 14.2c). A person looking at the moon when it is at right angles to the sun sees half of the sunlit moon (Figure 14.2b and d).

When the sun and the moon are at right angles, not in line (Figure 14.2b and d), their gravitational effects cancel somewhat. Being very close, the moon's effect is greater than the sun's effect and there is still a tide, but the range is smallest. This is called a *neap* tide. Since the moon goes around the Earth in a little less than a month, there are two times when there are spring tides and two times when there are neap tides (Figure 14.2).

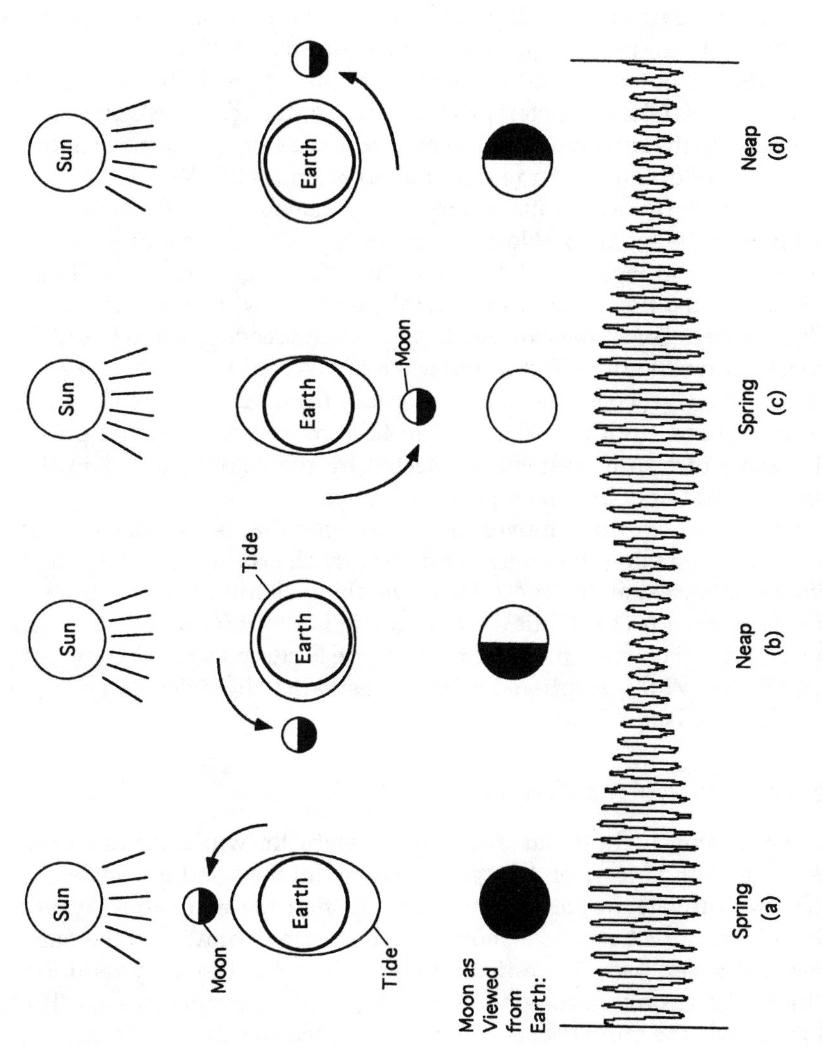

Figure 14.2 Movements of the Earth and moon that cause tides in the sea, with a graph of sea level through a monthly cycle of moon and tides. (a) Spring tide with new moon; (b) neap tide; (c) spring tide with full moon; (d) neap tide.

The pull of the sun and moon does not have much effect lifting the water straight up, because the Earth's gravity is too strong, but the moon and sun do pull water sideways, causing the surface to rise and form a broad mound many miles across, a few feet (about a meter) high. The mound of water moves around the Earth as a wave being dragged by the position of the sun and moon as the Earth rotates. If you are in a boat on the open sea and the wave of tide passes, the boat may go up several feet and down again in several hours without anyone being aware of it. But, at the shore, having the sea rise several feet (about a meter) produces tremendous changes on beaches and docks and in the river mouths. The pull from the side is greatest at about 45 degrees latitude, causing large tidal ranges at mid-latitudes.

A place on Earth, as it rotates once a day, usually passes through an elevated part of the sea twice. Note almost two ups and downs for each day in the tidal graph in Figure 14.2. Whereas the tidal range is large in a large ocean, in a lake the tide is almost negligible, only a few inches in the large Lake Okeechobee. Coastlines exposed to large open oceans (such as Florida's east coast) receive the full effect as the wave of elevated sea surface passes. The tides on the west coast are smaller, because the Gulf of Mexico is small and partly isolated from the whole ocean tidal movement. Because of the way the wave of tide movement is affected by the coast, the tide in the northern Gulf has only one main peak each day.

Where the wave of tide moves into a coastline that is cup-shaped, the wave converges, making the range of tide larger. The coastline just north of Jacksonville is an example. The tidal wave along Florida's fairly straight coastline is about 3 ft (1 m), but the tidal range is about 9 ft (3 m) north of Brunswick, Georgia, where the Atlantic coastline is curved so as to converge the tidal energy. With exceptional tides, the estuaries there develop exceptionally rich marshes.

14.1.2 Wind waves and currents

The largest currents in the ocean are partly driven by the winds above the sea surface. When wind blows on the sea, the energy of the wind is transferred to the energy of water waves and currents. At first there are small waves moving in many directions. Soon, however, those waves moving at the same direction and speed become unified into long lines, because they reinforce each other. The stronger the wind, the higher and longer the waves. The higher the wave, the more energy it carries and the faster it runs. In storms, waves may be about 25 ft (7.5 m) high or more and potentially destructive to boats and shore installations.

If the wind stops, the waves keep going because of their stored energy. The height and energy of the waves decrease gradually, but the length and time between each wave crest does not decrease. By counting the number of seconds between wave crests as they pass, you can tell what kind of wave it was when it was first formed. Ordinary waves that reach the east coast in summer, driven by normal easterly winds, have crests about 5 seconds apart. However, waves moving out from tropical storms and hurricanes may be ten seconds or more. Sometimes as you stand on a beach it is hard to make these counts if there is more than one kind of wave coming ashore from different directions. All the waves tend to turn parallel to the shore and mix as they break (Figure 13.7).

The waves on a particular beach maintain that beach. With stronger waves, the size of the sand grains is larger because smaller ones are washed away. In northern waters where breaking waves are often large, the beach can have a gravel texture. Where waves reaching the beach are small, the sediments are fine grained. If there are no waves breaking, then vegetation may take over the beach. The wave climate is part of Florida's natural resource that attracts tourists.

If waves are large and coming in perpendicular to the shore, the water that is thrust onto the beach when they break goes back out to sea as a current directly outward called a "rip" current. These are dangerous for ordinary swimmers who may be carried out. If they can stay afloat and on top of the waves, and if the wind is toward the shore, the wind and waves will eventually bring them ashore again. If the waves are coming ashore at an angle, the current they generate with their breaking is along the shore (*longshore current*) and not so hazardous to swimmers.

In addition to wind waves, there are long waves (longer than wind waves but shorter than the wave of tide) which run up the coast with several minutes between crests. Sometimes this is called *surfbeat*. Bathers can get in trouble if they are enjoying small waves in water where they can touch bottom one minute and then find that the sea level has risen several feet with the passage of the longer coastal waves.

For small boats, there is a very dangerous condition in inlets when the tide is running in opposite direction to the waves. If the tide and waves are coming into an inlet in the same direction, the surface may be smooth. Suddenly when the tide reverses direction and runs out to sea against the waves, the smooth sea can become dangerously chaotic with tall steep waves that can capsize small boats.

14.1.3 Gulf Stream and shelf currents

The largest currents affecting Florida are part of the general ocean circulation and are partly driven by the general wind circulation. When wind energy is transferred to make waves, not only do the waves move ahead, but the whole water mass is moved along. Figure 14.3 shows the pattern of wind and the resulting currents in the Atlantic Ocean and Gulf of Mexico. Because the winds over the open sea in summer at the latitude of south Florida are usually steady from the east (north and south of Cuba in Figure 14.3), the current is steady from the east. These easterly winds are called *trade winds*. In the winter when the cold fronts come to Florida (see Chapter 12), the easterly winds and currents are displaced southward.

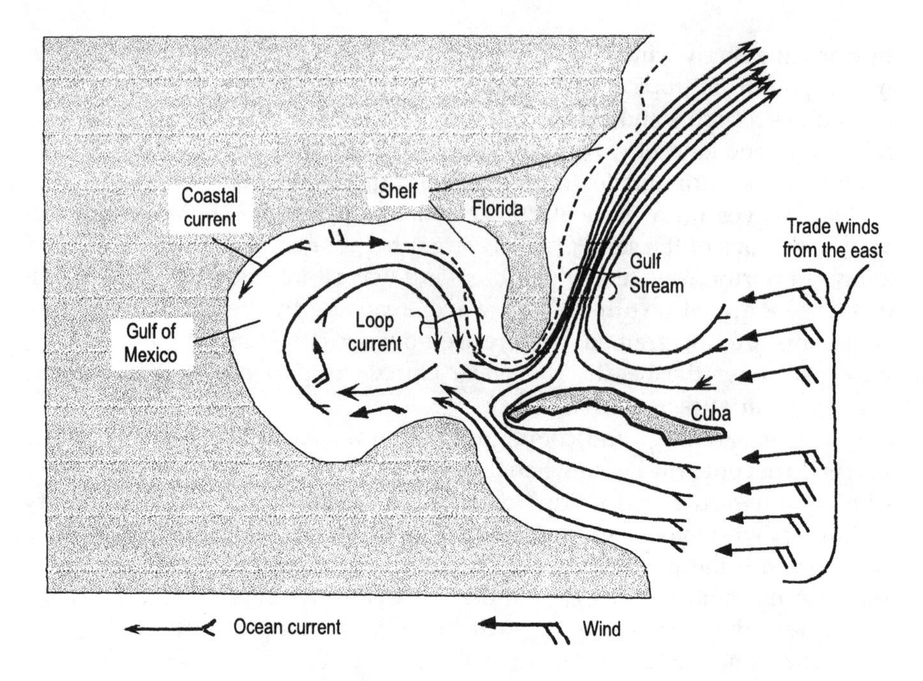

Figure 14.3 Winds and ocean currents influencing Florida.

The winds over the entire ocean to the east converge to produce the Gulf Stream, which swings north in the deep channel between Miami and the Bahamas (Figures 14.3 and 14.4). At times it may run as fast as 13 miles per hour (21 km per hour). North of North Carolina, this warm current moves out to sea, crosses the ocean, and gives northern Europe a milder climate than it would have without it.

The Gulf Stream is very blue because it does not have much sediment or organic matter in it. The water of the Gulf Stream has been moving across the tropical sea for many weeks during which plankton animals (zooplankton) have eaten much of what was in the surface and carried the organic matter to deeper water with their fecal pellets. Gulf Stream blue water has an ecosystem with very special organisms (Chapter 15).

Particularly when driven by the winds in summer, most of the open waters of the Gulf of Mexico have a clockwise circulation (loop current in Figure 14.3), so that there is a deep water current from north to south along the edge of the shelf on the Florida west coast which then flows east to join the Gulf Stream. Sometimes this current circulates on top of the shelf.

14.1.4 Seawater salinity and density currents

Density is the weight of a unit volume of liquid, solid, or gas. Density of liquid is often measured in grams per cubic centimeter, or milliliter (ml). Freshwater has a density of about 1.001 grams per milliliter (g/ml).

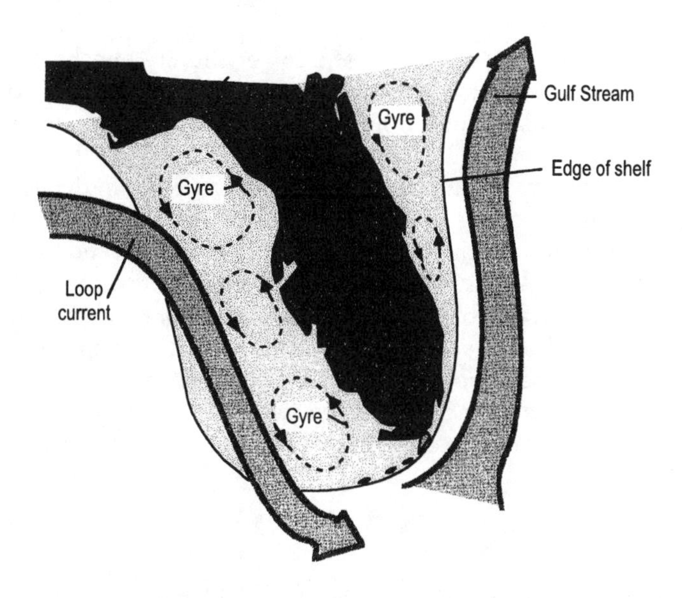

Figure 14.4 Coastal currents of Florida.

Average seawater, however, is heavier because of the salts it contains. Its density is about 1.025 g/ml. It contains 3.5% salt; this is its *salinity*. In other words, there are 3.5 parts salt per 100 parts seawater. In oceanography the salinity is often indicated by moving the decimal over one point, as 35.0 parts per thousand (ppt).

Water masses of different density cause currents. Waters of lower salinity or higher temperature have lower density. Waters of lower density stick up higher than waters of higher density. As a result, gravity causes the top water to flow downhill and to the right because of the Earth's rotation (see discussion of Coriolis force in Chapter 12).

14.1.5 Rivers and coastal currents

When rivers bring low-density freshwater to the sea, it tends to flow on top and to the right. The freshwater coming out of streams swings to the right because the Earth under it is rotating to the left. Light objects float on denser objects. In some estuaries, there are two layers: freshwater going out on top and saltwater coming in on the bottom (the St. Johns River mouth at Jacksonville is an example).

Reaching the sea, the freshwater turns down the coast to the right and mixes with the saltwater, giving coastal water a lower salinity than the open ocean. Being less dense, the coastal water is some inches higher than the denser saline waters farther out. Although the downhill pressure is directed towards the open sea, because of the Coriolis force the water current runs down the coast. Because there are many streams coming out of Florida on all its coasts, waters along shore are of a lower salinity and have the tendency to move to the right (southward on the east coast, northward on the west coast of peninsular Florida, and westward along the panhandle of Florida).

As we have already discussed, however, the currents in deeper waters farther offshore are in opposite directions (Gulf Stream going north; loop current coming east and south). Between the inshore current and the offshore current gyres develop. A *gyre* is water making a circular current (Figure 14.4). Gyres are very important to populations of marine organisms, which can release eggs into the waters of a gyre without the developing organisms being carried out of the region. By the time the organisms have grown up, many of them will have been circulated back to where they started. Many of these gyres are over the continental shelves and the circulation helps ecosystems develop there.

14.1.6 Storms

Descriptions of the winds and currents around Florida apply to ordinary times. When strong storms move in, the circulation is changed temporarily. For example, strong winter cold fronts often have cold winds from the north that drive water south and out of estuarine inlets, chilling what is left and sometimes killing many fish.

A hurricane, with its counterclockwise circulation (as you look down on its map; see Figure 12.8), causes the water underneath to circulate, too. The rotation-of-Earth effect tends to cause the currents under the storm to go to the right and away from the storm center. This causes the sea level to rise around the storm. Replacement waters come up from below (called *upwelling*). These deep waters have more nutrients, which fertilize the sea surface and cause extensive plankton growth after the storm passes.

14.2 Chemistry of seawater

Table salt is made of the crystals of two elements, sodium and chlorine, in the form of ions. An *ion* is a chemical unit with an electrical charge (+ or –). In table salt, the sodium ion has a plus charge and the chloride ion has a minus charge. The attraction of plus and minus charges helps to hold the ions together in a salt crystal.

When table salt is dissolved in water, the sodium and chloride ions separate and become dispersed in the water. You can get them back together again by evaporating the water down until it becomes supersaturated and the salt crystallizes out. Table salt is usually made from seawater, since sodium and chloride ions are the most abundant ions in seawater. Commercial salt is manufactured by taking advantage of the sun's evaporation of ponds in the Bahama islands, where rainfall is low; such salt was formerly made in the Florida Keys.

Table 14.1 Most Abundant Ions in the Ocean

Plus Charged	Minus Charged
Sodium (Na)	Chloride (Cl)
Magnesium (Mg)	Sulfate (SO_4)
Potassium (K)	Bicarbonate (HCO_3)
Calcium (Ca)	Bromide (Br)

Seawater is a mixture of other salt ions as well. In Table 14.1 are given the four most abundant ions with plus charges (cations) and the four most abundant ions with minus charges (anions). The amounts of these salts depend on the world cycles of the elements. For example, calcium from rivers accumulated in the sea until it reached its present concentration, which is enough to make it easily available to marine animals for building calcium carbonate skeletons. The calcium in the sea is now a balance between that going into the water from the world's rivers and that being sedimented from the sea as calcium carbonate in shells, reefs, and other skeletons.

14.3 *Ecological self-organization with coastal eutrophication*

Especially in recent years, runoff from farms and cities has washed high amounts of nutrients (fertilizer elements) into coastal waters, making them more fertile (more eutrophic). These conditions cause some marine life to prosper but are not favorable to most of those originally present. Included in the runoffs are toxic substances. Such conditions are still new, and we do not know yet what the outcome will be of the self-organizational process of nature in adapting the wide variety of possible species in the sea to the new conditions found in coastal Florida. Chapter 15 introduces the varied ecosystems found in Florida's marine waters.

Questions and activities for chapter fourteen

1. Define: (a) tide; (b) neap tide, spring tide; (c) full moon, new moon; (d) wave; (e) current; (f) salinity; (g) Coriolis force; (h) gyre.
2. Research Florida's most recent hurricane or the one that came closest to where you live. Where did it come from? How was it formed? Where did it land? How much destruction was caused?
3. If you are a surfer, find out what combination of ocean-bottom shape, waves, currents, and winds makes the best surfing waves. Which are dangerous? What kinds of ecosystems live in this environment?
4. What would be the direction of Florida's coastal currents if Florida were in the southern hemisphere with the Coriolis rotation effect to the left?

5. Make sketches (like Figure 14.2), showing the position of the Earth, sun, and moon in a solar eclipse and in a lunar eclipse.
6. On a world map, trace the Gulf Stream. What effects does it have on the east coast of the U.S. and on Great Britain?
7. If you have ever gone swimming off southern California, you know the water is very cold. Explain why that water is cold and Florida's water is warm.
8. Explain why there is more likely to be a "rip" current in a cup-shaped curved beach than in a straight one.
9. If there were an oil spill in Tampa Bay, where would it go?
10. In a river inlet on the east coast, which side — north or south — would have the highest salinity? Explain.

chapter fifteen

Marine ecosystems

Controlled by oceanographic processes (see Chapter 14), the marine ecosystems of Florida (see Figure 14.1) include the blue-water ecosystem offshore, the continental shelf ecosystem, coral reefs fringing the southeast edge of Florida along the Keys, beach ecosystems on the exposed coastline, and estuarine ecosystems (see Chapter 16).

15.1 Blue-water ecosystem

The blue-water ecosystem occurs in Florida outside of the continental shelves in the tropical surface waters of the deep sea (Figure 14.1) and includes the Gulf Stream. The waters are blue because they have very little nutrients and organic matter. Consequently, the short-wave light of sunlight being scattered around makes the sea blue. Because the blue ocean contains few particles, silt, or organic matter, light penetrates deeply and plants can photosynthesize down to 300 ft (100 m). Although life is sparse, it is diverse and interesting and includes floating *Sargassum* seaweed, blue jellyfishes, and flying fish (Figures 15.1 and 15.2).

Plankton are organisms suspended in water that move with the currents. The producers in blue waters are *phytoplankton* (microscopic algae, Figure 15.2). The small consumers are *zooplankton* (tiny invertebrate animals the size of pinheads and even smaller single-celled consumers). Many of the zooplankton migrate up at night to eat in upper waters and down in daytime.

There is a high diversity of animals in the food chain based on plankton (Figure 15.3). Small fish eat plankton, and the food chain continues up to fast-swimming fish, such as tunas, and very large game fish, such as sailfish and marlin, which are important attractions to tourists (Figure 15.1). Some of the smaller invertebrates and fishes move down in daytime 500 to 3000 ft (150 to 900 m) to depths where there is as much light from luminescence (light emitted from deep sea organisms) as there is from daylight penetrating down.

Turbulence, the circular motion of the water caused by wind, tides, and various currents, keeps the water constantly stirring. Driven by wind waves

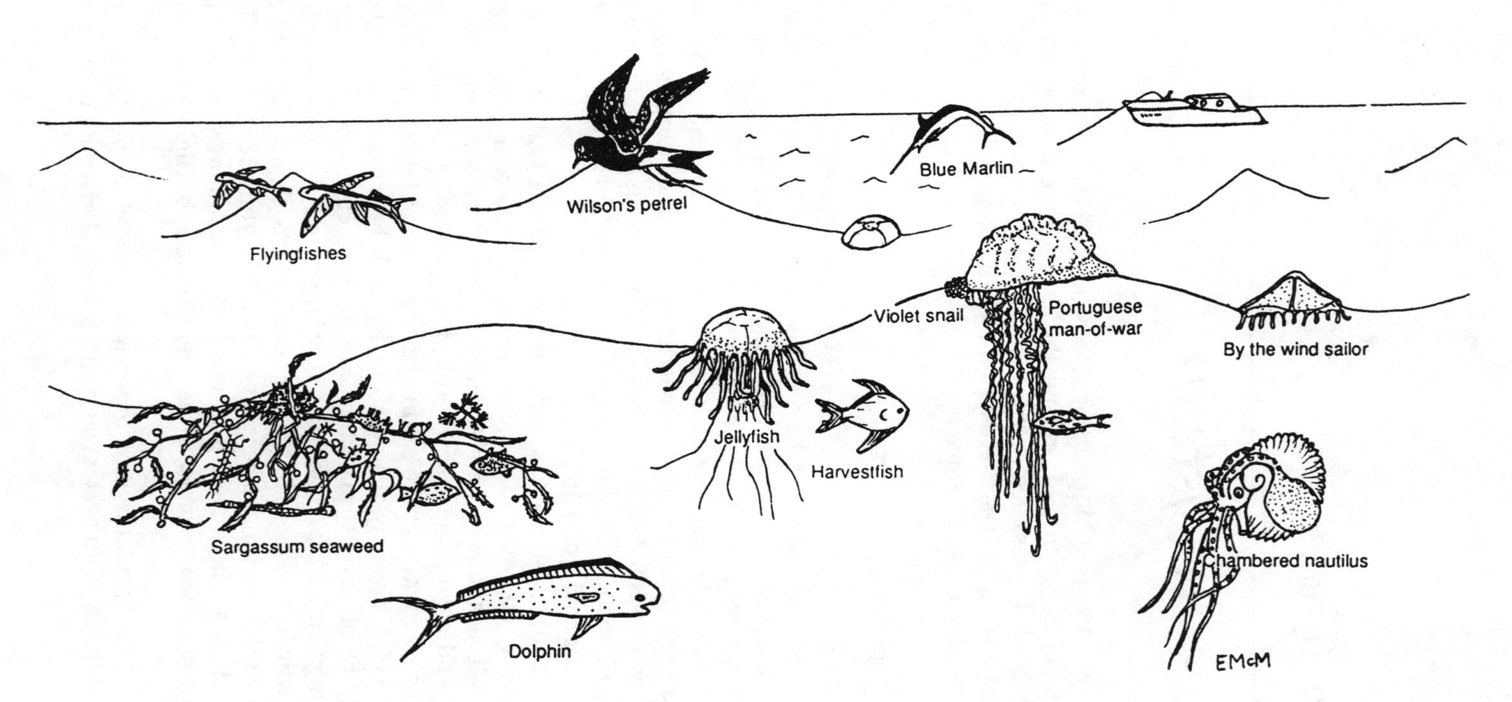

Figure 15.1 Organisms in the blue-water ocean system. (Illustration by Elizabeth A. McMahan.)

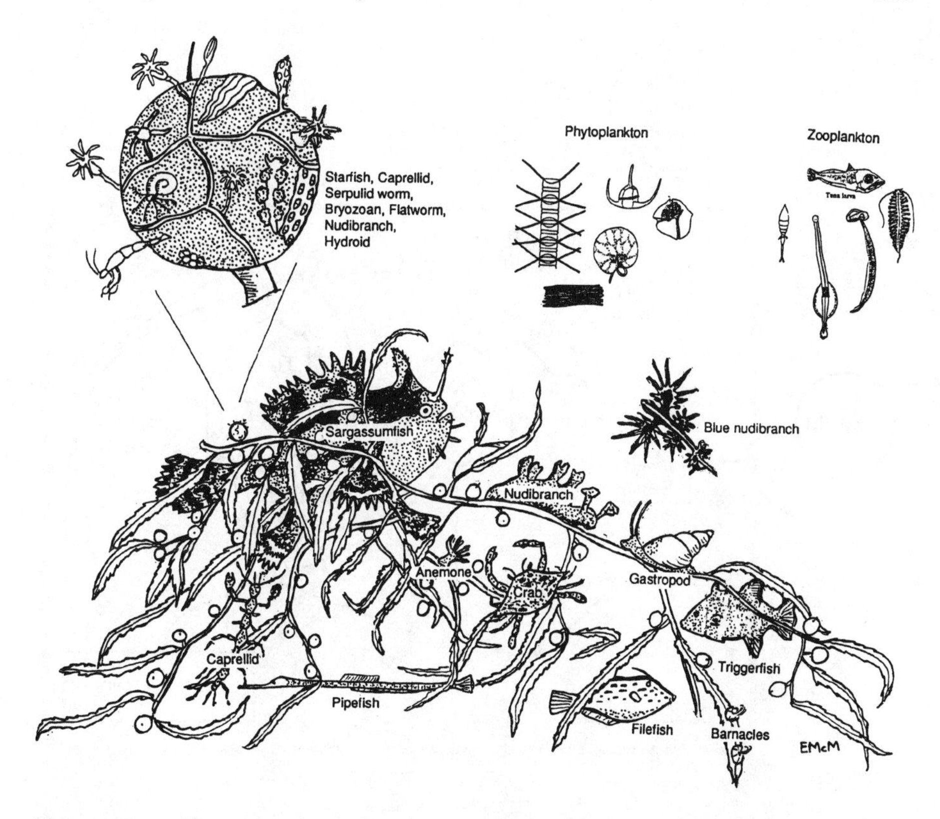

Figure 15.2 Close-up views of life in the blue waters: *Sargassum* seaweed with one of its floats enlarged; tropical phytoplankton, 0.001 to 0.1 mm in size; tropical zooplankton, including game fish larva, 0.1 to 10 mm in size. (Illustration by Elizabeth A. McMahan.)

and tides, water turbulence provides both vertical and horizontal mixing of nutrients, gases, and plankton. The systems diagram shows the action of the turbulent motion on phytoplankton and the transport of nutrients up from the depths (Figure 15.3).

The blue-water system includes a floating community. Brown *Sargassum* seaweed (Figure 15.2) is moved by turbulence into rows parallel to the direction of the wind and waves. Some floating animals are brilliant blue, such as the man-of-war jellyfish and a species of floating snail. Close-up views of *Sargassum* seaweed show a microecosystem of many associated animals (Figure 15.2).

15.2 Continental shelf ecosystem

Over the continental shelves (see Figure 14.1), the animals at the bottom of the ecosystem migrating up and down in the deep waters are replaced by

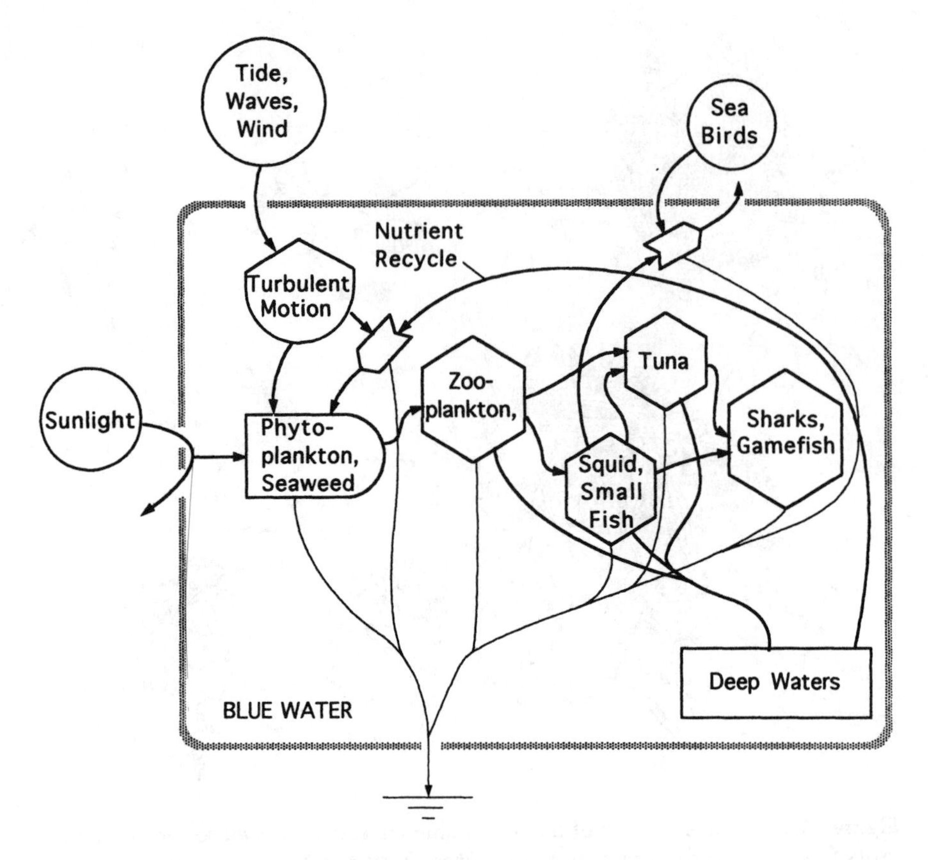

Figure 15.3 Energy systems diagram of the blue-water ecosystem of the open sea and its exchanges with deeper waters below.

many kinds of bottom animals that live within and above the muddy bottom. The coastal shelf areas are influenced by the sea breeze circulation, by cold winds from the land, and by sediments and nutrients in the runoff from shore. The continental shelf waters are greener and more turbid (cloudy) but have more photosynthesis by phytoplankton because of nutrient fertility. Plankton and larvae of shrimp, crabs, and fish can circulate but remain in the same general area over the shelf because of the gyre currents (Figure 14.4).

A systems diagram of the continental shelf ecosystem is similar to Figure 15.3, except the deep waters box is replaced with the bottom community. The ecosystem on the continental shelf has a plankton food chain in the upper waters and a food chain of bottom organisms and bottom-feeding fishes below. On the bottom, the food chain starts with bottom plants, detritus from the land, and a "rain of organic matter" from the plankton system above. Small bottom animals (called *benthos*) are consumed by many bottom fishes. During the daylight, zooplankton and many other smaller animals stay near the bottom, coming up to feed throughout the waters at night.

Many of the shelf species (crabs, shrimp, fishes) have juvenile stages which ride the wind waves and tides into the estuaries when food is temporarily abundant, migrating back out to the shelf waters for breeding and early life where conditions of salinity and temperature are uniform.

Commercial shrimp and many fish species are caught on the shelves of both coasts by shrimp trawlers pulling wedge-shaped nets along the bottom at night. As in many parts of the world, the fish, shrimp, and lobster catch per unit effort has been decreasing. This is because the number of boats and their catch are high in relation to the carrying capacity of the ecosystems, some waters are polluted, and habitat and nursery areas have been lost.

A somewhat erratic phenomenon in the high-salinity shelf ecosystem near shore is "red tide", a periodic "bloom" of a species of microscopic phytoplankton of the dinoflagellate type. Unlike many other algal blooms, this one makes a toxin that kills fish, which drift ashore along with pungent smells to make beaches unattractive for tourists. The periodic fish kills return nutrients to the system in which the fish grow again. This is a natural phenomenon that was observed by early settlers, but it has become more common, possibly due to runoffs from the land. It could be to the aquatic ecosystem what fires are to forests.

15.3 Shallow-water grass flats

Grass flat ecosystems prevail in clear shallow marine waters protected from heavy waves. The grass flat in Figure 15.4 from south Florida includes a high diversity of tropical fishes and invertebrates. In clear waters, where nutrients and currents are optimal, the turtle grass becomes dense and is extremely productive of organic matter.

15.4 Coral reefs

Along shores and on the bottom of shallow warm seas above 68°F (20°C), where sunlight, waves, and currents are strong, coral reef ecosystems develop. In Florida, there is a long coral reef bathed by tropical waters of the Gulf Stream along the edge of the Florida Keys facing the waves that run steadily from the east (Figure 15.5). Growing on the platform of old skeletons, corals organize in lines facing the waves (Figure 15.5). Keys are small low islands. Patch reefs also occur in Florida Bay and a few spots as far north as Ft. Lauderdale.

Plants and animals with calcareous skeletons build platforms of limestone made of their skeletons. The reef corals that build most of the reef are a kind of jellyfish that gets most of its food and energy from the photosynthesis of symbiotic (interdependent) algae called *zooxanthellae,* which live in their tissues (Figure 15.6). Coral animals also capture small animals with their stinging cells. The nutrients released from digesting these organisms are used for growth.

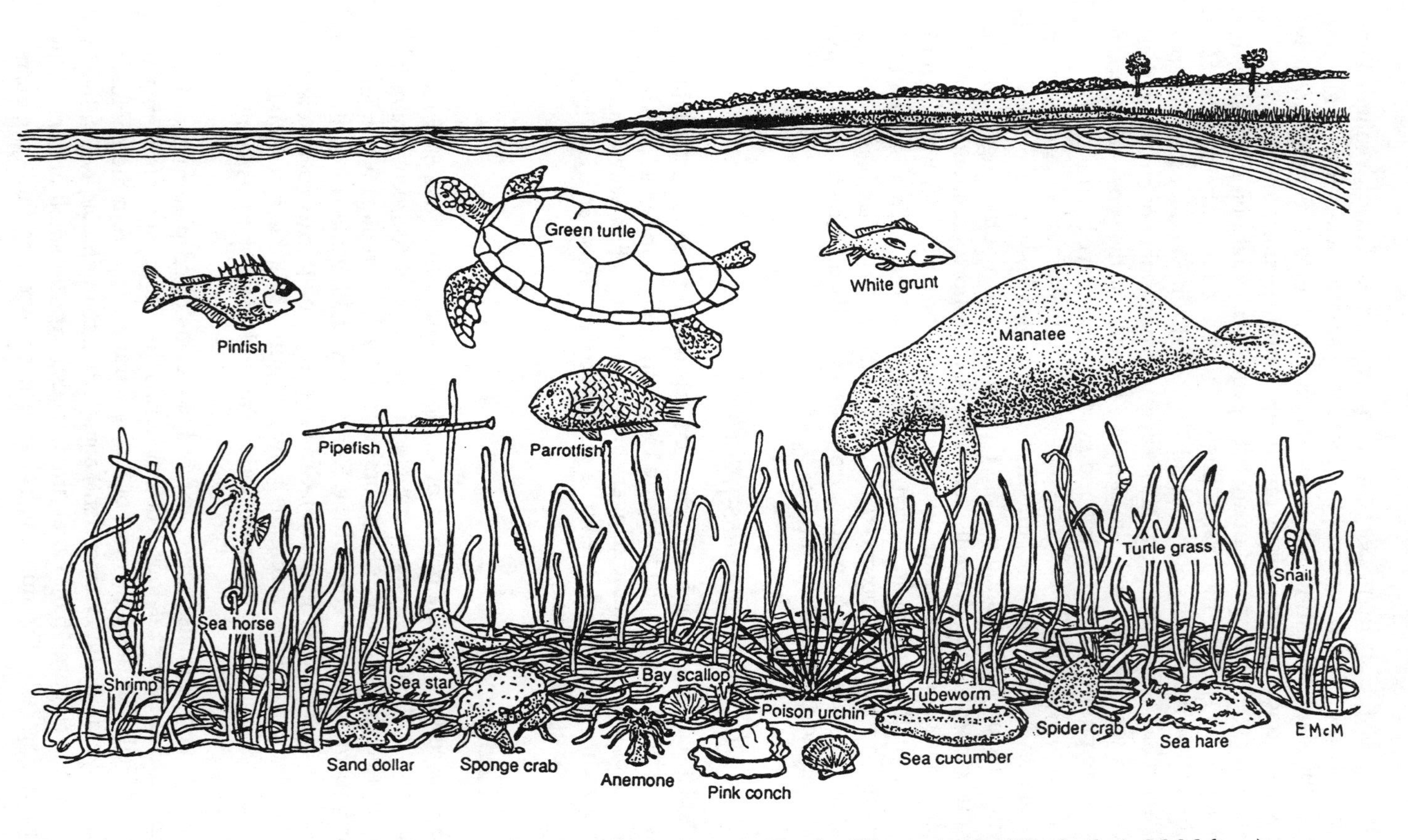

Figure 15.4 Organisms in a turtle-grass ecosystem in south Florida. (Illustration by Elizabeth A. McMahan.)

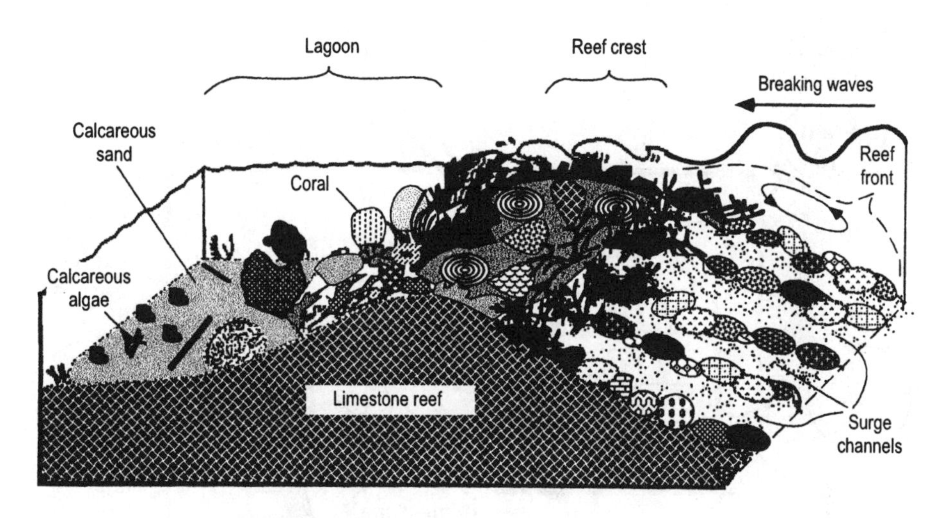

Figure 15.5 Zones in a Florida coral reef.

Photosynthesis by the symbiotic algae within coral tissues not only makes oxygen but raises the pH (making tissues chemically basic). Where the ocean waters and the jellyfish tissues are saturated with calcium and carbonates, raising the pH makes the calcium carbonate ready to precipitate. The coral and other skeleton animals can then deposit a layer of limestone underneath them (Figure 15.6) without much effort. The deposition of skeletal matter a fraction of an inch a year eventually builds a platform with zones

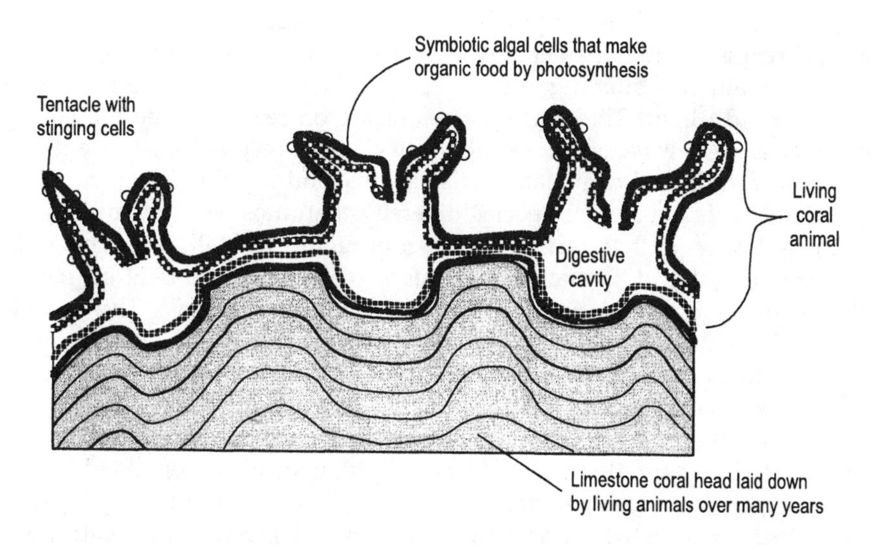

Figure 15.6 View of living coral and its skeleton in cross-section.

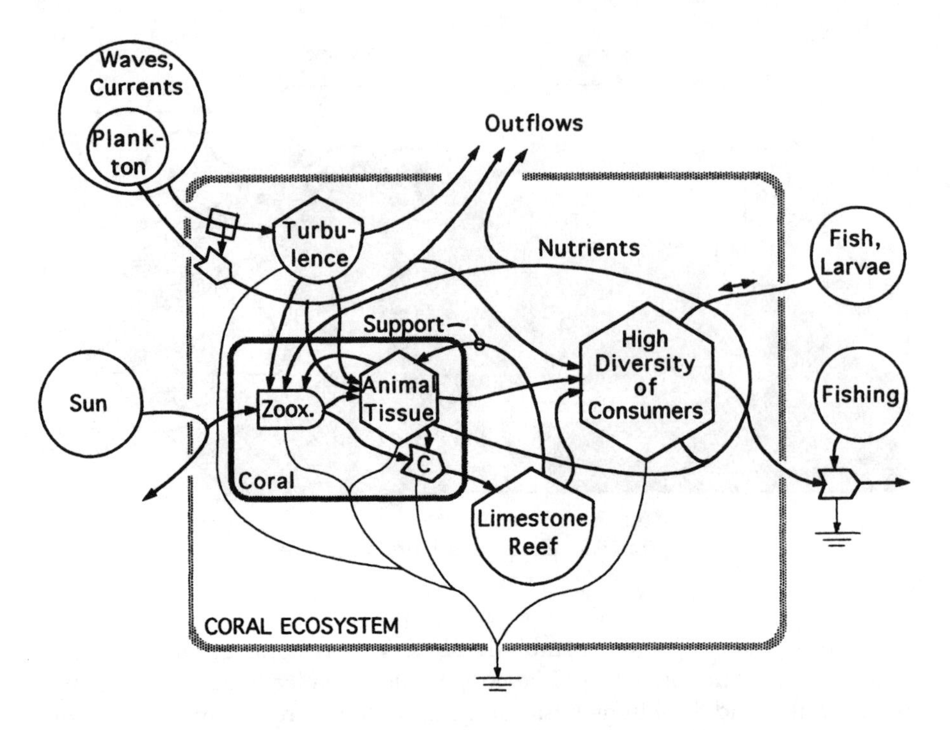

Figure 15.7 Summary of coral reef system. Zoox. = zooxanthellae undergoing photosynthesis; C = depositing of calcareous skeletons by coral tissue with the help of the zooxanthellae, making tissues more basic.

(Figure 15.6). The relationships of photosynthesis, respiration, and recycle in a coral reef are given in an energy systems diagram (Figure 15.7).

The main reef-building corals are near the sea surface because they require bright light. The dense masses of life on reefs also require strong currents and/or wave action to supply oxygen for respiration, nutrients for growth, calcium and carbonates for skeletons, and supplementary food.

Coral reefs have more species diversity than most ecosystems (Figure 15.7), and some of that variety is shown in Figure 15.8, although the great diversity of colorful animals and plants is difficult to represent on paper. There are many symbiotic relationships among reef animals. The various corals, shellfishes, sponges, and algae attach to each other to form complex reef structures in which other animals live. Large carnivores common on the reef include moray eels within the coral heads and barracuda and sharks at the edges of the reefs. Even though there are many kinds of organisms on the reef (high diversity), there are no large populations of any one kind.

Many reefs are losing their main photosynthetic corals, possibly because the transparent blue waters are being replaced with greener, more eutrophic,

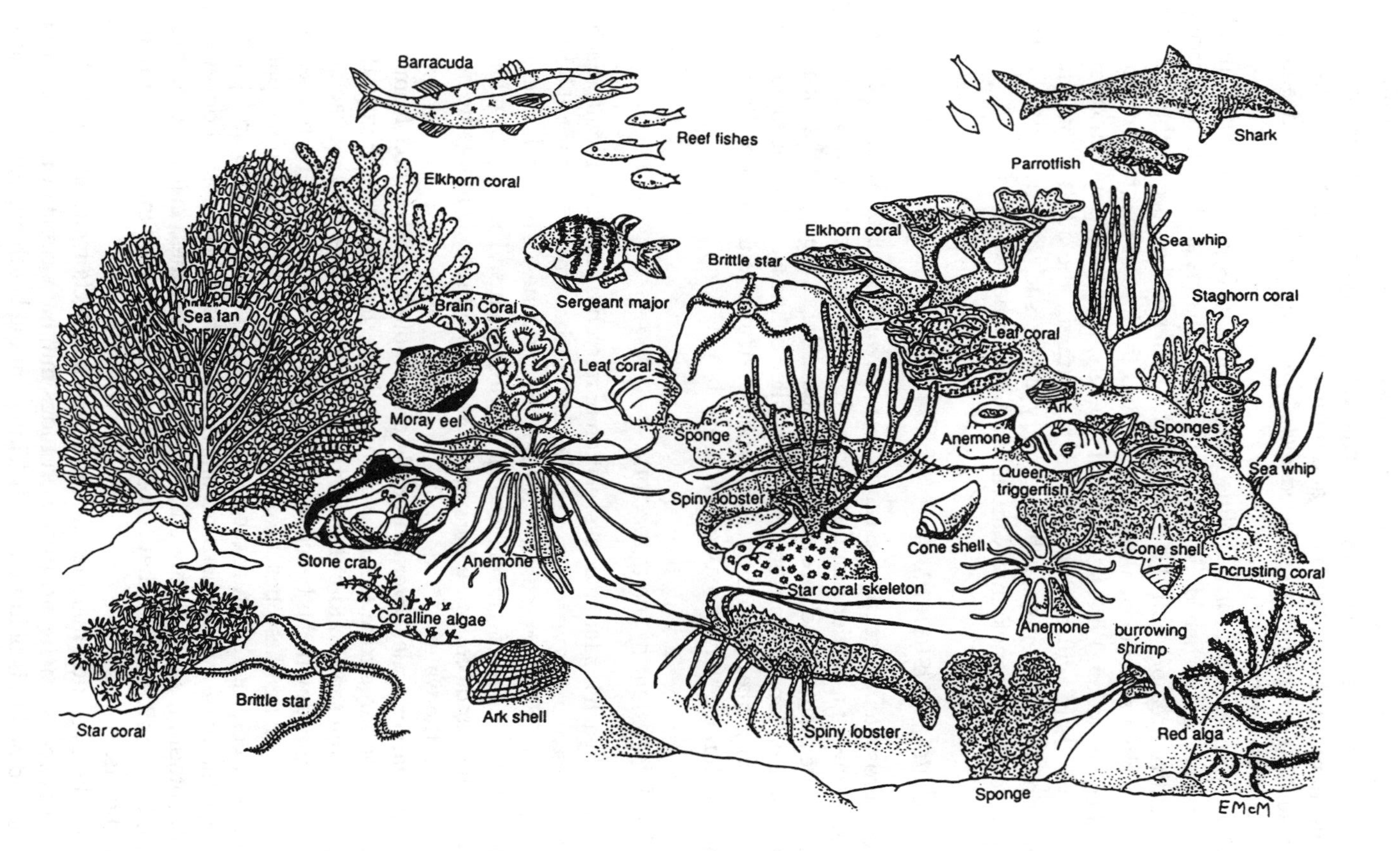

Figure 15.8 Life in a coral reef. (Illustration by Elizabeth A. McMahan.)

turbid waters which cause stress and disease. As the human population increases, more reefs are damaged mechanically by boats and divers, waters receive more wastes, and turbidity and sediments on the reefs from dredging increase. You can recognize stressed reefs by white bleached coral heads or heads covered over with algae.

The species of fish and shellfish from coral reefs which humans find edible are being overfished. For example, spiny lobsters are becoming scarce in the entire area of south Florida and the Bahamas because of the heavy fishing by divers, fishermen using traps, and the loss of reef habitat due to coastal construction. With fewer adult lobsters, the larval populations may get so low there will not be enough to maintain the adult populations. International agreements may be necessary to keep such fisheries alive. Reefs are protected in Pennekamp National Park and other reserves.

15.5 Beach ecosystems

Special ecosystems occupy the sandy, wave-swept beaches. Refer again to the overview of a barrier island in Figure 13.7. There is a terrestrial ecosystem of dune plants, which are able to withstand the salt spray on their leaves. Other life develops among the debris deposited by the waves at the top of the beach as a *drift line* community. A microecosystem is found among the sand grains within the wave-breaking zone of a beach.

15.5.1 Ecosystem of the wave zone

Within the beach is a fantastic association of microbes and tiny animals. Each breaking wave pours water through the sand so that when it comes out it has been filtered. Some water drains down from the dunes. The beach filter is a little like the filter of gravel used in sewage plants. The tiny organisms between the sands consume the organic matter and release inorganic nutrients back to the water.

On a larger scale, several animals are adapted to filter food from the waters in the surge zone. Coquina clams and *Emerita* sand crabs (Figure 15.8) pop out of the sands and burrow again with each wave, moving back and forth, up and down the beach with the tide. There used to be hordes of sandpipers along these beaches, but the numbers are fewer now, partly because of continual disturbances from people, cars, and dogs and partly because many of the northern nesting grounds have been lost to development. Although sea gulls are still found on the beaches, many more have moved to the cities where they have become scavengers of trash in the urban ecosystems.

At the high tide margins, floating debris collects including *Sargassum*, other seaweeds, shells, driftwood and logs that have come out to sea from rivers, and all kinds of human trash. Tiny jumping beach hoppers and many other organisms live in these high tide lines (Figure 15.9).

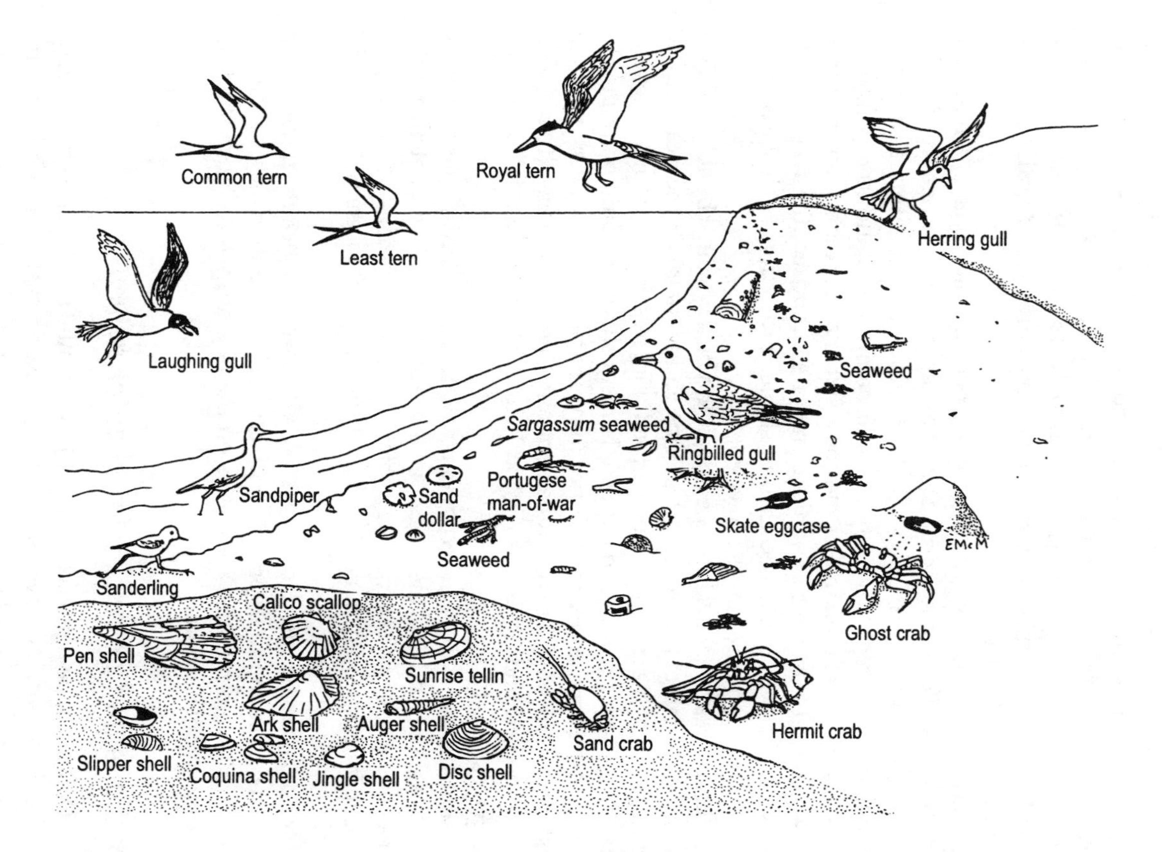

Figure 15.9 Creatures and debris found at high tide lines. (Illustration by Elizabeth A. McMahan.)

15.5.2 Sand dune ecosystems

Above the beach drift line in undisturbed areas, sand dune ecosystems occur (Figure 15.10). The dune vegetation and the sand dunes develop together, built up from the blowing beach sand. The succession of a sand dune starts with pioneer plants, such as the grasses and sea oats. Their seeds are easily carried by birds and small mammals. These tall grasses catch the sand carried by the wind. Their long fibrous roots reach down to the groundwater that collects from rain percolating through the very porous sand grains (observe groundwater in Figure 13.7).

Where the dunes have not been disturbed for many years, a maritime forest develops. The spray from sea storms tends to kill the leaves. But, the forest develops a tight canopy so that the leaves underneath are protected from salt spray. A high diversity of sable palms, oaks, and saw palmettos develops. This vegetation makes the coast stable and secure against storms. If the vegetation is removed and the dunes are bulldozed, the sand starts to move with the wind and becomes unstable. Protection against seawater breakthrough in hurricanes is removed.

To keep the system healthy, the dunes and beaches must be able to adjust to the rise and fall of the sea and to storms and to maintain their capacity to reform. The beach zone should be a broad one, free from paving and exotic kinds of vegetation, for this to happen. Houses should be built on pilings so the sands can move under them. Since the dune plants are not very resistant to vehicles, dune buggies should be kept out. The natural vegetation makes a beautiful surrounding and is a good habitat for many animals including rattlesnakes.

Sea turtles are part of the ecosystems of the coastal ocean, but they have to come ashore and lay their eggs in the beach sands. The hatched baby turtles have to get back to the surf, which can be very difficult when beaches are occupied by people, dogs, and cars. There may not be enough protected beaches for the turtles.

Florida's beaches used to be covered with carpets of beautiful shells, and some still can be found in protected places such as the state parks. Unfortunately, many communities allow cars on the beaches which crush shells as fast as they come ashore.

Questions and activities for chapter fifteen

1. Define: (a) pelagic ecosystem, (b) continental shelf ecosystem, (c) coral reef, (d) beach ecosystem, (e) sand dune ecosystem, (f) Gulf Stream.
2. On a map of Florida, locate its continental shelf, pelagic regions, coral reefs, beach ecosystems, estuaries, and a coastal county.
3. Give three examples of diversity in a coral reef.
4. What kinds of services do you think people can feed back to the marine environment to restore the fish yields?

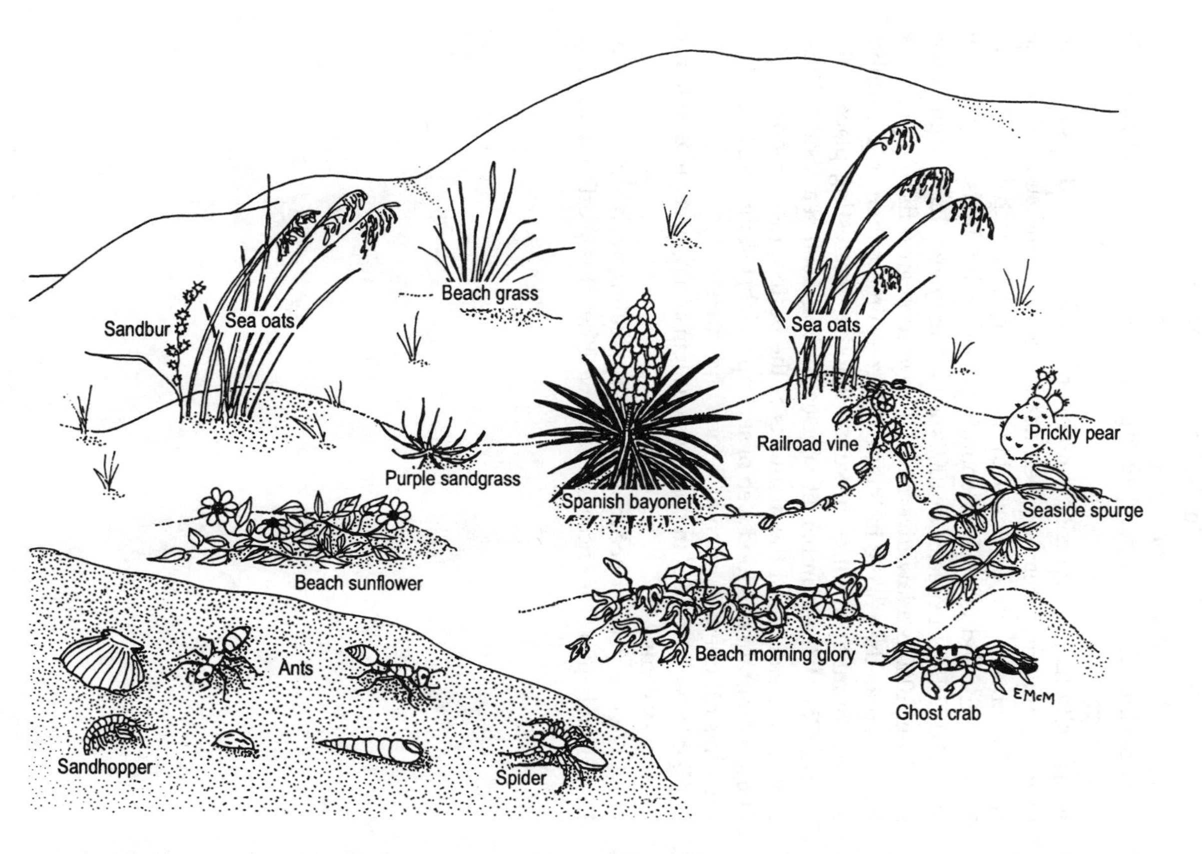

Figure 15.10 Organisms of upper beach and sand dunes. (Illustration by Elizabeth A. McMahan.)

5. Describe how beaches protect the land.
6. Describe the roles of sand in Florida's coastal ecosystems.
7. Draw a model of the ecosystem of the continental shelf, by replacing the box in Figure 15.2 with the bottom community.
8. Describe dune succession.
9. How do the reef corals get their organic fuel for growth? Their nutrients for growth?
10. A diversity index (diversity is number of different kinds) is a useful way to compare systems — their health, stresses, and stage of succession. To obtain one diversity index, count the number of different kinds of individuals when counting 1000 individuals. A count of about 30 different species in 1000 means the system has a high diversity; 6 per 1000 is a low diversity. Go to your nearest fishing port. Ask fishermen as they bring in their catch to estimate their total number and the number of different kinds. Another interesting place is the beach. Count the number of species among 1000 sea shells. These represent a sample of the mollusks at the bottom just offshore.
11. Draw a diagram of a food web of shrimp or lobster. Be sure to include the physical energies such as winds and currents.
12. Explain how "red tide" in the ocean system could be compared to fire in land systems.
13. How is your nearest ocean community dealing with shore erosion? Discuss whether it is successful and whether it is worth the cost.

chapter sixteen

Estuaries

Along the sea coast where land and freshwaters join the sea are the estuaries, salty waters with fertile ecosystems important to coastal fisheries. An *estuary* is an arm of the sea mostly surrounded by land. In most estuaries, freshwater streams mix with the saltwaters that the tide exchanges with the open sea. The main ecosystem of many estuaries is composed of plankton and bottom organisms (benthos). We call this ecosystem *plankton-benthos.* Other ecosystems in estuaries are grass flats in shallows where bright light can reach the bottom, oyster reefs, and wetlands around the edge.

Ranging from almost freshwater conditions to saltier water than in the open sea, salinity is the most important factor controlling the species. Expanding economic developments increasingly impact estuaries, releasing wastes, dredging navigation channels, and changing the circulation. Some of the largest estuaries in Florida are labeled in Figure 16.1, which also shows the connecting streams. This chapter ends with problems in managing some of Florida's estuaries.

Figure 16.2 has the main features of many Florida estuaries — a freshwater stream flowing in from the left and the sea on the right exchanging waters through an inlet in the outer barrier island of beaches. The estuarine plankton-benthos ecosystem occupies the deeper waters in the middle. Bottom grass communities are in shallower and clearer zones. Oyster reefs develop bars across the currents. Marshes are on the shallow shores.

16.1 Salinity

Salinity may range from 0.01 parts per thousand (ppt) salt in freshwater to 36 ppt for full seawater to 45 ppt in Florida Bay, where evaporation exceeds rainfall. Where freshwaters join saltwater in depths of 5 ft (1.5 m) or more, two layers may develop with the lighter freshwater on top and the heavier saltwater underneath.

The zones of salinity that slide in and out of the estuary with the tide are also dependent on the velocity of the freshwater streams. Some estuarine organisms are especially adapted to the large variations in salinity. For

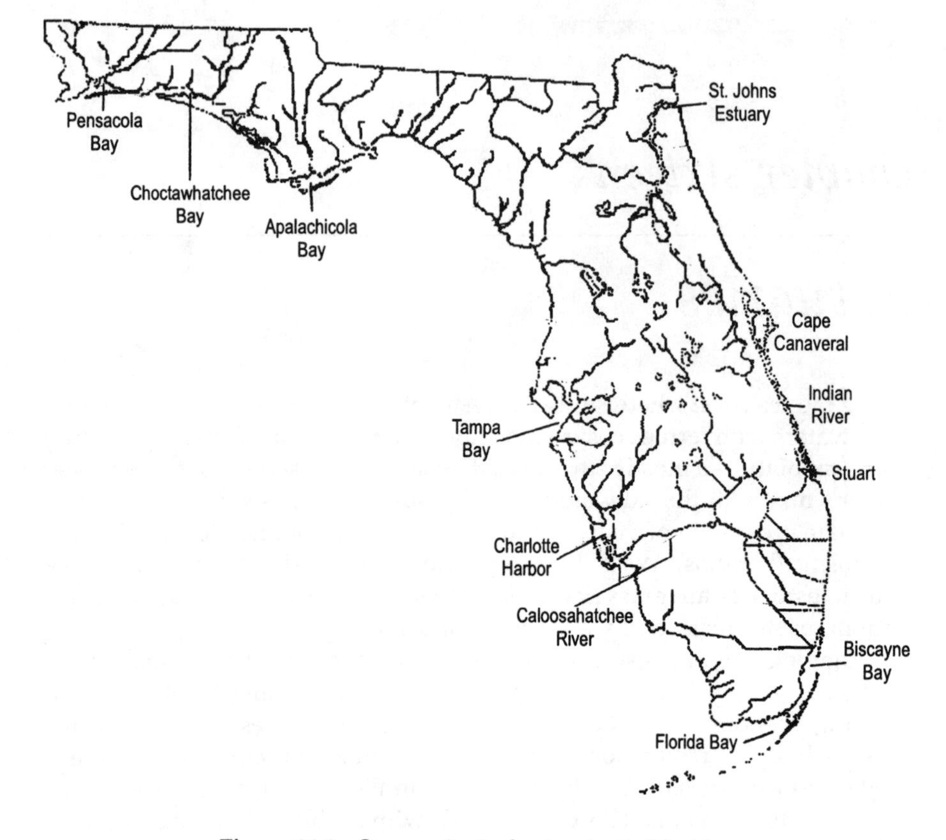

Figure 16.1 Some principal estuaries in Florida.

example, commercial oysters can live in water with salinities from 2 to 30 ppt. In estuarine organisms, energy must be used for kidneys and other special adaptations to varying salinities rather than for developing diversity. There are fewer species in an estuary than in rivers or the open sea, but, because fertility is high, there is high production of the few species present.

16.2 Circulation

Some of the high productivity of estuaries is due to their good physical circulation (high kinetic energy). Some of this energy comes from the discharge of rivers, some from the tides, and some from coastal winds. Phytoplankton cells are kept suspended by the motion which also helps their photosynthesis and that of attached plants by bringing necessary materials such as carbon dioxide, nitrogen, and phosphorus to them. The circulation helps the recycle process. The stirring keeps organic matter particles in suspension to be eaten by zooplankton. Where sediments are loose, the

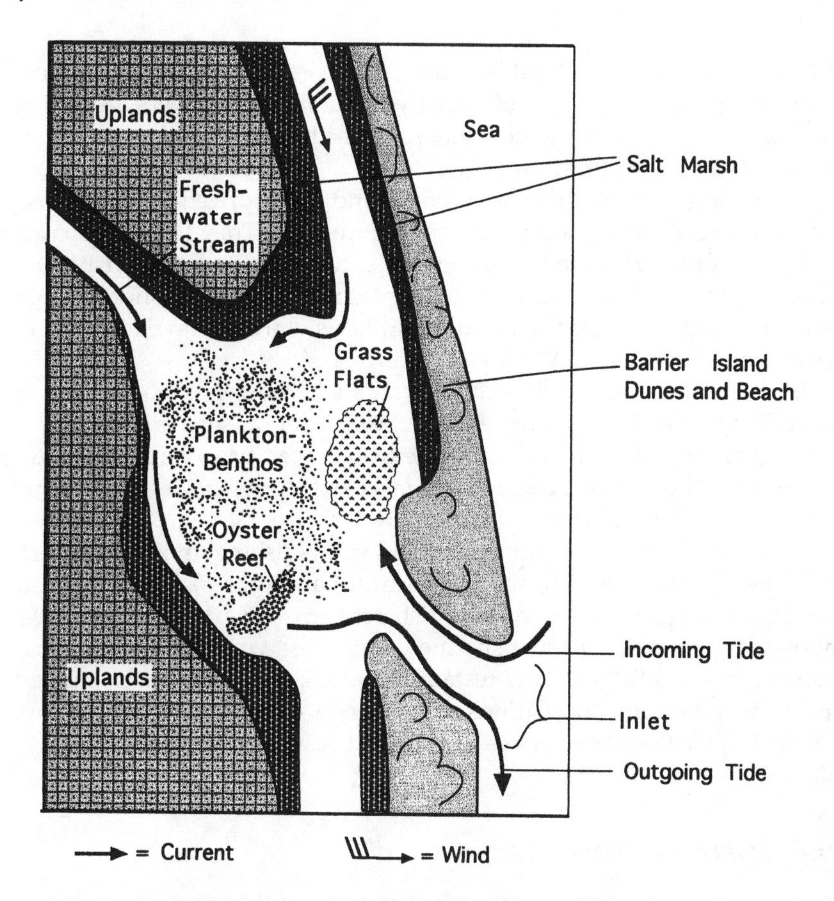

Figure 16.2 General features of an estuary including freshwater stream, currents, inlet to the sea, and areas occupied by ecosystems.

circulation makes the water turbid, reducing the light for the phytoplankton and bottom plants.

Where rivers flow into the estuary, the Coriolis force turns the stream to the right (Chapter 14). When the seawater flows in, and the Coriolis force bends it to the right, the right side of the estuary has higher salinities. Winds develop waves and currents bending to the right. Differences in density between freshwater and saltwater result in currents to the right (see Chapter 14). As viewed from above, the resulting circulation is counterclockwise (Figure 16.2).

16.3 Food-rich estuarine nursery

High levels of organic food develop in estuaries. Some comes down the rivers, and some is produced by the plants and algae aided by the circulation and nutrients. As the sunlight increases, estuaries have a burst of productivity in spring and a high rate of growth in summer. The rivers also bring in sediments such as sand and clay that form the bottom ooze in and on which ecological bottom communities live.

Many of the species of oysters, crabs, and fishes prized by sports people and commercial fishing are estuarine (Figure 16.3). They live and spawn as adults offshore, and come into the estuaries as larvae and small fishes as the waters warm up (Figure 16.3). Mullet, for example, spawn offshore and grow up in the estuaries. Then, as adults, mullet swim in schools up rivers to freshwater lakes all over Florida (Figure 16.4).

Some fishes have a cycle with migrations in the opposite direction (Figure 16.5). Shad and sturgeons breed in headwater streams and have young that pass through the estuary on their way to the sea, growing rapidly during their time in the estuary. Because the larvae of many species mature in the estuary, it is often referred to as a *nursery*.

Because of too much commercial and sports fishing, many of the estuarine fishes (mullet, redfish, snook, sea trout, and others) were depleted in Florida. Consequently, the marine fish resource was diminished, a value important to tourists, sports fishermen, and those who expect inexpensive, fresh commercial fish on their restaurant tables. Net fishing was banned in Florida to give fish populations a chance to come back. Seafood prices became higher than beef prices when most restaurants began serving fish from out of state.

16.4 Systems overview

An energy systems diagram summarizes some of the main processes in an estuary (Figure 16.6). In addition to the sun and wind, the outside energy sources of the estuary system are the freshwaters of the rivers and the saltwaters of the ocean coming in with the tides and the larger species that migrate in from the river and from the sea. The river also contributes nutrients and organic detritus. The wind, the river, and the tide keep the waters circulating and exchanging smaller organisms such as zooplankton, invertebrates, and eggs and larvae of various animals with the sea. This constant inflow of species provides organisms suitable to a wide range of conditions and makes it easier for the system to adapt to change.

Where the water flows are large, the small organisms are mostly those of the inflows (river or sea); however, when the exchange of water is less, the estuary develops a distinctive community of organisms different from that in the outside waters (Figure 16.3). These species can live with a wide range of salinity.

Many species of birds, flying in and out, are part of the estuarine ecosystem. The gulls feed on animals in the estuarine mud and on the beach at low tide. Wading birds, such as herons, willets, and rails, feed in the marshes, and diving birds, such as pelicans and cormorants, fish in the water (Figure 16.3).

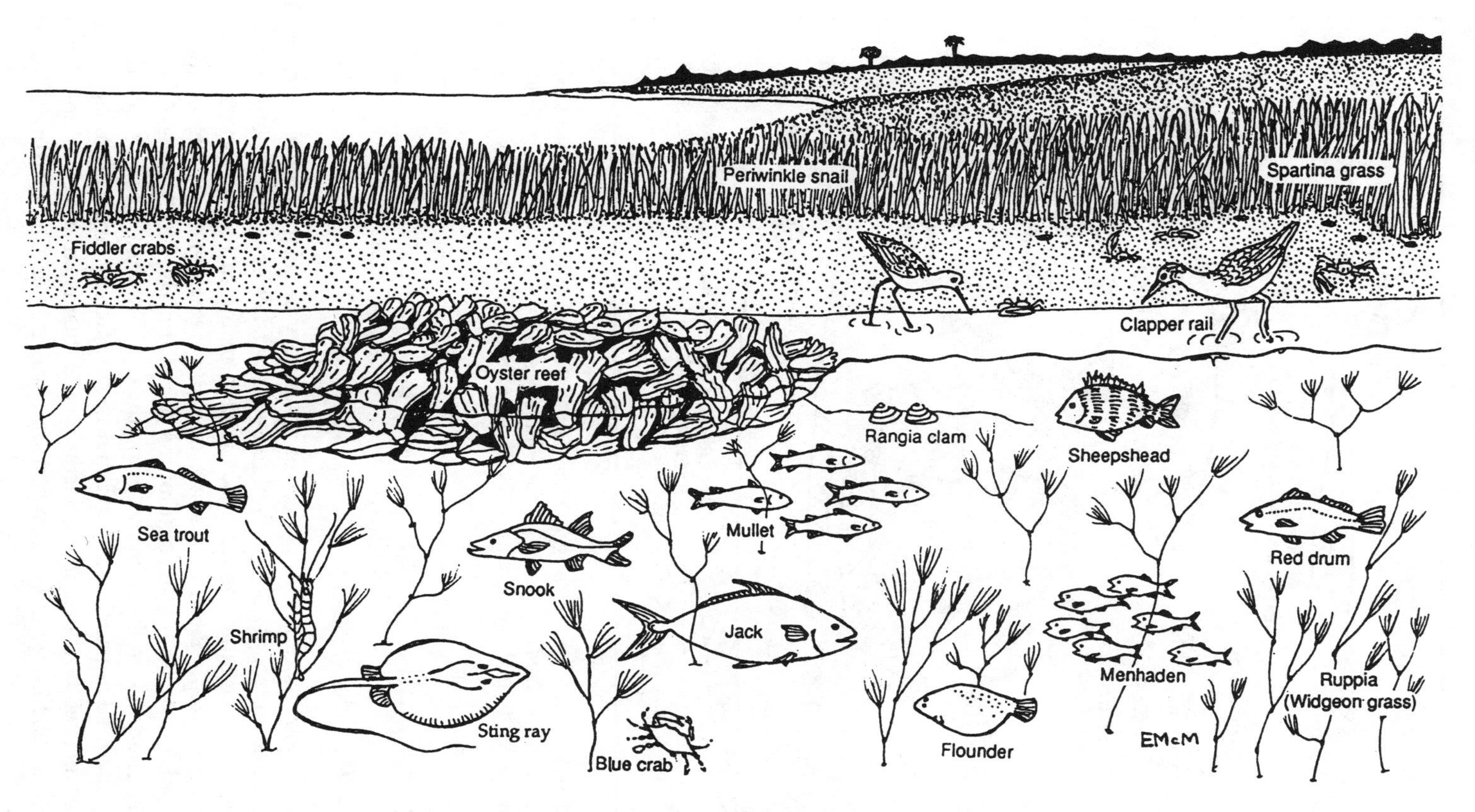

Figure 16.3 Life in a Florida estuary. (Illustration by Elizabeth A. McMahan.)

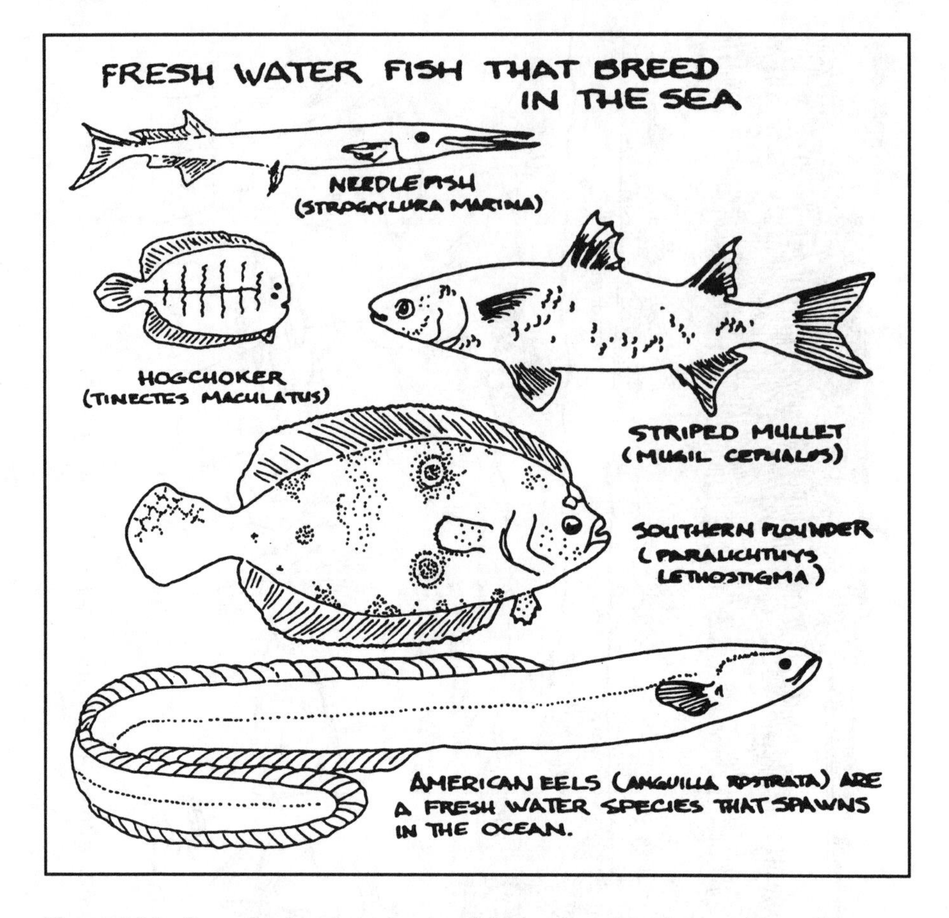

Figure 16.4 Some freshwater-estuarine fish that breed in the sea. (From Wharton, C.H., *Forested Wetlands of Florida — Their Management and Use,* Center for Wetlands, University of Florida, Gainesville, 1976. Illustrated by Joy Bartholomew.)

16.5 Estuarine ecosystems

16.5.1 Plankton-benthos

In the deeper, more turbid waters of an estuary develops an ecosystem of phytoplankton, zooplankton, plankton-eating fishes, and bottom organisms (Figure 16.7). Organic matter is suspended in the turbulent waters, some from the phytoplankton and some from the fecal pellets of the animals. Zooplankton are found throughout the water at night, but they tend to sink into the shadows on the bottom during the day. They eat the phytoplankton and suspended organic matter and in turn are eaten by larger animals. Other organic matter flows in from decomposing plants in marshes and from upstream. Microorganisms then consume this matter. The mixture of dead

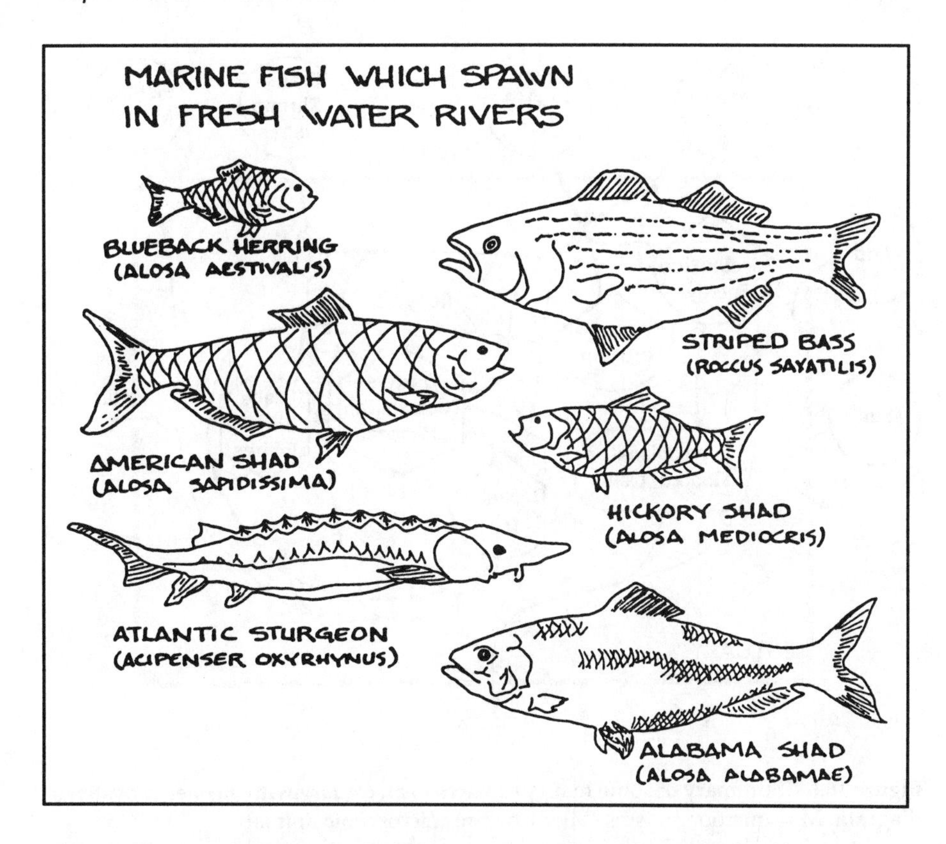

Figure 16.5 Some saltwater fish that spawn in freshwater rivers. (From Wharton, C.H., *Forested Wetlands of Florida — Their Management and Use,* Center for Wetlands, University of Florida, Gainesville, 1976. Illustrated by Joy Bartholomew.)

organic particles and the microbes that are decomposing it is called *detritus.* Detritus is a rich food for the rest of the estuarine food chain. Some bottom animals (benthos) consume the sedimentary detritus; some live in the sediments with tubes to pump in waters for respiration and for filtering out food. Many of the zooplankton are larval stages of bottom dwellers.

16.5.2 Grass flats

Where waters are clear and shallow, 1 to 6 ft (0.3 to 1.8 m) deep, they receive enough light to produce dense beds of bottom plants and animals. The species depend on the salinity. In Florida there are freshwater species in the upper reaches, widgeon grass in low salinity zones, and turtle grass in higher and little-varying salinity zones. Turtle grass ecosystems within estuaries are like those in more tropical waters of south Florida (Figure 15.4) but with fewer species. Species of small, filter-feeding clams and scallops among the

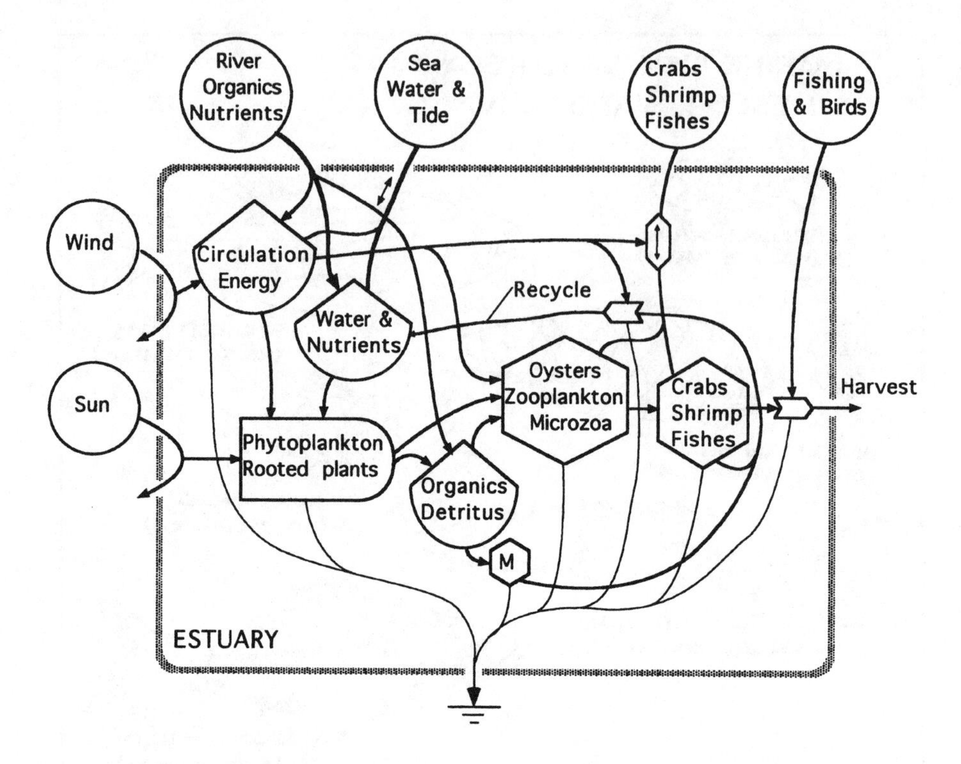

Figure 16.6 Summary of some main processes in an estuary with an energy systems diagram. M = microorganisms. Microzoa are microscopic animals.

plants help keep the waters there clear. Shoal grass has adapted where the salinity is changeable, sometimes higher than seawater.

16.5.3 Oyster reefs

Oyster reef ecosystems develop in some estuaries where the salinity varies widely. Oysters attach to each other, building up mounds of shells. As bottom oysters die, larvae attach to old shells, increasing the size of the reef. As the mounds build up, the oysters have increased access to currents which bring in food and carry away wastes. Oysters filter algae and organic particles from the water for food. Oysters also remove the suspended sediment particles, depositing them with mucous by-products on the bottom around the reef.

Oyster populations are reduced by drills, disease, and harvesting. *Drills* are snails that drill a small hole through the shell of the oyster, eating the insides. Drills and other predators and competitors increase at high salinity when conditions are fairly constant. Where there are large variations in salinity, drills and competitors are reduced and oysters prosper. However,

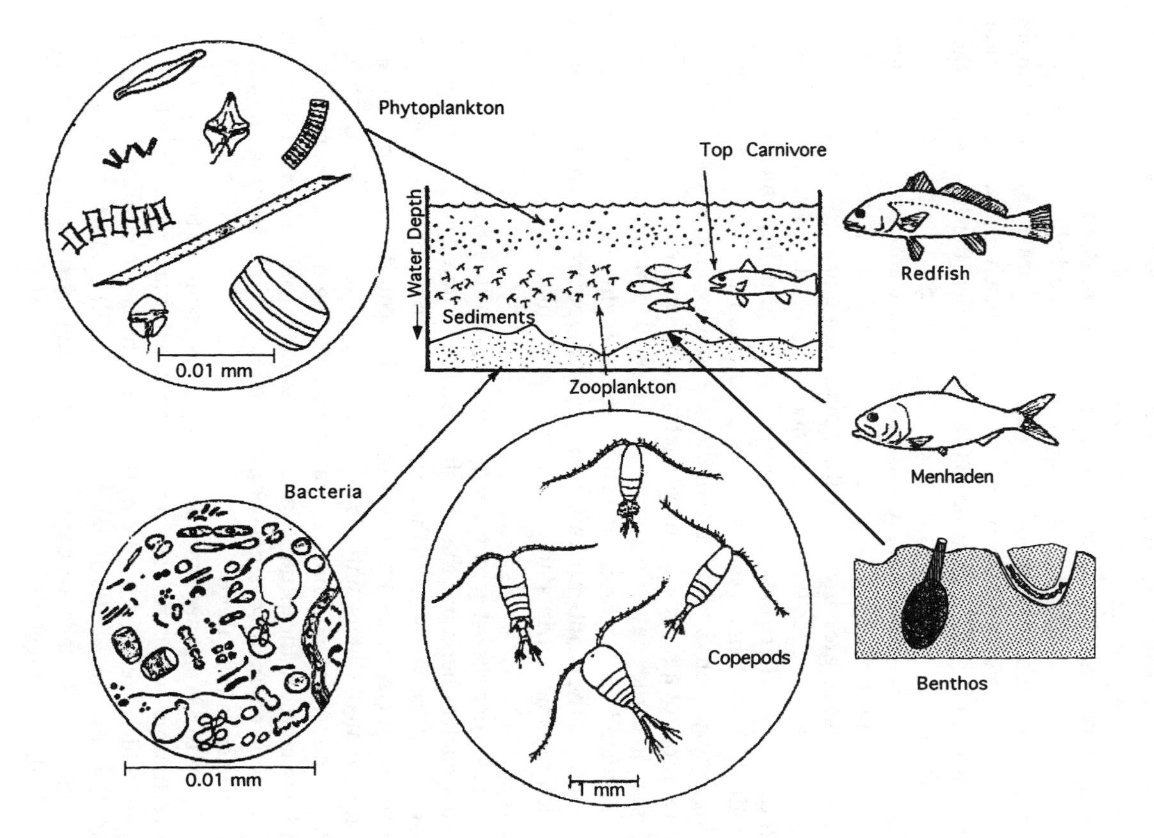

Figure 16.7 Plankton (suspended organisms) and benthos (bottom organisms) in an estuary.

oysters are also killed if the estuarine waters become completely fresh in times of flood. When conditions of food and salinity become unfavorable, oysters develop diseases that consume their bodies quickly and cleanly. Without these diseases, dead animals decompose slowly, with toxic by-products delaying the growth of replacements.

Even in high salinity, oyster reefs can form in the *intertidal zone* between high and low tide where their shells protect them and alternating exposure to heating by the sun keeps other species out (Figure 16.3). For example, intertidal reefs occur in the estuary at Crescent Beach in northeast Florida. Since intertidal oyster reefs cannot filter-feed when the tide is out, these oysters do not grow as fast as those on deeper reefs.

Whereas coral reefs have uniform conditions and high biodiversity, estuarine oyster reefs have varying conditions and low diversity of species. Species of oysters growing on drilling platforms and reefs in the open sea have to share resources with many other kinds of filter-feeding organisms (high diversity). There are not enough of one kind to be commercial.

Industries that harvest oysters help maintain the size of the reefs by putting empty shells back on top of the reef to which the larvae attach. This is an example of a fishing industry that returns services to help the natural processes; in earlier days, the reefs were removed to make roads. Oysters are also cultivated on floating rafts with racks hanging below on which the oysters grow. The number of rafts that can be supported depends on the rate at which organic detritus and suitable phytoplankton food appear in that estuary.

Some oyster reefs become contaminated with human disease microorganisms, such as hepatitis, that the shellfish have filtered from contaminated estuarine waters. Many areas that used to supply oysters to be eaten raw are intermittently closed because of high bacterial counts. Oysters moved to uncontaminated waters will clean themselves in a few days. Oysters on contaminated reefs off-limits to harvesting can be a useful sanctuary for supplying larvae to the estuary. In 1962, after radioactive fall-out from atomic testing occurred, many oysters became radioactive from feeding on radioactive particles in the fallout.

Figure 16.8 is an energy systems overview of some of the main processes on an estuarine oyster reef. The waters from the tide and river bring food and oxygen, which are used by the oysters. The shells accumulate as a reef on which the new larvae attach, later becoming adult oysters. The oysters are removed by drill snails, disease, and harvesting by fishermen, who return some shells to the reef.

16.5.4 Marshes and mangroves

An estuary is usually edged with wetlands: swamps with mangrove trees or marshes with salt-tolerant grasses standing in water for part of the day. As the waters flow in and out of the marshes, there are exchanges of nutrients,

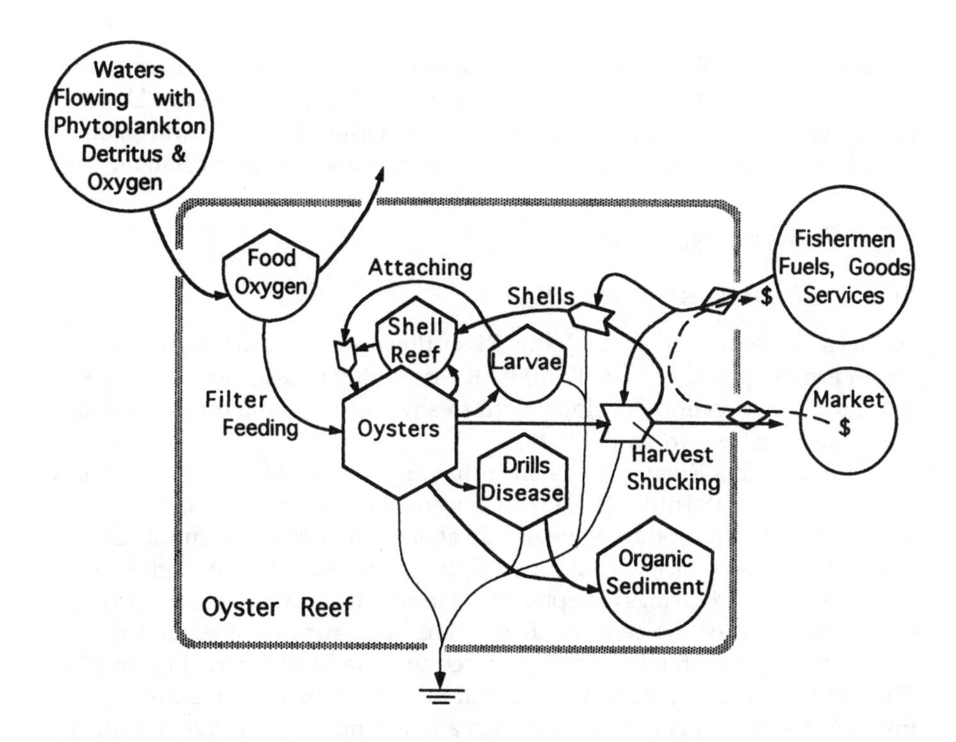

Figure 16.8 Summary of some main processes in an oyster reef ecosystem with an energy systems diagram.

organic matter, and organisms. Tidal channels develop. Many invertebrates live in the marsh mud. The marsh offers excellent protection for larvae and small fish which come in and out with the tides. With their leaves up in the air and their roots deep in rich sediments, marsh plants and mangroves are highly productive, especially when the salinity is not too high to inhibit their transpiration. Where salinities are very high, mangroves become dwarfed. For more on saltwater wetlands, see Chapter 18.

Salt marshes have a high diversity of terrestrial insects which may keep populations of plants and animals stable. Spraying marshes with insecticide to eliminate mosquitoes kills much of the rest of the animal life and is both expensive and ineffective, as mosquitoes develop resistance. On the drier, land side of salt marshes and mangroves where water is intermittent, hordes of saltwater mosquitoes develop that make areas within several miles almost uninhabitable.

To get rid of these mosquitoes, many of the shallows were turned into shallow freshwater ponds by the construction of small earthen dikes. The small mosquito fish in these freshwater pools and wetlands eat mosquito larvae, gulp air when oxygen is low, and reproduce quickly after a drought has temporarily reduced their populations and habitat. These shallow pools grow freshwater plants used by ducks and other waterfowl. In other places, the zone of intermittent saltwater has been filled (e.g., with sand). Unfortunately, both the fill and the dikes cut off the marine larvae of shrimp, crabs, and fishes from marshes, eliminating the role of marshes as marine nursery.

16.6 Florida estuaries

16.6.1 St. Johns River estuary

At Jacksonville, where the St. Johns River flows slowly into the Atlantic, the depth is maintained at 40 ft (12 m) by the dredging of deep ship channels. A two-level stratification develops, with freshwater outgoing on top and saltwater being drawn in on the bottom.

The St. Johns River is unusual in having such a gradual slope of only about a meter in 180 miles (300 km). Consequently, the rise and fall of the sea level at the mouth sends a wave of tide all the way upriver without sending any sea saltwater past Green Cove Springs. Because of the slightly salty groundwaters and springs (Chapter 13), marine fishes go many miles upstream.

At the turn of the century there were huge runs of shad that moved southward up the St. Johns River to breed in fresh headwaters (Figure 16.1). The mullet and many other marine fish moved in the opposite direction to breed in the sea. This estuary was also a major nursery for juvenile shrimp and marine fish. The St. Johns River was always somewhat fertile from the nutrients inflowing from the springs and had abundant grass flats in places, but it was black with swamp water in other places. Later, sewage, paper wastes, and other industrial wastes caused a zone of the river at Jacksonville to remain at such a low level of dissolved oxygen that the normal movements of fishes, shrimp, and larvae were blocked.

Dams and water control weirs stop migrating fishes. A dam that formed the Rodman pool cut off most mullet and other migratory fish from the Oklawaha River, although a few passed through with boats going through the locks. Fish migrations can be restored by removing dams and weirs. Or, small channels for fish can be arranged around dams. Allowing these channels to be filled with vegetation keeps the waters flowing slowly, and lakes above are not drained.

16.6.2 Indian River and Cape Canaveral estuaries

Stretching half the length of the state behind the front beach of the Atlantic Ocean is a shallow natural estuary, the Indian River, which is connected with the sea through several inlets. A 12-ft (4-m) deep "intracoastal" channel was constructed down the estuary and through various islands to form the inland waterway for small boats here and around the rest of Florida. The spoil (dredgings) was left as sandy island bars that became covered with the *exotic* (not native) tree, Australian pine, which had been introduced earlier.

Where connections with the sea are small, the regular ups and downs of the 3-ft (0.9-m) tide twice a day do not penetrate far. But, when the sea level is raised for several days or more by wind stresses over the open sea, then saltwaters have time to flood into the Indian River and the various shallow estuaries that used to cover Cape Canaveral.

In order to keep water levels in Lake Okeechobee from getting too high (relative to the levee around it), waters are discharged in some years into the Indian River estuaries at Stuart on the east coast (Figure 16.1) and through the Caloosahatchee River on the west. During times of the freshwater release, these estuaries turn fresh, killing the normal estuarine life. Many believe ways must be found to keep freshwaters for inland use.

16.6.3 Miami estuaries

With the development of Miami and Miami Beach, the nearby estuaries became eutrophic (high nutrients), with sewage and other wastes passing down boat channels. In some places, the plankton shaded out the grass flats, but in other places huge densities of new tropical invertebrates developed. Dumping these waters into the sea is wasteful, when instead they could enrich wetlands inland and be restored to good quality (Chapter 18).

When nuclear power plants were built at Turkey Point with large needs for cooling waters to remove the excess heat, initial use of the estuary was proposed. This was stopped by public opinion. Instead, about 5.6 square miles (3600 acres, or 1400 hectares) of dwarf mangroves were dredged to make a network of canals to recirculate the hot water.

16.6.4 Florida Bay

Between the line of the Keys and the mainland is Florida Bay, where rainfall is much less than elsewhere in Florida. The bay is a shallow estuary with grass flats. It is cut off from circulation energy so that the saltwater evaporates in greater quantities than the amount received in rain and runoff. Thus, the salinity in many years rises from the usual 36 ppt in the sea off south Florida to 45 ppt or higher. Part of this area is in the Everglades National Park. Because high-powered fishing boats are allowed, many manatees are killed by outboard motor blades. The western part of this bay is one of the principal areas that produce pink shrimp.

16.6.5 Charlotte Harbor and Tampa Bay

On the west coast of Florida are estuaries that receive high phosphate from rivers that drain the rich phosphate rocks around Bartow. Even before mining started, the Alafia and Peace Rivers had higher than usual phosphate contents so that their estuaries were fertile. Mining increased the outflow of phosphate.

One nutrient in excess does not make a eutrophic problem if other nutrients are limiting; however, Tampa Bay receives many other kinds of wastes, sewages, and runoffs, so that it has become one of the most fertile estuaries in the world. In inner areas, nutrients are clearly pollution in the sense of reducing oxygen levels that can cause fish kills and reduce diversity. Further seaward, this nutrition may increase general levels of productivity.

16.6.6 Apalachicola Bay

In the panhandle, the Apalachicola River, draining much of Georgia, reaches the Gulf of Mexico carrying heavy loads of clay sediments and is thus quite different from other Florida rivers. The estuary received surges of freshwater and developed magnificent oyster reefs that have supported stable commercial operations for a century. Seasonal flooding has given the estuary ranges of salinity from 2 ppt salt to nearly marine 25 ppt salt and has concentrated the biological resources in those few species that can live with sharply varying salinity. Lately, however, dams upstream have reduced the amount of organic matter coming down the river and modulated the pulse of alternating salinity. Less sediment is deposited in wetlands downstream in Florida. In order to give shrimp trawlers and party boats better access to the outside waters, 12-ft (4-m) channels were dredged through the barrier beach which allowed more saltwater to come in on the bottom. This higher salinity has allowed a diversity of other species to come in, reducing the production of commercial oysters.

Questions and activities for chapter sixteen

1. Define: (a) estuary, (b) nursery, (c) microzoa, (d) filter feeders, (e) sediment, (f) salinity, (g) parts per thousand (ppt), (h) fertility, (i) reef, (j) intertidal zone, (k) detritus.
2. Give examples of the types of plants and animals found in an estuary.
3. Discuss the physical characteristics of an estuary.
4. How is an oyster reef formed?
5. Find out more about oyster farming operations and report your findings to the class.
6. Give two reasons for the high productivity of underwater grass flats.
7. Discuss the methods of control of saltwater mosquitoes and the advantages and disadvantages of these controls.
8. Describe the effect of phosphate mining and sewage disposal on the Tampa Bay estuary.
9. Discuss conflicting uses of estuaries.
10. Develop a photo essay of Florida's estuaries. Your photographs may come from magazines, newspapers, or your personal collection.
11. If possible, take a field trip to an estuary and collect as much information about it as you can. Divide into groups to:

(a) Investigate outside sources — sun, wind, tide, rivers, pollution.
(b) Observe zones from low to high tide.
(c) Observe producers such as algae and seaweed; look for patches of seaweed.
(d) Look for the presence of a marsh. Is there a marsh? If there is, how big an area does it cover and what kinds of plants or animals are present? If there is no marsh, what is at the edge of the estuary?
(e) Observe bottom organisms; at low tide, dig in the mud. Look for oyster mounds.
(f) Observe fish — look, and ask any people around what they catch.
(g) Observe birds — be quiet and watch; field glasses and a bird identification book are helpful.

Put all the information together. Draw a diagram of your estuary on the blackboard. How does it compare to Figure 16.6? Explain the differences.

12. Discuss diversity in estuaries and its relation to salinity and salinity variation.
13. Explain why the migration of fishes between freshwater and saltwater is an advantage for the fish, increases ecosystem functions, and is an advantage to human society.
14. How would you decide what human activities would be best for an estuary like the one you studied? Consider alternatives as viewed with a map.
15. Check on seafoods for sale in your area and where they come from. Which of them spend some of their life in an estuary?

chapter seventeen

Freshwater ecosystems

Because of its abundant rainfall and its flat, irregular surface, Florida has many freshwater ponds, lakes, and streams, each containing interesting ecosystems. Over millions of years, rainwaters percolating through wetlands and becoming acid dissolved the underlying limestone, creating water-filled depressions and sinkholes. Lakes also formed in old stream beds and among dunes left when the sea retreated. Human activity added reservoirs, quarry pits, ponds to delay storm runoff, and basins from phosphate mining.

17.1 Ponds and lakes

Shown in Figure 17.1 is a sketch of a Florida pond ecosystem. Because many Florida lakes are as shallow as ponds, some lake ecosystems are similar. Figure 17.2 is a systems diagram of the main components and processes. You can trace the action of light on the left driving photosynthetic production of the organic matter that may accumulate to cover the bottom and help support the rest of the consumer food chain: *zooplankton*, small bottom invertebrates, fishes, and microbes. The carbon dioxide needed for photosynthesis comes into the water from the air and from decomposing organic matter. Ponds contain three groups of photosynthetic producers: *phytoplankton* (small, suspended algae), bottom plants and algae, and emergent plants. Other algae are attached to the leaves and stems of the plants. Runoff from the surrounding area brings in organic matter and nutrients.

Because of the spring dry season and the summer rains, water levels naturally rise and fall, producing a broad margin with wetland plants where the land is only submerged part of the year. The time when a wetland is covered by water is its *hydroperiod*. The alternating wet and dry conditions are important in many organisms' life cycles. The emergent plants along the shores add to diversity and absorb nutrients. This zone makes good habitat for wildlife.

The natural pattern is a seasonal high in late summer after thunderstorms and tropical storms, and a low at the end of the dry winter-spring

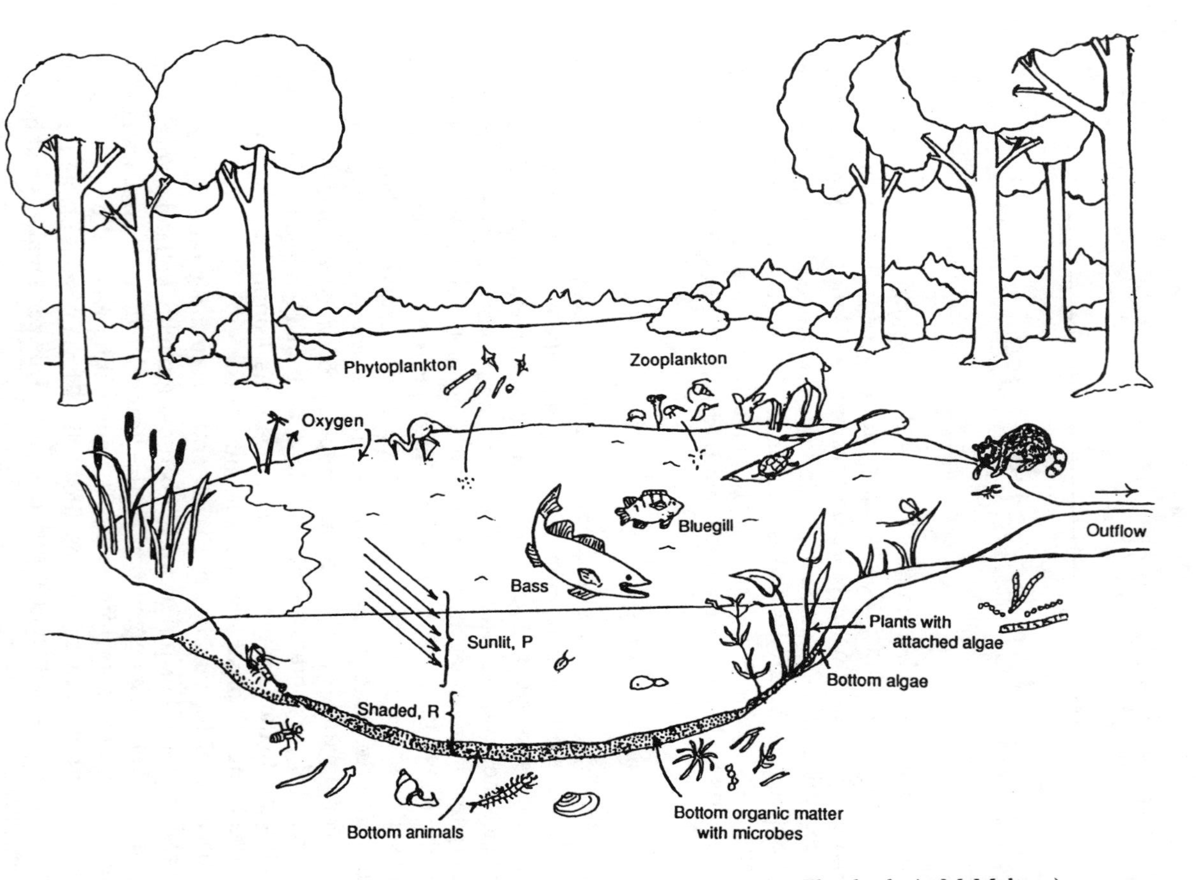

Figure 17.1 Components of a freshwater pond. (Illustration by Elizabeth A. McMahan.)

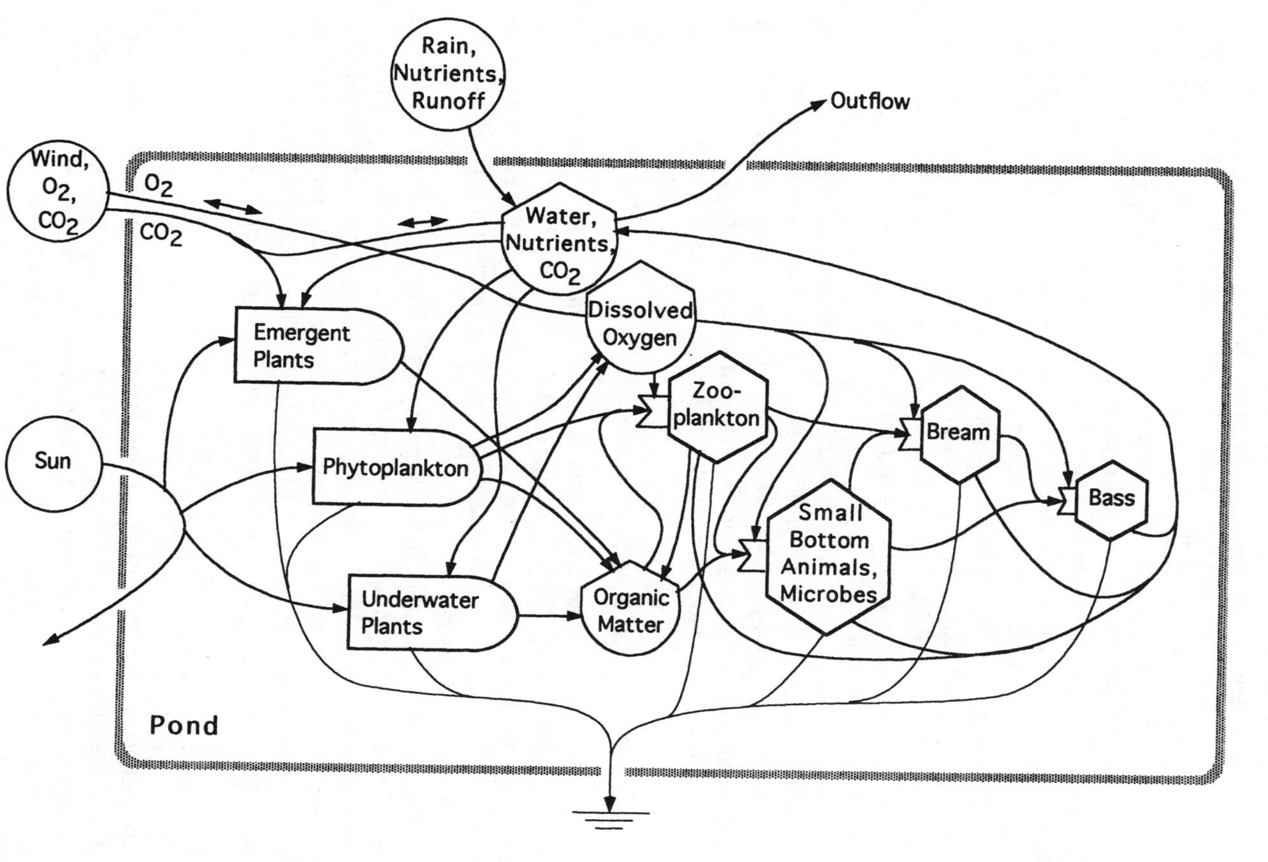

Figure 17.2 Systems model of a pond.

period. Year-to-year cycles run 7 years or more, with ponds and lakes standing 10 ft higher in the wet years than in the driest times.

As people develop homes around lakes, they often want to keep the water level constant so their piers and boats will be close at hand. Many lake levels are stabilized with small outflow dams. The effect has been to eliminate much of the emergent and wetland plants and wildlife. Restoring natural water level fluctuations and hydroperiods is needed in many situations.

Some ponds and lakes are *softwater* because they are located in sands and receive primarily rain, which has few dissolved chemicals. In limestone areas, calcium and carbonates are added to water from the dissolving limestone. Carbon dioxide and carbonates react to form bicarbonates. Many lakes that receive groundwaters through limestones and marl are *basic,* above 7 on the pH acid-base scale (1 to 14) because of their concentrations of carbonates, and *hard* because of their high concentrations of calcium and magnesium. Many of these are relatively clear.

Other lakes are brown (100 to 400 parts per million on the black-water color scale) because of stream inflows through peat or seepages from wetlands around the shores. Most lakes in Florida are shallow and differ from ponds in having longer *turnover times* (the time it takes for the water to be replaced) and a larger area which allows stronger wind waves to develop.

17.1.1 Eutrophic and oligotrophic waters

Waters with high nutrient levels are called *eutrophic*. Those low in nutrients are called *oligotrophic*. Most ponds and lakes in Florida are very fertile (eutrophic) with the nutrients for photosynthesis because they drain phosphate rocks or receive nutrients from stormwater, farms, runoffs, and wastes from cities and industries.

Oligotrophic lakes still exist in some places in Florida where the drainage of water includes only rain water or runoff from poor superficial sandy soils — for example, sandhill lakes. The sandhill lakes in north central Florida are oligotrophic. Although their fertility is not so great and their growth rates less, the variety and diversity of life are large. These lakes are surrounded with maidencane grass and rushes and tend to be clear for swimming.

17.1.2 Productivity

As explained in Chapter 4, photosynthesis uses light energy, carbon dioxide, water, and nutrients to make organic matter and oxygen. Respiration (animals, many microbes, and parts of plants) does the reverse, using organic matter and oxygen to do the work, releasing as by-products carbon dioxide, water, and nutrients. You can trace these processes in the pond systems diagram (Figure 17.2).

In the sunlit day, the pond metabolism first generates dissolved oxygen and organic matter in the water, then, at night, it consumes the oxygen again.

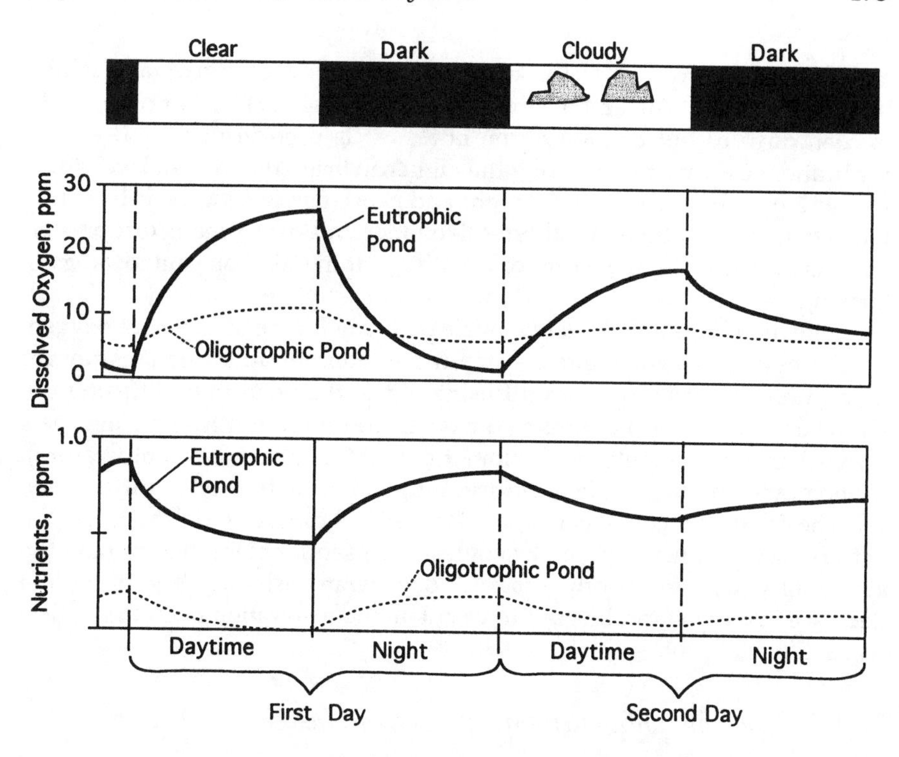

Figure 17.3 Changes in oxygen and nutrients in an oligotrophic pond (dashed lines) and a eutrophic pond (solid lines) through two days and nights.

The rise and fall of dissolved oxygen are shown in Figure 17.3. Also shown is the daytime fall of nutrients in the water as the plants take it up by photosynthesis to be released again with night respiration. The rise in oxygen is greater on a sunny day. The respiration following a sunny day is also greater because there is more organic matter and oxygen at the end of the day to stimulate consumption. After a night of respiration, there are more nutrients in the water to stimulate the morning photosynthesis. More metabolism is shown for the eutrophic condition than for the oligotrophic condition. Because the second day was cloudy, less sunshine fell on the pond, and photosynthesis was less. This caused less oxygen and organic matter to be produced.

The maximum amount of any gas that can dissolve in water (the saturation level) depends on temperature. For example, freshwater saturated with oxygen at 70°F (21°C) contains about 9 parts per million (ppm) dissolved oxygen. When the temperature increases, the amount of dissolved oxygen that can be held in the water decreases, causing the surplus to diffuse out of the water. If the temperature decreases, the amount of oxygen the water can hold increases and oxygen from the air diffuses into the water.

During a sunny day in eutrophic water, both oxygen and organic matter build up quickly. The amount of oxygen may rise to 30 or 40 ppm. Some oxygen diffuses out of the system, but most is used in plant and animal respiration. Consumption during the night can bring the oxygen level down to 1 or 2 ppm by the end of the night and cause fish kills. Most fish require more than 4 ppm. Some with lung-like blood vessels in the throat and air bladders can live in almost no oxygen by gulping air (top minnows, gars, tarpon).

Shown in Figure 17.2 is a reversible pathway for the diffusion of oxygen. It diffuses from the water into the air when its concentration and pressure are high (more oxygen molecules diffusing out of the pond than diffusing in). Diffusion reverses if the dissolved oxygen concentration in the water gets lower than that in equilibrium with air (about 9 ppm at room temperature). In the daytime, oxygen often diffuses out, and at night it often diffuses in.

The daytime increase in oxygen (Figure 17.3) can be used to calculate the net primary productivity (photosynthesis), a useful tool for determining the ability of a lake to absorb wastes and generate fisheries. The nighttime decrease in oxygen can be used to calculate the respiration of all the underwater life.

17.1.3 Problems with eutrophic ponds and lakes

As human settlements developed, enormous quantities of sewage, agricultural wastes, and highway debris have been running off the lands, making lakes very eutrophic. With these new rich conditions for plant growth, the conditions are ready for new species to take advantage of the new opportunities. Introduced plants, called *exotics*, such as water hyacinths and Asiatic milfoil, spread wherever the nutrient conditions are extremely high. These have been regarded as pests because they block the movements of boats and interfere with fishing and swimming.

Attempts to remove the plants have not been very successful. Adding herbicides puts decomposing plant matter in the bottom of the pond. The decomposers then release nutrients and stimulate the regrowth of the same plants. This poisoning also disrupts many other aspects of the ecosystem. Adding plant-eating fish also accelerates the cycle of regenerating the nutrients and plants. A method of catching these nutrients has been developed using natural wetlands such as marshes and swamps. By arranging these wetlands between the wastewaters and lakes and rivers, the nutrients can be filtered out to grow swamp trees or maintain greenbelts and wildlife areas (Chapter 18).

The best solution is "simple" — keep the extra nutrients out of navigational and recreational waters. As fertilizer gets more and more expensive, there will be more efficient use and less waste. More efforts will be made to conserve and recycle nutrients. Eventually, most of the sewage and agricultural waste waters will be recycled to fertilize forests, crops, and pastures.

Water hyacinths in eutrophic waters live on top of the water in solid rafts that block boat access and cause the waters underneath to become *anaerobic* (without oxygen). As an ecosystem, the water hyacinth beds are not so different from similar ones in the Amazon. They are full of life, with many species of air-breathing fishes and amphibians. Although several attempts have been made to harvest the water hyacinths for biomass or cattle food, the product is nutritionally poor and too full of air and water to be valuable enough to cover the costs of collection. When nutrient levels are low, water hyacinths are not a problem as they become one of the many diversified aquatic plants growing sparsely along the edges.

One real value of the water hyacinth community may be as a nutrient filter that lowers the eutrophic state of the waters. The way to eliminate the hyacinth problem may be to use hyacinths in one body of water to clean up the water for the next. Hyacinths often fill the canals and sloughs that contain eutrophic waters flowing from farms or towns. These might be arranged in strips through the peatlands. While the peat is being used up on the farmed peaty areas, it can be depositing under the hyacinths. Then the land use can be reversed, farming where the hyacinths were, letting the other areas build up peat again.

Lake Apopka is a well-known example where inflows of nutrients from agriculture growing on peatlands generated an extremely eutrophic lake, often green as pea soup and without many hook-and-line fish. A restoration project pumped the lake waters back over some of the peatlands, removing many nutrients.

17.1.4 Lake Okeechobee: a eutrophic lake

Large Lake Okeechobee is like a giant water tank for south Florida (Figure 1.2). It receives water from higher grounds to the north through the Kissimmee River. The waters are highly colored, but there is a regular fisheries food chain. The large area allows winds to keep the waters stirred and aerated. The lake originally had its overflow over the south shore, seeping through custard apple swamp forests and moving slowly southward through the Everglades ("river of grass"). There were miles of shoreline for excess water to flow out gently. Early settlers, beginning to farm the peaty lands below the lake, put up flimsy levees (dikes) to drain these lands. Then, when hurricanes came in 1926 and 1928, the winds built the water levels up because of the levees' resistance. When the levees finally broke, the wall of water drowned 400 people in 1926 and 2000 more in 1928. After that, the Federal Corp of Engineers built a massive levee around the lake, probably heavy enough and tall enough to hold a hurricane surge. With increasing wastes in inflow streams, Lake Okeechobee has been getting more eutrophic and its value as a water supply is threatened.

The level of the lake used to be about 20 ft (9 m) above sea level, but due to the many demands on water in the basin, the lake is usually at 13 ft (4 m) or less. When the levee was built, lake shallows were included within the levee, and with the level maintained at lower depths, a very large new marsh was generated. Many water birds have adapted to this marsh, compensating somewhat for losses of habitat elsewhere.

17.2 Stream ecosystems

The ecosystems that develop in streams have many of the same components as pond and lake ecosystems but are organized by the flowing water. The current brings nutrients to plants and food and oxygen to animals, and it carries wastes away. In other words, the ecosystem uses the physical energy of the water's downhill flow, often developing high concentrations of life.

Because Florida is low and flat, it has fewer streams than do some states, and for this reason its streams are highly valued as recreation areas for canoe trips, as corridors for wildlife to move about in an urban landscape, and for the good fishing. Figure 1.3 shows the rivers and larger streams of Florida. Some streams called *strands* are broad, full of vegetation, with waters flowing slowly.

In streams, the food web starts with the algae and with organic matter from the land. Nutrients are absorbed by algae for photosynthesis and are also used directly by microbes. Many streams draining rocky or sandy areas are oligotrophic. They can become eutrophic if they receive enough nutrients from mineral deposits, sewage, or pasture runoff. Organic matter from the land is an important source of stream organic matter (detritus). Some detritus is decomposed by microbes, and some flows on downstream. In small, shaded, woodland streams, organic detritus is the primary support for the rest of the food chain.

Freshwater insects spend most of their lives in the water as *larvae*. They mature, emerge from the water briefly, and fly in a large swarm over the water, depositing their eggs in the water. They feed on the organic slime of detritus and microbes and may in turn be eaten by carnivorous fish. Figure 17.4 is a picture of the food web of a stream flowing through a swamp.

Streams that drain swamps have dark water due to the decomposing organic matter flowing directly through sandy soil into the stream. (The dark water does not indicate they are dirty or polluted.) Many are oligotrophic. Examples are the Suwannee and St. Marys Rivers draining the Okefenokee Swamp of southern Georgia, and the Santa Fe River. Figure 17.4 shows typical animals of a black-water swamp stream.

Streams that cross phosphate rocks pick up high concentrations of phosphorus that are further increased by runoffs from phosphate mining and processing. Photosynthetic production may be stimulated, but fish kills occur. Branches of the Peace and Alafia Rivers are examples.

Turbid, sediment-loaded streams occur where rivers drain areas of clay soils. The rivers tend to become turbid with suspended clay, often yellow at times of high runoff. Generally, the fishes of these rivers are adapted to the turbidity. Where the river waters slow down, the sediment deposits add to

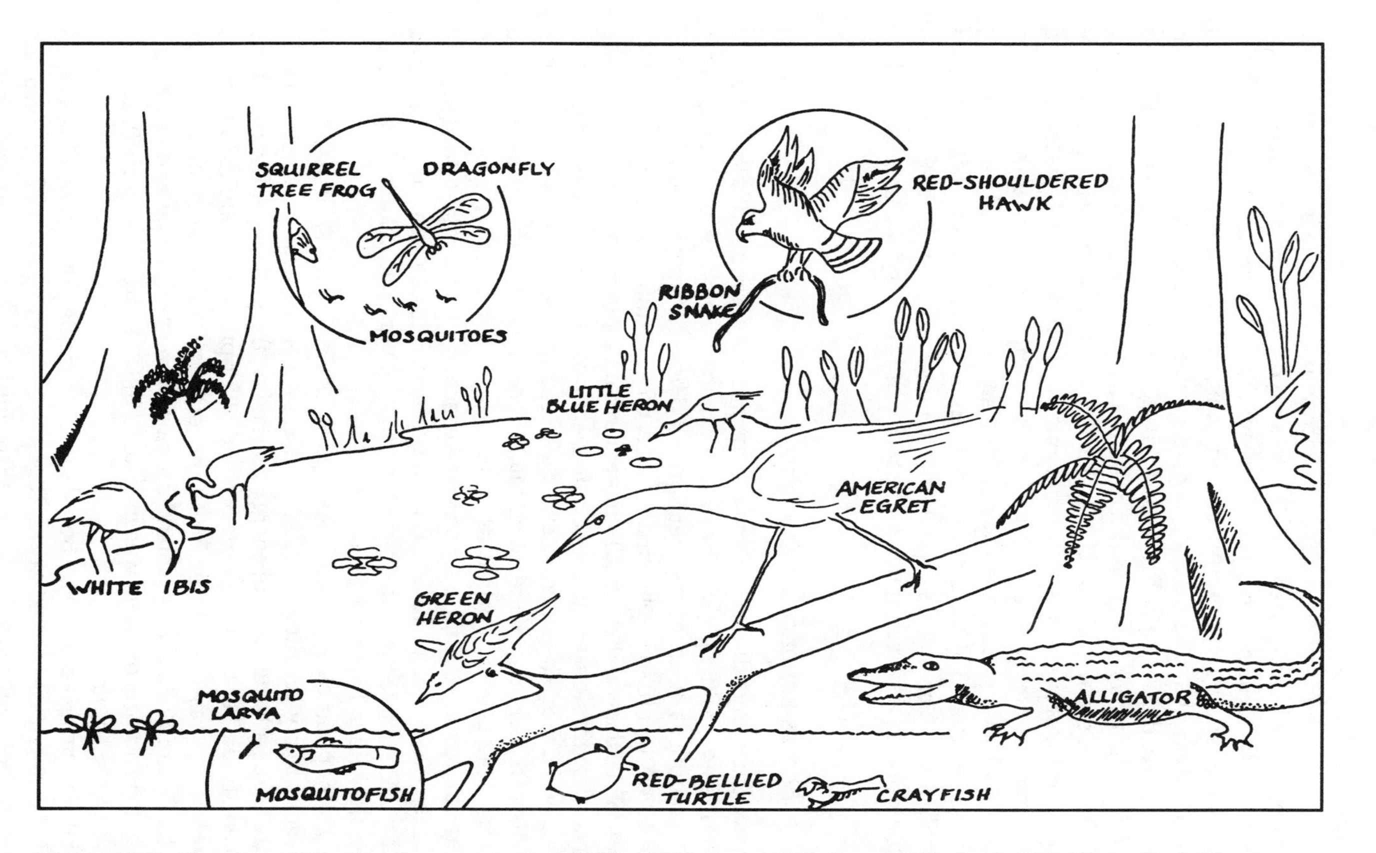

Figure 17.4 Dominant animals of a swamp stream. (From Wharton, C.H., *Forested Wetlands of Florida — Their Management and Use*, Center for Wetlands, University of Florida, Gainesville, 1976. Illustrated by Joy Bartholomew.)

the fertility of local soils. An example is the Apalachicola River that drains the clay hills of Georgia.

The lower parts of rivers that connect with the sea are influenced by the tidal seawater. Because of the salt the water is dense and goes underneath the freshwater to form a two-layered stream. The tide raises and lowers the water levels at the river mouth, causing a period of river hold-up followed by rapid discharge. This tidal effect goes upstream far beyond the saltwater. Examples are the Wakulla River, St. Johns River, and Spruce Creek. The rise and fall of the water levels produces both freshwater and saltwater marshes.

When the sea was about 25 ft (7.5 m) higher than at present, during past ice ages, the porous sands and limestones of Florida's lands became filled with saltwater. This salt has not washed out completely; thus, groundwaters, springs, and streams that draw water from areas below 25 ft (7.5 m) above sea level (see Figure 13.5) have more salt than do other streams. Many marine animals, such as blue crabs, find it easy to move into such streams. Examples are the Homosassa River, much of the St. Johns River, and the Caloosahatchee River.

17.2.1 Ecosystem of Silver Springs

Some of the waters that percolate into the porous upland sands and swamps of Florida go into the groundwater and eventually emerge at lower levels in large volume as clear, hardwater springs. These flowing waters have moderate levels of nitrates, phosphates, and other requirements for growth of plants. Since the water is clear, light penetration is good, and very productive ecosystems develop with full complements of algae, rooted plants, insect larvae, and fishes. After the streams flow several miles, they pick up detritus and dissolved organic matter, eventually becoming like other streams. These streams have been important as water supplies, recreation spots, favorite places for scuba diving, and tourist attractions where glass-bottom boat excursions have been successful for a century. The Chipola River, Wakulla Springs, Weekiwachee Springs, Rainbow River, Ichtucknee River, and Chassahowitzka River are examples.

One popular spring, Silver Springs at Ocala (Figure 17.5), has been the subject of many scientific studies. A diagram of the ecosystem is given in Figure 17.6, with the flows of energy marked on the pathways. The dissolved oxygen added between the outflow boil (2.2 ppm) and a station 3/4 miles downstream was used to estimate the primary productivity (see, for example, Figure 17.7). The seasonal variation of the production and consumption of the spring ecosystem was measured in 1954 and again in 1980, and the figures were similar.

Silver Springs had large populations of mullet and giant blue catfish until the downstream Oklawaha River was dammed in 1967 to form the Rodman Pool (without a fish ladder). The mullet no longer had free access to the sea to reproduce and return, and only a few pass through the locks for

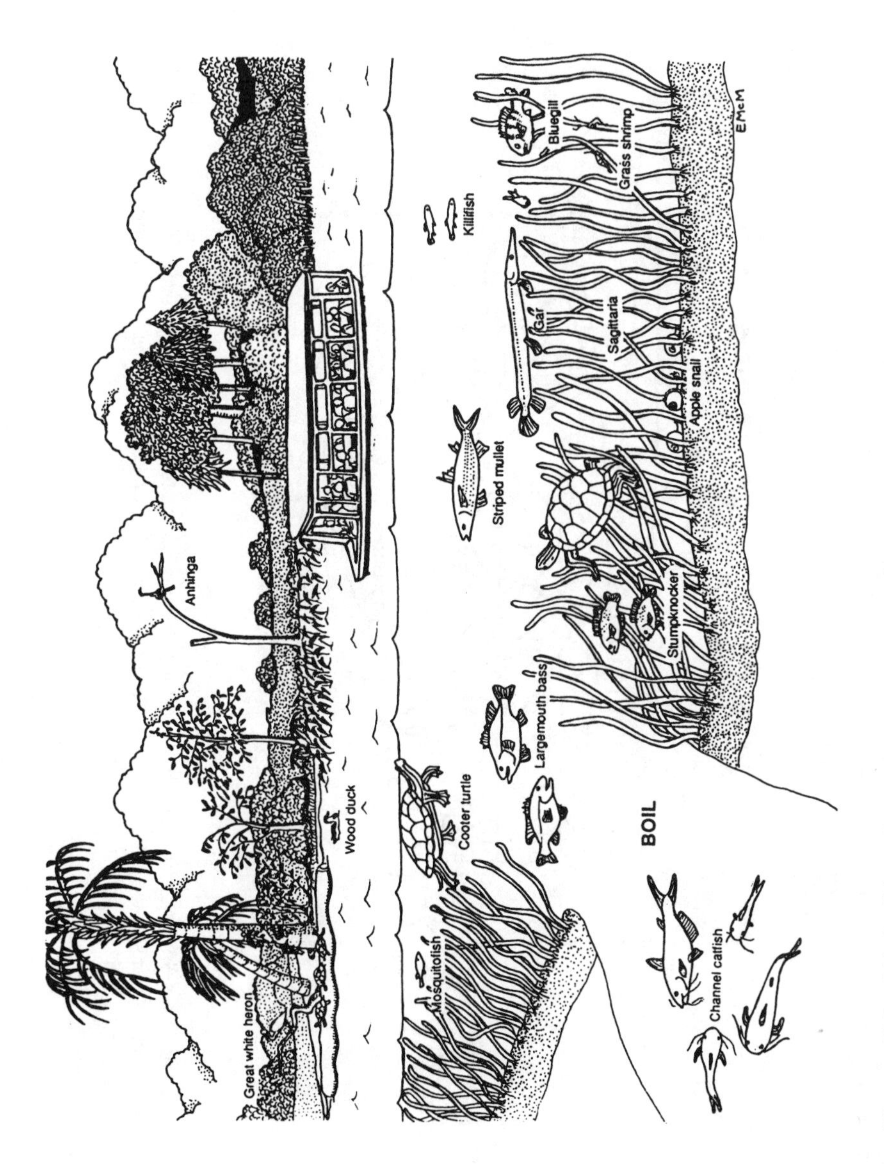

Figure 17.5 Life in Silver Springs before 1967. (Illustration by Elizabeth A. McMahan.)

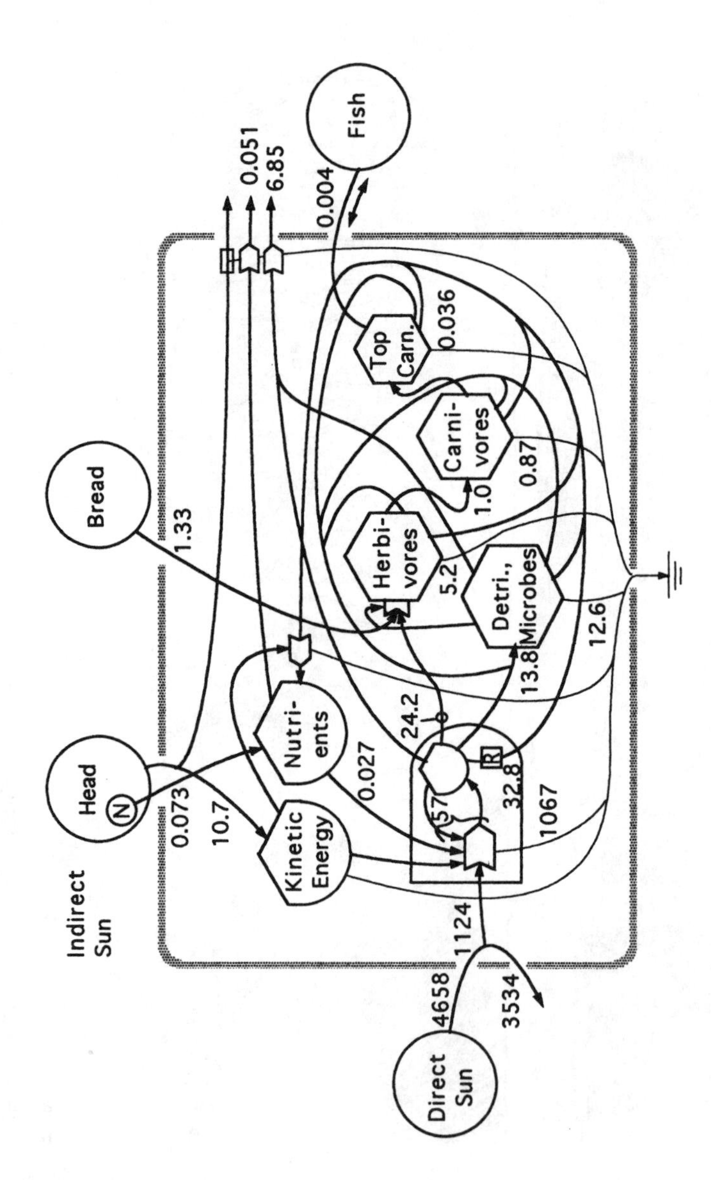

Figure 17.6 Summary model of the Silver Springs ecosystem including average flows of energy in kilocalories per square meter per day. (Adapted from Odum, H.T., *Systems Ecology*, John Wiley & Sons, New York, 1983.)

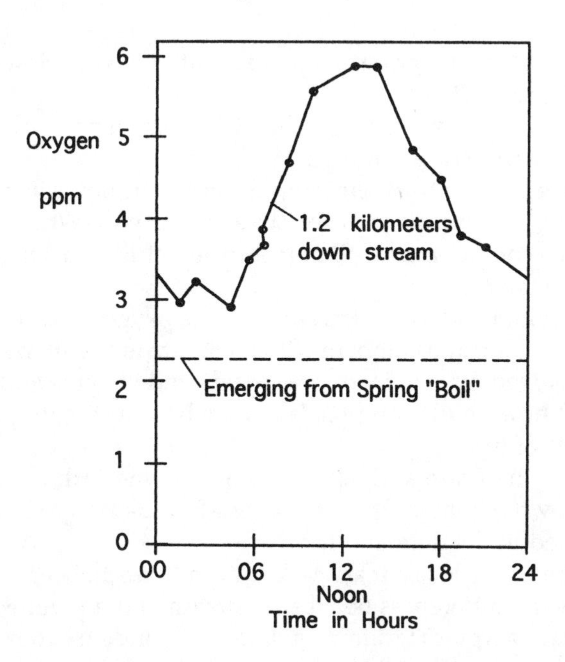

Figure 17.7 Typical graph of the rise and fall of dissolved oxygen over 24 hours in Silver Springs, May 1955. (From Odum, H.T., *Ecol. Monogr.*, 27, 55, 1957.)

boats. Another herbivorous algal-eating fish, the gizzard shad, replaced it as a prominent member of the spring system. The large blue catfish also disappeared, apparently because its place of breeding was below the Rodman pool.

Other springs have different chemical components. Some springs come out of the ground without oxygen and support interesting blue-green algae and white sulfur bacteria, without any fish except those that can get air from gulping bubbles. In Warm Salt Springs, which connects with the sea and is as salty as an estuary, marine organisms occur, including the tarpon, which can gulp air in the night hours when the dissolved oxygen is low.

Since springs depend on the pressure of groundwater in surrounding lands, the springs are easily stopped where they are unwisely dammed to make pools or when the groundwater level is lowered by strong pumping of water from wells nearby. Large popular recreational springs in Tampa and near Bartow were stopped by large groundwater users diverting the waters. Clear springs are becoming turbid as the groundwaters become polluted.

Questions and activities for chapter seventeen

1. Define: (a) softwater, (b) hardwater, (c) groundwater, (d) benthic, (e) hydroperiod, (f) aquatic, (g) eutrophic, (h) oligotrophic, (i) saturation level, (j) turbid, (k) percolate, (l) sediment, (m) larvae.

2. Draw a systems diagram of a pond familiar to you. How does it differ from Figure 17.2?
3. Why are oxygen and carbon dioxide levels in a pond different during the day than they are at night?
4. Compare and contrast oligotrophic and eutrophic lakes in Florida.
5. Consider a eutrophic pond or stream near you. Where are the extra nutrients coming from? What measures could you suggest to reduce the nutrients?
6. With a Winkler kit or oxygen probe, take oxygen measurements early in the morning and compare with those late in the afternoon. Measure the dissolved oxygen in several ponds and streams and compare.
7. Discuss how an oligotrophic lake may become a eutrophic lake over a period of time.
8. Research the Rodman Pool (reservoir) on the Oklawaha River, produced by the dam built for the Cross-Florida Barge Canal. What has happened to the dam and lake?
9. How are water levels managed in Lake Okeechobee?
10. Discuss the differences between a stream and a pond ecosystem.
11. Draw an energy diagram of a stream. Include in your diagram the energy sources, producers, consumers, energy flows, and important storages.
12. Why does Florida have fewer rivers than many other states?
13. In what ways do spring runs differ from other streams?
14. Visit a spring near you. What are some of the problems caused by people? Discuss how both its flow and beauty can be protected.
15. Compare a general freshwater ecosystem (Figure 17.2) with a land ecosystem (Figure 7.2). Identify producers, primary consumer, carnivores.
16. To study succession in a freshwater ecosystem, put 10 glass microscope slides in a pond or stream. Take 2 out after 1 day, another 2 out after 1 week, and 1 out each week to the end of the term. Look under a microscope at what has grown on the slides. Keep a record of the quantities and kinds of organisms you see and how they change over time. Compare your results with the characteristics of succession in Table 8.1.
17. If you have access to a computer and the BioQUEST module, Environmental Decision Making with EXTEND, run the pond ecosystem simulation (see Appendix A for source information).

chapter eighteen

Wetland ecosystems

Where lands are flooded intermittently, *wetland* ecosystems develop. Because water displaces the air from the soil, only plants and animals adapted to low oxygen conditions can live. Depending on the water regime, many kinds of wetland ecosystems develop. In this chapter, we consider the wetland processes, the kinds of wetland ecosystems that result, and the benefits and problems of fitting wetlands with human society.

Wetlands occupy 15% of Florida's area (Figure 18.1). Forested wetlands are called *swamps*, and grassy wetlands are called *marshes*. Wetlands are found bordering lakes and in the floodplains along streams. Many of the wetland areas of Florida are isolated depressions on flat tablelands higher than surrounding areas. Headwaters of streams such as the Suwanee River and the Withlacoochee River start from elevated swamp areas. A wetland where waters drain slowly through vegetation without a stream channel is called a *strand*. Strands drain into streams. At lower elevations where groundwaters seep to the surface there are wetlands. The margins of estuaries are lined with saltwater wetlands: salt marshes and mangroves. Figure 18.2 is a bird's-eye view of wetlands in central Florida.

Once wetlands were drained, but now they are preserved because of their role in maintaining the quantity and quality of freshwaters and protecting wildlife, as well as for their aesthetic beauty. Wetlands develop wood products and peat. Because wetlands receive water from higher ground, they act as natural filters that absorb nutrients, turbidities, toxic substances, and disease microbes. A cypress-gum swamp west of Tallahassee absorbed lead from years of washing car batteries with acid, preventing serious health hazards downstream. Wet season rains are stored in wetlands for groundwater recharge and are slowly released to strands and streams to maintain river flows during drier times of the year. Famous wetlands such as the Everglades and the Okefenokee Swamp (just north of Florida in Georgia) are tourist attractions.

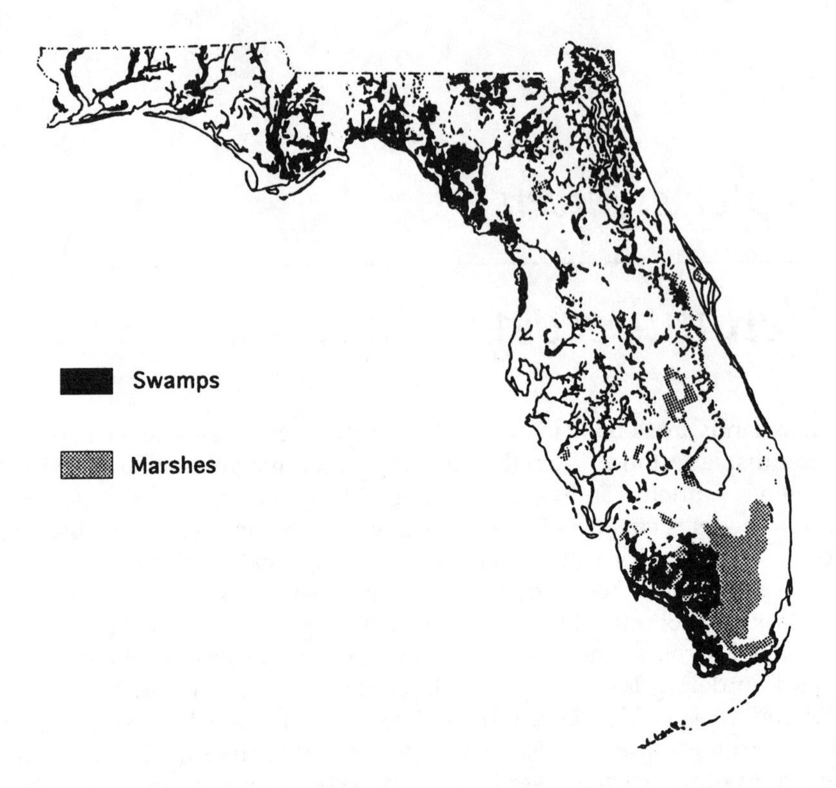

Figure 18.1 Marshes (grassy wetlands) and swamps (forested wetlands) of Florida.

18.1 Water regime and wetlands

When land is flooded, the air spaces in soils fill with water. With access to atmospheric oxygen limited, soils become *anaerobic* (without air). Roots of ordinary plants cannot breathe. The only species of plants that survive have developed special means of obtaining energy that are different from those in unflooded areas. For example, roots of the swamp black gum trees derive their energy from anaerobic respiration. A by-product is alcohol, which is sucked up the trunk by transpiration. The half-completed respiration process starting in the roots is completed in the trunk and leaves. Cypress trees grow special roots above ground, called knees, through which they exchange some carbon dioxide and oxygen.

A wetland can be defined as land with anaerobic soils for at least part of the year. In practice, the presence of wetland plant species is used to locate and define the boundaries of wetland ecosystems.

The length of time of flooding is called the *hydroperiod*. Swamps flooded most of the year tend to grow pond cypress and gum. Swamps flooded for only a few weeks tend to have *bottomland hardwoods* with wetland species of red maple, oaks, ash, and hickory.

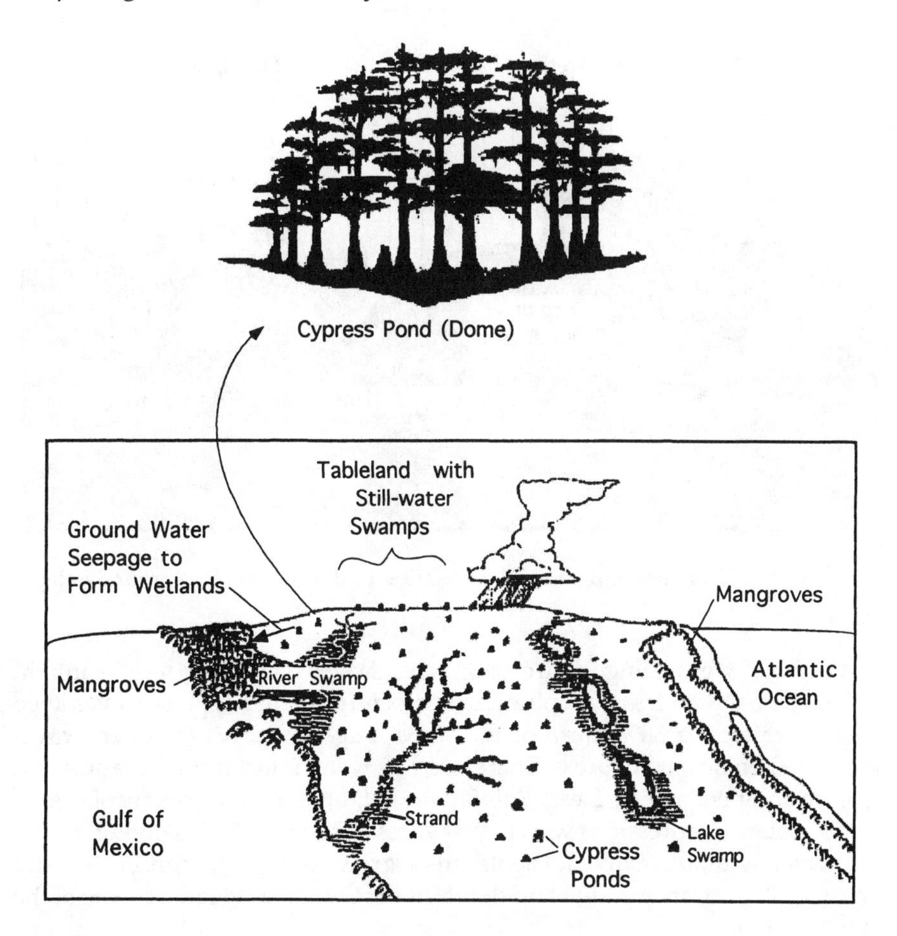

Figure 18.2 Wetlands of central Florida showing cypress domes on the uplands, lake-border swamps, strands, swamps, and marshes on river floodplains and at the edge of estuaries.

In Florida, rains percolating into the land maintain groundwaters close to the surface. The groundwater controls the kinds of wetland (Figure 18.3). The water stored in elevated swamps supplies outflowing streams and flows down into the ground, recharging the groundwater. Wetlands at lower elevation receive groundwater seepage, keeping them wet. For example, the Okefenokee wetland area in southern Georgia supplies water to the Suwanee and St. Marys rivers of Florida. The area known as the Big Cypress east of Naples is a flat elevated area that receives and stores rainwater in small basins that recharge the groundwaters and outflowing strands of the region (Figure 18.4).

Although wetland plants must transpire water vapor from their tissues into the wind, some transpire less than others. Transpiration helps keep

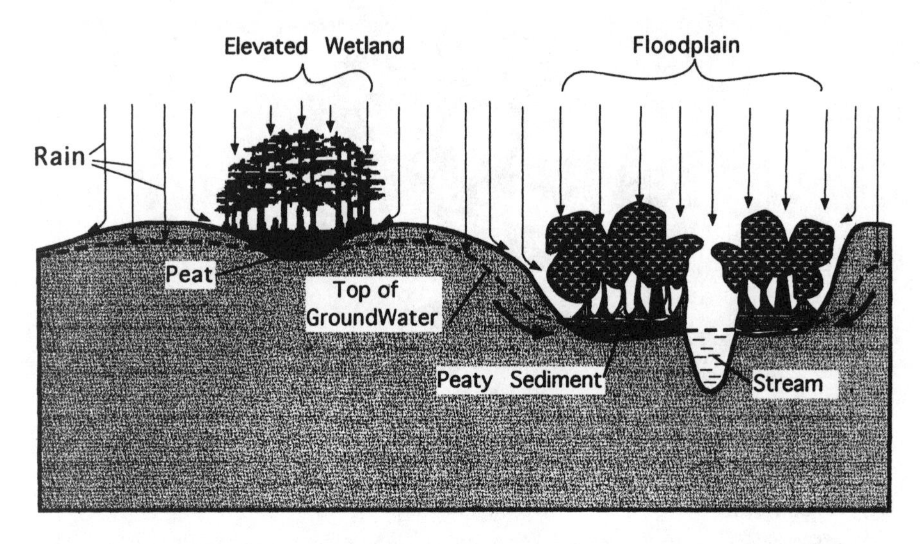

Figure 18.3 Marshes and swamps in relation to the near-surface groundwater.

leaves from overheating in direct sunlight. Leaves that reflect more sunlight can keep cool with less transpiration. For example, pond cypress leaves keep cool by reflecting 30% more of the sun's long-wave rays (infrared wavelengths) than do bald cypress. Therefore, less water is lost from these swamps than from open lake surfaces. Rainfall on flat, upland areas often collects in small, isolated depressions with pond cypress (Figure 18.2). Receiving mostly rainwater without many nutrients, trees grow slowly but conserve water because of low transpiration and the ability of the leaves to reflect some of the sun's energy.

Pond cypress tablelands form the headwaters of many streams in Florida — for example, the Santa Fe and Withlacoochee Rivers. The Big Cypress area (Figure 18.4) has low nutrients and dwarfed pond cypress trees, each more than a hundred years old. In contrast, where water is flowing through cypress, as in the Fakahatchee Strand (Figure 18.4), trees became tall and luxuriant. After logging left few seed trees, a scrubby maple forest developed.

If inflowing waters are high in nutrients, wetlands produce excess organic matter and bind the nutrients into *peat*. Peat is the brown organic sediments made from plant decomposition. It has the ability to absorb and bind inorganic and organic chemicals. It helps retain water. When dried it is a good fuel. Peat is the first step in formation of coal.

Conversely, wetlands that are decomposing release their bound nutrients back into the waters. Waters draining farms on peat around Lake Apopka gave the lake continuous algal blooms. Now, lake waters are being pumped back into wetlands to rebind nutrients into peat again. Thus, wetlands can serve as a buffer, tending to keep a balance of nutrients and organic matter in the waters around them.

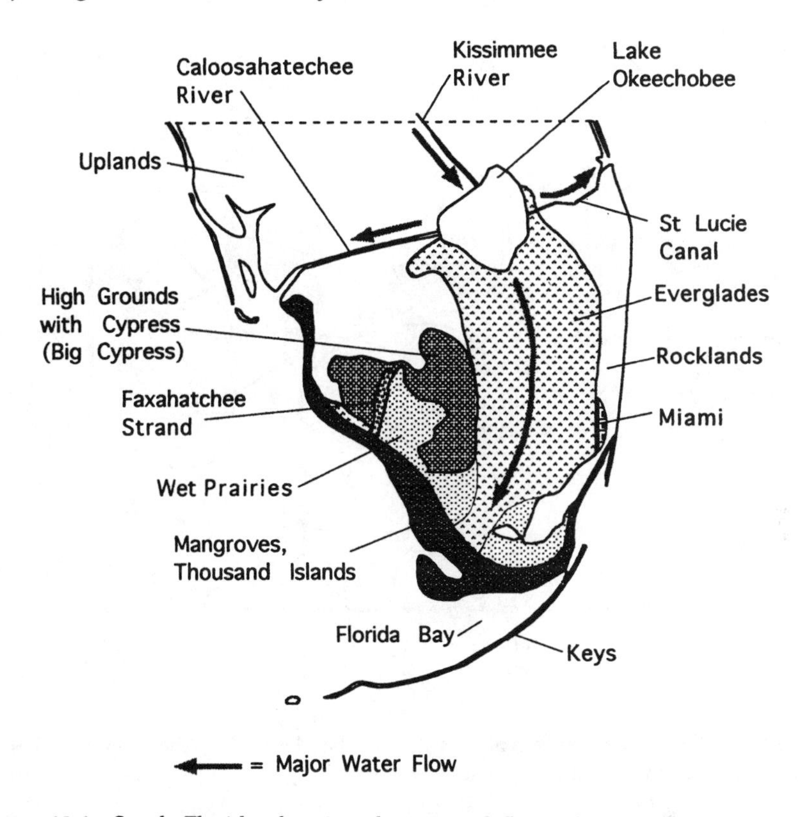

Figure 18.4 South Florida showing the original flow of water starting with the Kissimmee River and Lake Okeechobee, through the Everglades into the saltwater mangroves, the Big Cypress plateau, and the Miami rocklands. Also shown are the channels constructed later for discharging Lake Okeechobee east and west.

Because energies must go into special adaptations for life in and out of water, the diversity of plant species is less than in ordinary forests. Some wetlands, such as salt marshes and mangrove swamps, have only a couple of plant species, while floodplain forests have a much greater diversity of trees. The variety of small insects, birds, and other animals is often higher in wetlands than in upland forests because animals of both upland areas and aquatic environments are found there. Many of Florida's endangered animals, such as the panther, find sanctuary in wetlands.

The now extinct ivory-billed woodpecker lived on insects in giant floodplain trees. The now extinct Carolina parakeet ate the seeds in cypress tree balls. The extinction of these species occurred when most of the virgin wetland forests were cut early in this century. You can see remnants of these virgin cypress strands with giant, widely spaced trees at the Audubon Corkscrew Swamp Sanctuary near Naples and the Fakahatchee State Park on the Tamiami Trail (Route 41).

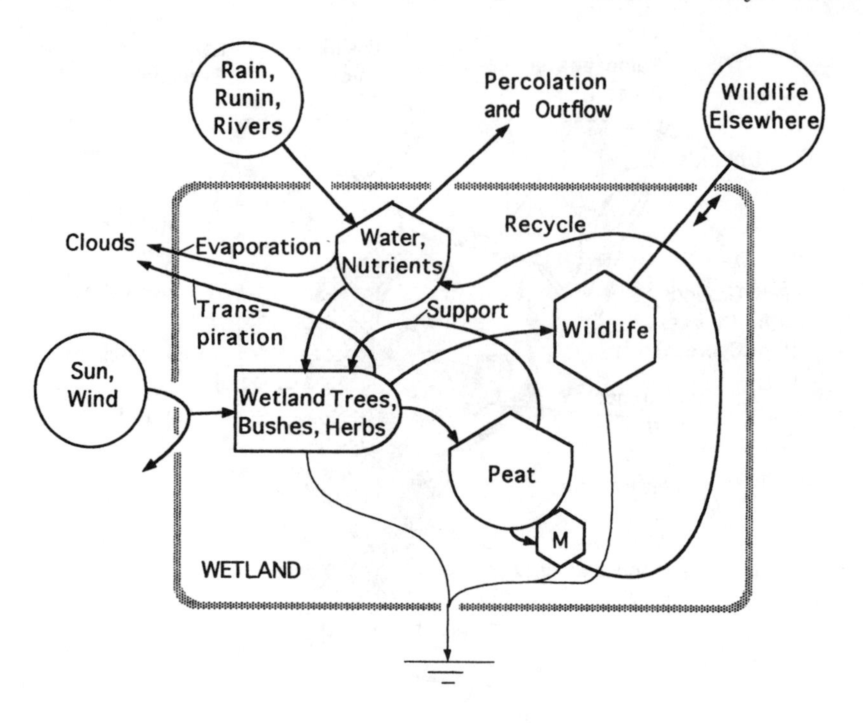

Figure 18.5 A systems diagram of a swamp (forested wetland) showing the processing of water and the buildup of peat. M = microbes.

The systems diagram in Figure 18.5 summarizes some of the distinctive characteristics of wetlands. Standing water within wetlands slows down decomposition, and organic matter is buried as peat. The pathways show the inflowing waters going out as evaporation, transpiration, and outflow in streams, strands, or underground. An elevated swamp often contributes to the groundwater; if it is at a lower level, it usually receives water from groundwater.

18.2 *Types of wetland ecosystems*

Disturbed areas with wetland water regimes start with marsh plants. Normally marshes go through succession (Chapter 8) and are eventually displaced by trees, with seeds supplied by floods and birds. Eventually, most wetland ecosystems in Florida's climate develop wetland forests (swamps), unless the succession is interrupted by fire or extreme drought. Types of wetland ecosystems in Florida are described in detail in Appendix D, Table D1.

18.2.1 *Freshwater marshes*

Marshes are dominated by herbaceous and grass-like vegetation. They vary in size from small marsh ponds of several meters in diameter to extensive marshes such as the headwaters of the St. Johns River or the Everglades. In Florida, semi-permanent marshes apparently occur wherever succession is arrested by frequent fires or grazing.

18.2.2 Lake margin wetlands

Swamps and marshes may grow around the edge of a lake if its shore has a gentle slope. Unless interfered with by people, lakes in Florida have water levels that go up and down several feet, with wet and dry seasons and wet and dry years. The larger the water range, the more land is covered and the wider the zone of wetlands. When the lake levels are regulated for the convenience of piers and boats, wetlands are diminished and with them the filtering of excess nutrients and toxic substances.

18.2.3 Floodplain forest

By depositing sediments during floods, many rivers develop a flat area on either side called a *flood plain* (bottomland, river swamp; see Figure 18.3). Hardwood forests develop. With frequent deposit of nutrient materials, tree growth is rapid. During dry weather, water flows in a small channel that meanders back and forth across the floodplain. During wet weather, water levels rise and overflow the banks. By spreading out among the tree trunks, the waters flow slowly and are not destructive downstream. While over the floodplain, the waters drop their valuable sediments, and fishes have a large feeding area.

If people build houses in floodplains, they are frequently flooded out unless they build on well braced pilings so that water can flow underneath the house. To encourage development in wetlands, some deep straight canals have been built (channelization). Because the channelized floodwaters flow quickly, they create greater floods downstream, and their sediments, nutrients, and toxic substances are not filtered out.

18.2.4 Saltmarsh and mangroves

Coastal wetlands that are covered with saltwater part of the time have characteristic vegetation adapted to the saline conditions. In Florida, north of Tampa where winter freezes kill mangroves, salt marshes predominate and are composed primarily of *Spartina* and black *Juncus*. In the more tropical areas of Florida without frost, mangrove swamps predominate, composed of red, black, and white mangroves.

Mangrove swamps have highly branched, above-ground roots (Figure 18.6). Freshwaters draining the land bring freshwater, sediments, organic matter, and nutrients. Tides move saltwater in and organic matter, pollutants, and sediments out. The tide also helps exchanges of fish, plankton, and larval stages of animals with the open waters of the estuary and ocean. Tidal energy

Figure 18.6 Mangrove swamp with red mangroves (prop roots) near the channel and black mangroves away from the channel on the left. (From Wharton, C.H., *Forested Wetlands of Florida — Their Management and Use,* Center for Wetlands, University of Florida, Gainesville, 1976. Illustrated by Joy Bartholomew.)

interacting with the plants and topography of the land surface forms a network of channels (tidal creeks).

Marsh vegetation and mangrove trees (Figure 18.6) have special adaptations to obtain freshwater from the saltwater that covers their roots. Some use the sun's energy to transpire water, and the pull within their stems draws freshwater into their roots, leaving some salt behind. Others secrete salt from their leaves. With energy going into salt adaptations, there is less for biodiversity. Most saltwater wetlands are relatively low in diversity.

Coastal wetlands provide protection from hurricane tides and protect the coastline from erosion. The organic matter dropped from mangrove and salt marsh vegetation is a rich source of food for estuarine nurseries (Chapter 16). Many of Florida's sport and commercial fish depend on this source of food and would disappear if these wetlands were destroyed. The disturbance of saltwater wetlands for housing developments and for mosquito control has interfered with their contributions to fisheries (see Chapter 24).

18.3 Kissimmee River and the Everglades

Originally, the Kissimmee River supplying Lake Okeechobee (Figure 18.4) was a slowly flowing river that meandered through a wide floodplain dominated by marshes. Summer runoff from central Florida reached the lake in the autumn with nutrients filtered out. In the early 1950s, the river was replaced with a large straight canal, allowing cattle to use the flood plain for pasture. With summer rains, waters reached the lake sooner, raising water levels during hurricane season and contributing to eutrophication. To protect levees, excess waters were dumped into the sea through the Caloosahatchee River and St. Lucie Canal. Now government agencies are spending millions of dollars to "undo" the ditching and to allow the river to once again meander through the floodplain marshes.

Before canals and highways were developed, excess water in Lake Okeechobee used to overflow the southern lake margin through a tangled swamp of custard apple trees and then flow south and southwestward to the Gulf of Mexico, producing the Everglades, a hundred miles of broad marsh wetland (Figure 18.4). The Everglades, the "river of grass", is an oligotrophic sawgrass marsh with small, oval-shaped tree islands elongated in the direction of water flow (Figures 18.7 and 18.8).

Water evapotranspires each day from the hot, wet surface of the marsh and contributes to cloud formation (Figure 18.5), showers, and thunderstorms that release rains from tropical air masses flowing in from the Caribbean Sea. The water of the Everglades turns over many times through additions and losses.

The scattered tree islands with bay trees that occur within the "river of grass" (Figures 18.7 and 18.8) have peat mounds above the surrounding water surface. The wet peat helps retard fire in dry years and shelters wildlife in wet years.

Alligators dig deep holes in the peat of the Everglades to protect their nests. Standing water remains in the holes during the dry season, offering a refuge for fish and other aquatic life when the marsh dries out.

Because of the limestone marls and groundwaters in south Florida, the water in the "river of grass" and the wet prairies is hardwater (high in calcium and carbonates) favoring animals with skeletons. Large apple snails grow and support the unique bird, the Everglades kite (Figure 18.7), now endangered as the result of the loss of snail habitat.

When the waters flowing down the "river of grass" reach the saltwaters of the Gulf and its tidal actions, mangroves replace the freshwater marshes. The coastal zone has hundreds of islands and channels covered by mangroves (Figure 18.6), some growing over oyster reefs. The lower part of the "river of grass", the Thousand Islands and Florida Bay, became the Everglades National Park.

18.3.1 Everglades canals, dikes, and highway fill

The flow of the Everglades was partly blocked by the construction of highways and dikes to make water reservoirs for agriculture and cities. Much of the water that originally flowed into the Gulf was diverted. With less freshwater reaching the park, the saltwaters moved farther inland so that much of the Everglades National Park developed scrub mangroves with their accompanying mosquitoes.

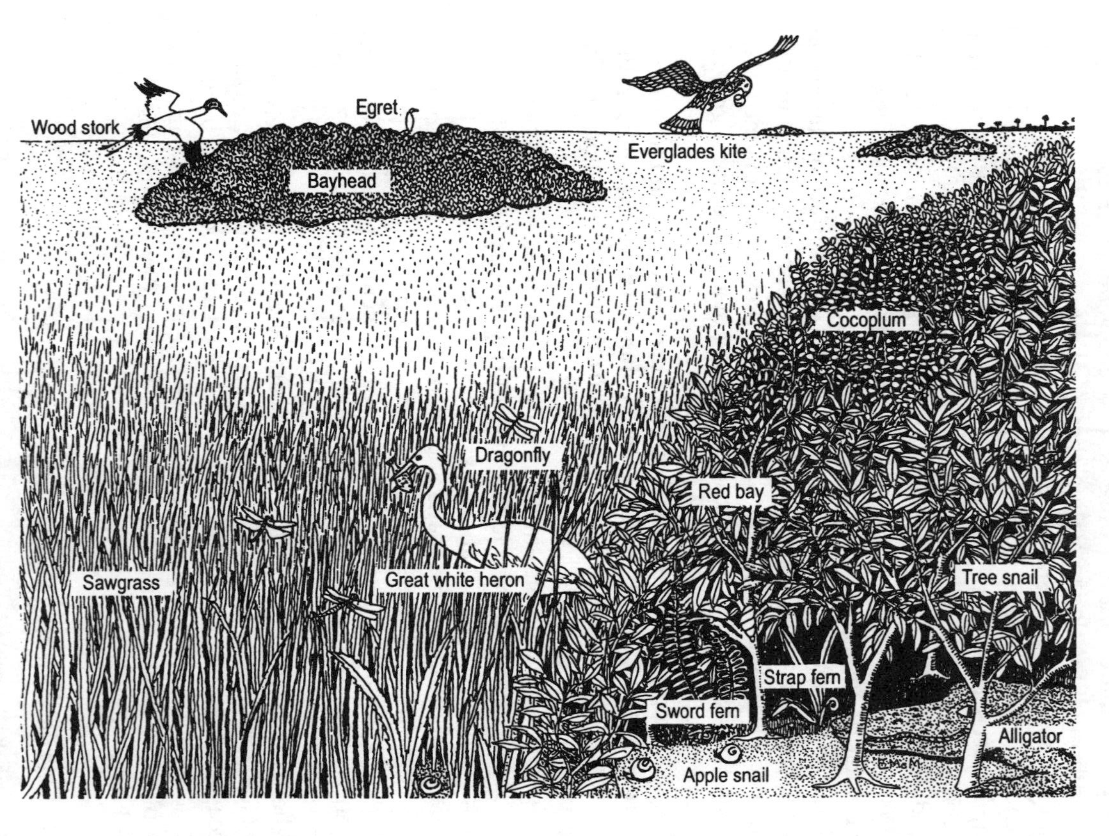

Figure 18.7 Everglades marshes, tree island bayheads, and associated organisms. The endangered Everglades kite eats apple snails. (Illustration by Elizabeth A. McMahan.)

Figure 18.8 "River of grass" alligator holes and nests. Alligators dig deep holes in the peat of the Everglades to protect their nests. Standing water remains in the holes during the dry season, offering a refuge for fish and other aquatic life when the marsh dries out.

Now, in wet years, the diked areas get too much water; in dry years, they go bone dry and much of the aquatic life is lost. Highways have become dams, and water has piled up too deep on the north side, while on the south side of the highways the marsh has become too dry.

With such severe fluctuations where there used to be less extremes, many of the aquatic birds have been lost. Water tables have generally been lowered more than 4 ft (1.2 m). Exotic Brazilian pepper and *Melaleuca* (an exotic tree from Australia) began to displace native vegetation because of the different lowered water table and higher nutrients. *Melaleuca* transpires more water and dries out the areas it occupies.

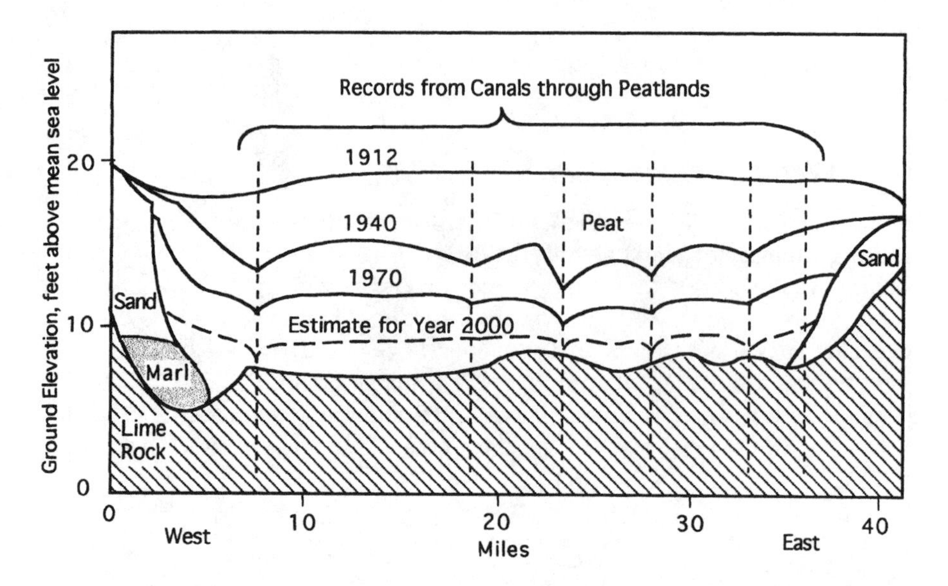

Figure 18.9 West to east cross-section of the peatland just south of Lake Okeechobee showing loss of peat by oxidation since 1912 due to farming. (From Browder, J. et al., *South Florida: Seeking a Balance of Man and Nature*, Bureau of Comprehensive Planning, Division of State Planning, Tallahassee, 1977.)

The old "river of grass" is still operating somewhat, just south of the Tamiami Trail in Everglades National Park and in the Loxahatchee National Wildlife Refuge in western Palm Beach County. Because the highways were not built with enough underpasses, wildlife has been killed by cars unnecessarily. However, highway I-75, built across south Florida later, has 36 underpasses for water and wildlife. Night photography and tracks show panthers and other wildlife are using them.

18.3.2 Peatland oxidation

Under the wetlands of the upper "river of grass" formerly occupied by custard apple swamp just south of Lake Okeechobee, peat beds 20 ft thick were deposited, a process requiring hundreds of years. The wetland vegetation on top was replaced by sugar farming, with soils dry in winter (see Chapter 23). In the Florida climate, peats oxidize and lose their nutrient storages if they dry out. Peat layers have been oxidizing away at an average rate of 1 inch (2.5 cm) per year. Most of the peat of the sugar cane area has already oxidized away (see a cross-section of the peatland in Figure 18.9). Without these soils, many agricultural operations will no longer be profitable. Now projects are getting underway to revert some of the farmed areas to wetland and to rebuild peaty soils as part of a long-range rotation with agriculture.

18.4 Wetlands management

In order to restore waters to the Florida landscape and because of the nutrient filtering capacity of wetlands, all runoff from agricultural lands, cities, and sewage wastewater should be directed through a wetland before reaching public lakes, streams, estuaries, or groundwaters. In this way, groundwater levels are sustained, waters and nutrients are recycled back to the land, and open waters are protected from excessive nutrients. Studies show that salt marshes and mangroves, like freshwater marshes and swamps, are effective filters. The use of wetlands for tertiary treatment of city wastes is discussed in Chapter 34.

Because the temperature of air over water cannot get below freezing until all of the water has frozen, wetlands help keep the crops from freezing. A landscape composed of wetlands is a warmer landscape in winter and a cooler one in summer. The drainage of wetlands may have caused increased frost damage to citrus and farms in central and south Florida.

In Miami and elsewhere, treated city sewage has been ejected into the ocean, causing a loss of freshwater. An alternative plan returns the treated wastewater to an Everglades area to have its nutrients removed by the marsh into peat so that waters can be reused. After a few miles of cattails, the low-nutrient sawgrass regime returns and the water becomes suitable to recharge groundwaters again.

18.4.1 Constructed wetlands and self-organization

Where it is undesirable to disturb a natural swamp, new wetlands are being constructed to receive the high-nutrient waters of street runoff and the treated sewages of towns. If there is a suitable surface, controlled hydroperiod, and source of seeding of suitable organisms, a wetland ecosystem adequate for improving water quality develops in several years.

After phosphate mining in areas of central and north Florida, wetlands have been restored by arranging water regimes with an appropriate hydroperiod, seeding, and planting seedlings to re-establish the species. Because of the wide extent of the vegetation disruption in these areas, not enough natural seeding occurs to restore wetlands in a reasonable time without human help.

Figure 18.10 shows a storm water pond allowed to develop as a swamp and a drainage ditch that has been allowed to vegetate as a swamp. These are examples of how humans can work with nature's self design during development to increase value and symbiotic relationships with the natural environment of Florida. This kind of management is called *ecological engineering.*

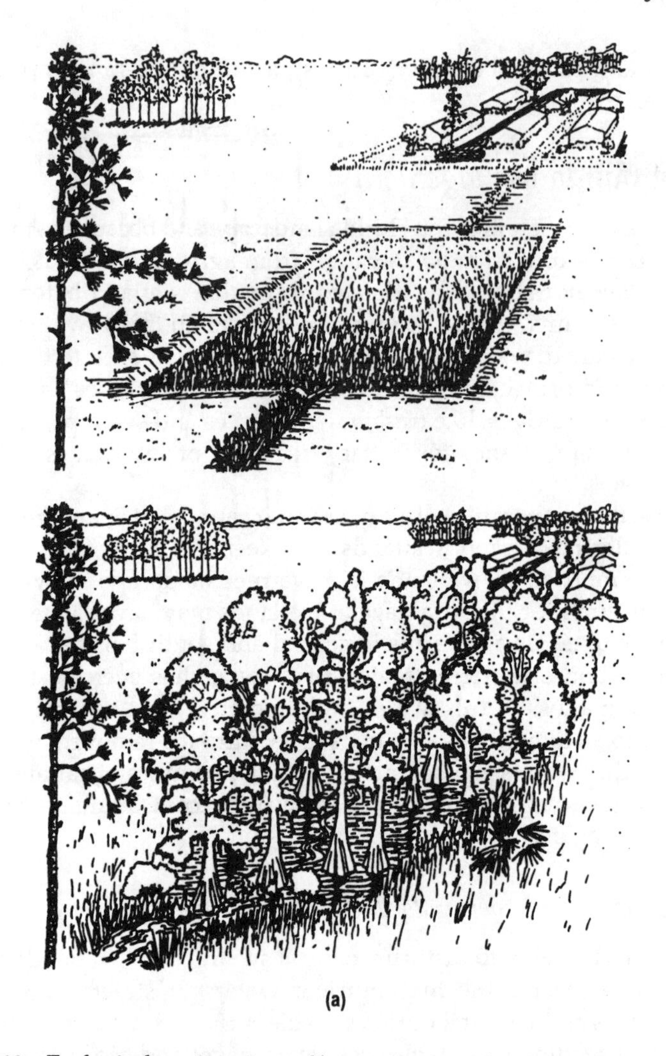

Figure 18.10 Ecological engineering self-organization of water courses around developments to reduce flooding and to increase water quality, wildlife, and aesthetics. (a) Stormwater retention ponds; (b) drainage ditches.

Questions and activities for chapter eighteen

1. Define the following types of wetlands and give some characteristic plant species: (a) swamp, (b) marsh, (c) cypress "dome", (d) mangrove swamp, (e) salt marsh.
2. Discuss the differences between marshes and swamps and the forces that may cause marshes to persist instead of undergoing succession to swamps.

(b)

Figure 18.10 (continued)

3. Look for articles in your local newspaper about wetlands, and discuss the issues in class.
4. Discuss how the development of south Florida's east coast has impacted the Everglades.

5. Visit a nearby wetland. Determine what type of a wetland it is. Take along a field guide and identify the major plants. Determine what water sources keep it wet. Draw an energy diagram that includes the food web.
6. Discuss the difference between aerobic and anaerobic respiration, using a wetland as an example.
7. Discuss the characteristics of Florida (climate, location, and topography) that produce the different wetlands.
8. List several common animals found in the Everglades. How does this animal population differ from the population of the African savannah?
9. Explain the role of fire in maintaining the Everglades.
10. Consider the possibility of restoring a continuous strip of the "river of grass" one half mile wide from Lake Okeechobee to the Thousand Islands, bordered by dikes on either side with bridging at all road crossings. Draw this on a map and discuss its possible relations to agriculture, to wildlife, to the Everglades Park, etc.

chapter nineteen

Upland ecosystems

Before European colonization, the uplands of Florida were covered with pine and hardwood forests, and remnants of these forests are scattered over the state. Pines are found where succession is restoring forests or where fires are frequent and pines are semi-permanent (called a *fire climax*). Where fires and other disturbances are excluded, climax hardwood forests develop with oaks and hickory trees predominant. Various pines and hardwoods are native in different parts of the state, depending upon the temperature and water in soils. Natural upland forests in Florida are described in this chapter, starting with tree growth, succession, and soil formation. Forests managed for commercial harvest are described in Chapter 22.

19.1 Forest succession

Much of Florida's land areas are in various stages of succession between cleared land and fully developed forest. Succession in forests, as described in Chapter 8, accumulates soil storages, grows populations of organisms, and develops the properties found in mature forest. The characteristic tall, old trees of mature forests have large, widely spaced trunks, and on the ground there is a complex structure of fallen logs and limbs useful as wildlife habitats, as stores of organic matter and nutrients, and as platforms for epiphytes and seedlings. Except where a tree has fallen, there is a dense canopy of leaves that, together with understory and ground plants, use almost all of the light and maintain a moist microclimate. There is a great variety of plants and animals (biodiversity). The type of mature forest, which tends to replace itself, is the climax for that area.

Figure 19.1 shows typical stages in forest succession in north Florida. On bare land that has been farmed, fast-growing weeds appear, such as grasses, herbs, and dogfennel. The first weeds cover the ground quickly; help catch sunlight, rain, and nutrients; and start accumulating organic matter in the soil as they die and decay. In a year or two, these early weeds are replaced by longer-lasting grassy and shrubby plants, such as broomsedge and blackberry.

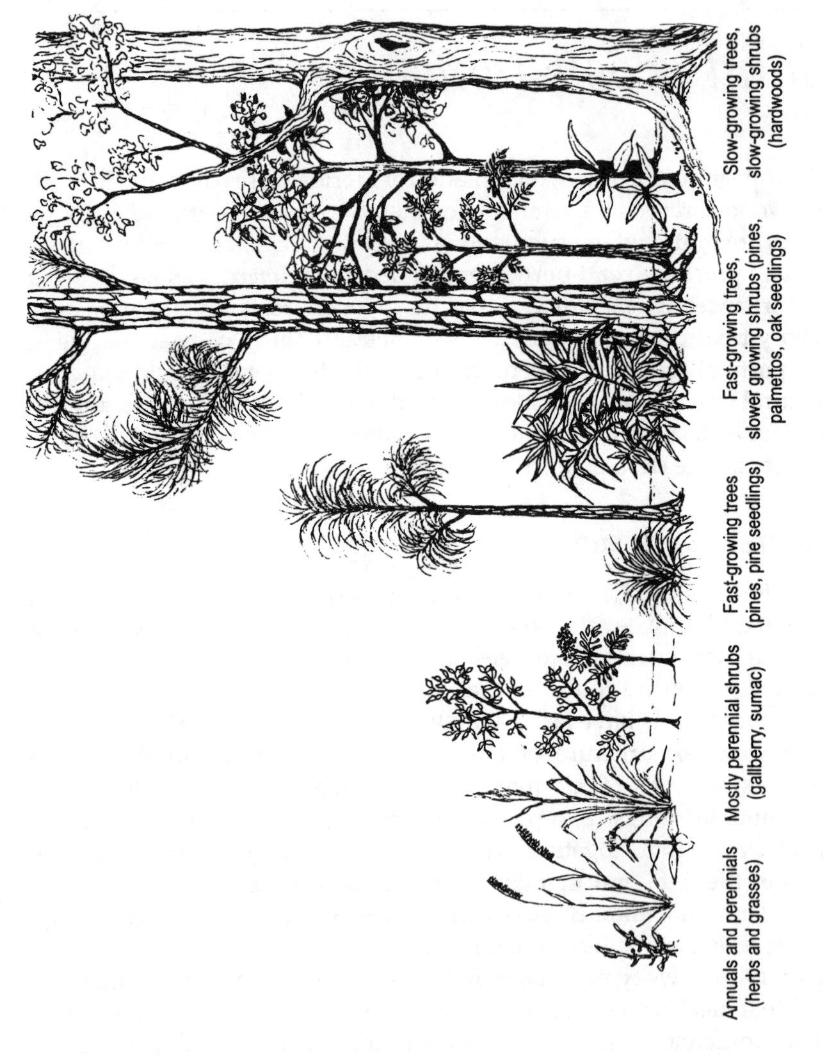

Figure 19.1 Sketch of stages in succession in an evergreen hardwood forest (hammock) in north central Florida. (Illustration by Denise Guerin.)

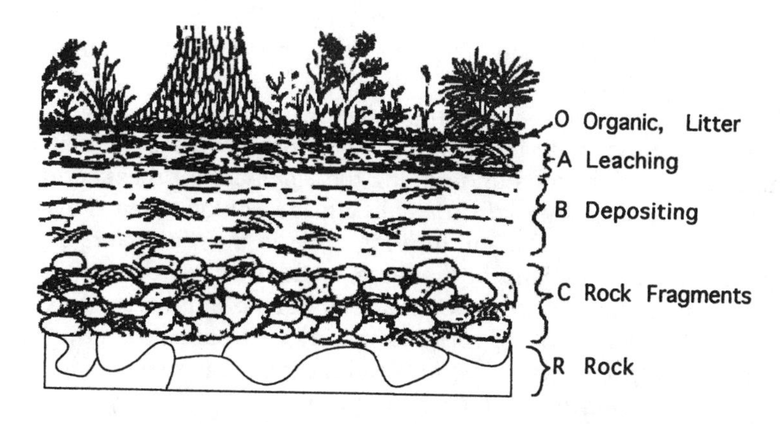

Figure 19.2 Cross-section of layers in soil.

If seed sources are nearby, evergreen pines come in after several years, soon overtopping the grassy and shrubby vegetation and shading it out. At this stage, with pine-needles and sun keeping the ground dry, conditions favor fire. As the pine trees increase the shade, hardwood seedlings come in. Later, when the hardwood forest develops, conditions for fire decrease. Old pines often remain in hardwood forests after succession, in a good position to reseed succession if forest areas are destroyed.

19.2 Soil formation

The succession process also builds soil, a major part of the forest ecosystem. Half of the organic matter of an ecosystem is underground in roots and organic matter that give the soil its structure. Organic matter from photosynthesis comes down tubes in the trunk to the roots and outside the trunk in falling litter. Earthworms process the material through their stomachs. Soil is a storehouse of the materials used by the ecosystem for living and nonliving processes. The actions of rain and the spread of roots break apart rocks into their mineral fragments, adding nutrients and helping the process of nutrient recycle.

Figure 19.2 is a typical cross-section of a forest soil with the letter labels usually assigned to the five main levels. The "O" level contains freshly fallen organic matter on top. These dead leaves, twigs, and limbs are the natural litter. As litter decomposes, its organic matter becomes mixed with ground particles to form the "A" level. This layer is full of consumers that release carbon dioxide, which makes this top part of the soil slightly acid.

Some nutrient substances are dissolved by the acid waters and percolate downward when it rains to interact with the sand or rock fragments below. Since this lower area is less acid, some materials such as calcium, phosphorus, and iron tend to precipitate and accumulate in this layer. The deposition layer is sometimes called the "B" layer. It is often reddish because of the deposition of the iron.

Below the deposition layer are the rocks or sand that geologic history provides as ground material. The "C" layer has the fragments of partly decomposed rocks or sands undergoing the weathering process and being acted on and dissolved by all the many living and nonliving reactions taking place in the soil. Below the soil-forming zone is the "R" layer, the main geologic material of the land in that area, such as limestone rock in many areas of Florida. The most common material of land in Florida is just sand. You can think of the sand and rocks as coming up from below slowly as an input to the forest ecosystem. "R" materials become fragmented and weathered as the "C" layer, which then becomes part of the "B" layer. In storm rains, topsoils wash away, and new soil is formed by materials from below.

19.3 Pine forests and fire climax

In Chapters 3 through 7, we used the pine forest ecosystem to illustrate principles of ecology and the systems modeling discussed in Part I (Figures 4.5, 7.2, and 7.4). The switch symbol was used for fire (Figures 4.4 and 4.5) because it is an on-off process turned on by lightning or people, and turned off when most of the dry organic matter is used up.

If fire is regular every few years, the ecosystem remains in the pine stage because the hardwoods that tend to come up in a shaded environment are killed by ground fire. Fire that is not too hot — such as a ground fire that moves slowly because the vegetation is not very dry — eliminates only the hardwood seedlings which have thin bark sensitive to fire.

Pine trees drop needles and cones which form combustible litter, making the pine forest more susceptible to fire. This kind of continuing pine forest is called a *fire climax* because it replaces itself. Lightning and humans start the fires. Two centuries ago, much of Florida was covered with pine forests maintained by regularly occurring periodic fires (started by lightning and Native Americans).

When the undergrowth burns, phosphorus and potash are released to the soil, but nitrogen blows away as a gas. To replace this nitrogen in the soil, frequently burned pine forests have a rich cover of leguminous herbs that convert nitrogen gas from the air to nitrates, replenishing the storages of nitrogen in the soil (see Figure 5.5 and accompanying text). Fire often stimulates plants in the understory to flower and develop seeds to take advantage of the open space and released nutrients.

Life in a fire climax pine forest shown in Figure 19.3 includes ground plants that fix nitrogen, gopher turtles that dig burrows, a ground cover of grasses or palmetto, and rattlesnakes. Pines are conifers with cones and no flowers, but they release enormous quantities of wind-blown pollen in the spring.

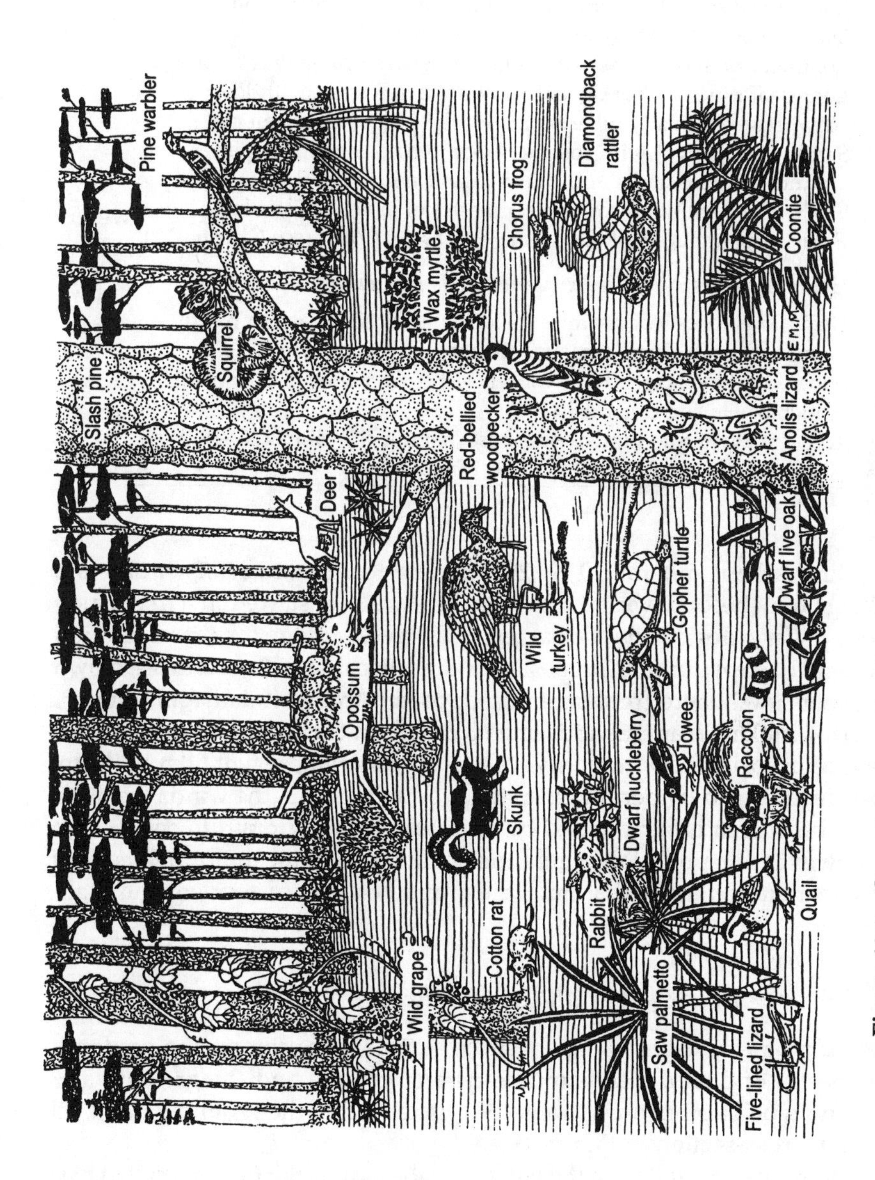

Figure 19.3 Organisms in a pine forest. (Illustration by Elizabeth A. McMahan.)

Table 19.1 Types of Upland Forest Ecosystems in Florida

Pine Forests — Successional Stages or in Fire Climax
Loblolly or short-leaf pines, inland in northwest Florida on clay soils
Slash pines with wiregrass, in wet "flatwoods"
Long-leaf or slash pines with palmetto where drainage is moderate
Long-leaf pine and turkey oak, on dry barren sites, with low diversity, especially where many of the pines have been cut
Sand pines (scrub) with rosemary, scrub oaks, endemic species; found on deep sands and dunes inland with very dry soils
Slash pine variety on rockland (limestone) in south Florida
Forestry plantations of slash pine or long-leaf pine (see Chapter 22)
Hardwood Forests — Climax Where Fire Is Excluded
Deciduous forests of northwest Florida, typical of eastern North America, with sharp seasonal changes and winter freezing
Subtropical evergreen hardwood forest (hammock), climax over much of Florida uplands
Tropical evergreen hardwood forest (hammock), climax in south Florida with many species of West Indian trees

If fire is kept out for 10 or 20 years, the underbrush of shrubs and seedlings may become very thick. Then, if a fire gets started in a dry windy period, a very destructive *crown fire* kills trees, understory, and ground plants and destroys human homes in the area. Because of fire controls, crown fires occur now where frequent burning used to prevent damaging fire.

The "Smokey Bear" advertisements designed to prevent all fires are not always appropriate. *Control burns* every few years can prevent litter buildup. These are arranged by permit when the ground is damp, the winds are light, and the burn area is surrounded by strips of plowed ground to contain the fire; however, air pollution regulations make it difficult to arrange control burns near cities. Smoke that hugs the ground can cause accidents on highways.

19.3.1 *Type of pine forest and groundwater*

Whether a pine forest is a stage in succession or a fire climax, the predominant species of pine depends on the type of ground and the groundwater level. The type of pine forest (Table 19.1) is related to the groundwater level shown in cross-section in Figure 19.4.

Conditions are wetter on the left where the land is flat (an area called *pine flatwoods*). Slash pines are dominant, and the ground under the trees is covered with wire grass and occasional shrubs such as gallberry and wax myrtle. Drainage is poor, and these areas have patches of wetlands. Where limestone is at the land surface in southeast Florida, a variety of slash pine

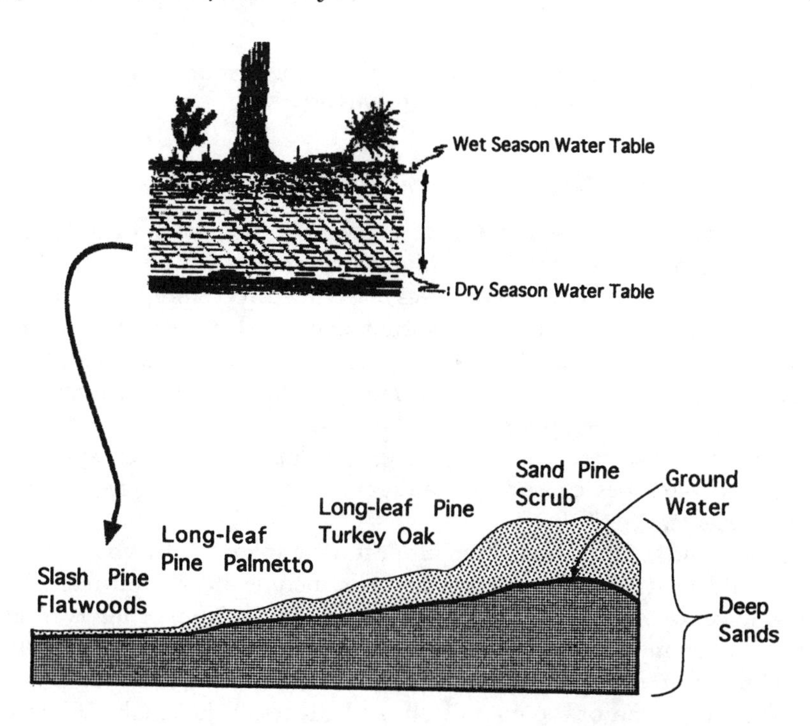

Figure 19.4 Types of natural pine ecosystems with fire according to the topography and superficial groundwater.

forest has developed called *rockland pines*. Most of these areas are now occupied by the cities of the Gold Coast.

Where there is more relief and water can run off better, the ground between the trees may be covered with saw palmetto, and the forest is called *pine-palmetto*. Like the pines, palmetto is adapted to fire with a thick, well-insulated stem that lies along the ground and has plenty of stored resources to put out new palm leaves after a fire.

In north and west Florida, the *long-leaf pine* and its associated community of organisms is the fire climax on typical uplands. Long-leaf pine has thick, fire-resistant bark and long protective needles around the growing point of young seedlings and saplings. On the drier lands, long-leaf pines have an understory of turkey oak (an area called *long-leaf pine–turkey oak*). Many of these original long-leaf pine forests have been converted to commercial slash pine plantations.

On old sand dunes, waters drain away from the surface quickly, producing the very dry sands for which *sand pines* are adapted, with rosemary and myrtle oak as understory. Sand pine can only germinate from its seeds after fire has caused the cone to open up and the seeds spread. Large areas of *sand pine scrub* grow on the ancient sand dunes of the Ocala National Forest and on modern dunes along the coast. *Loblolly pines* grow in north Florida on soils with more clay to hold water and bind nutrients.

19.4 Hardwood forests

Dominant hardwood trees are able to grow up in the shade of older pine trees. Trees grow in several levels. Maple, oak, magnolia, and hickory are very common in the *canopy* (the tall trees receiving direct sunlight). A diverse understory develops that contains shrubs and shorter trees such as dogwoods. On the forest floor there are herbaceous plants, ferns, mosses, and seedlings. With passing storms sticks fall from former trunks, limbs, and twigs. The litter builds up a storage of organic matter that includes seeds ready to repair the forest if an opening develops. A complex system of animals and microbes consumes the organic matter and recycles the contained nutrients.

Leaves, buds, seeds, and litter support a complex food web of insects, spiders, and other invertebrates. Mammals include deer, squirrels, chipmunks, mice, opossums, foxes, and raccoons. Before human intervention, there were also more bobcats and black bears, and even wolves. Figure 19.5 is a picture of an evergreen hardwood forest. Figure 19.6 is a systems diagram showing some of the complexity and diversity that make the forest productive and efficient.

19.4.1 Temperature and type of hardwood forests

Three kinds of hardwood forests are described for Florida (Table 19.1). The transition from deciduous forest in the northern counties of the state to subtropical evergreen forest in central Florida and patches of tropical forest in south Florida is gradual from north to south and is caused by increasing temperature and decreasing frosts.

19.4.1.1 Deciduous hardwood forest

The predominant upland forest over the eastern U.S. is the *deciduous hardwood forest*, found in Florida along its northern border where winters are cold. Hardwood species have flowers and broad leaves, mostly dropping their leaves in autumn. All life in the deciduous forest is dominated by the seasonal cycle of leaf fall in autumn and regrowth of leaves in spring.

Deciduous hardwoods are able to maximize the use of the sunlight available during the spring and summer and to store food in their fruits and nuts, roots, and underground stems. The trees form annual rings by depositing wood during summer photosynthesis, a process interrupted by the annual leaf fall. The forest produces an abundance of leaves, fruits, nuts, and seeds which provide food for a wide variety of animals in the summer and autumn.

When the days become shorter, and the temperatures drop in autumn, the leaves of deciduous trees change color to reds, yellows, and orange. This

Figure 19.5 Dominant trees, shrubs, and animals of an evergreen hardwood forest (hammock) in north central Florida. From left to right: live oak, water oak, hickory, smilax, red-shouldered hawk, palmetto, armadillo, Spanish moss, green dragon, hop-hornbeam, poison ivy, sweet gum, squirrel, magnolia, mulberry, raccoon. (Illustration by Denise Guerin.)

color change is caused by the plants re-absorbing the green chlorophyll from the leaves so that previously hidden colored plant pigments show up. In deciduous forests, a few hardwood species such as magnolia and holly are evergreen.

In the winter, dead leaves become litter which decomposes partially to make humus, giving soil a fertile texture. When tree limbs are bare, there is no transpiration and water is conserved. Some animals hibernate, while others develop heavier coats.

In the spring, the trees use the organics stored the previous summer to produce new leaves and flowers. With new buds, the insect populations increases rapidly. Birds, including the returning migratory species, nest, singing most during this part of their life cycles.

19.4.1.2 Subtropical evergreen hardwood forest

In central Florida, the climax ecosystem on uplands where fire is excluded is the evergreen hardwood forest, also called *hardwood hammock* (Figure 19.5).

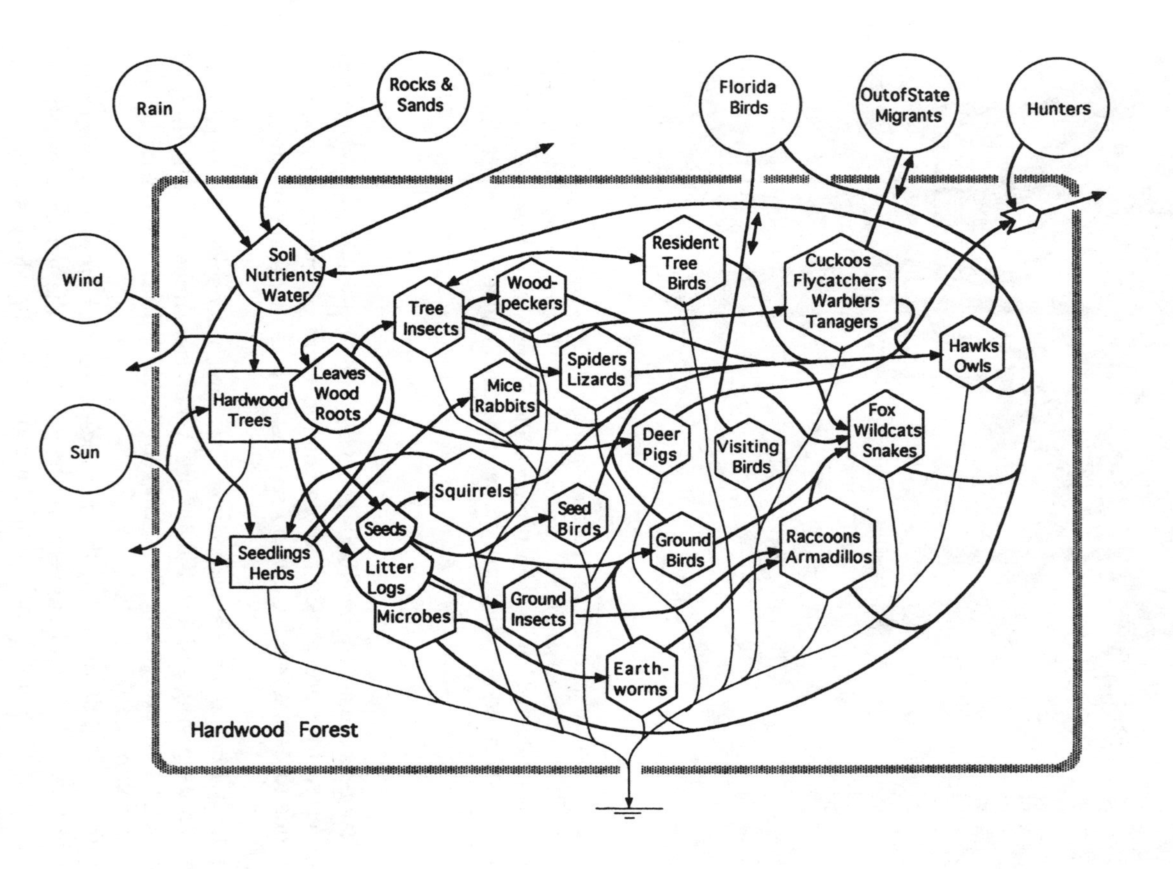

Figure 19.6 Systems diagram of an evergreen hardwood forest (hammock).

Hammock means a shady place. Hardwood hammocks are still found in state reserves and parks and in protected tracts within towns — for example, the San Felasco Hammock in Gainesville.

Dominant trees are the evergreen live oak and laurel oak, while a few trees are deciduous with limbs bare in the winter. The top leaves of the crown trees adapting to bright sun are small, thick, and heavy, whereas the shade-adapted leaves of the understory trees are broader and thinner. Evergreen leaves are thick and leathery and conserve moisture during the occasional long dry periods in winter and spring. The live oaks drop their leaves all at once in the spring, just as the new leaves come out. Other species are deciduous, dropping their leaves in the autumn. Measurements indicate that leaf-falls in autumn and spring are often equal. The understory has a diversity of small trees such as dogwood and holly.

Tree limbs are coated with mosses, lichens, orchids, and "Spanish moss", a flowering plant that hangs down in graceful, gray-green drapes, giving these forests a special appearance. Plants that grow on other plants are called *epiphytes*.

A great variety of insects form a food-chain network that includes herbivores, sucking insects, carnivorous insects, bees which pollinate flowers, termites which break down wood, and ants which consume detritus. Birds help regulate insect populations and transport seeds, as do mice and squirrels. Tanagers, cuckoos, and flycatchers nest in summer and migrate south in winter. Woodpeckers, titmice, chickadees, and wrens are all-year residents. Carnivores include lizards, coral snakes, and hawks.

19.4.1.3 Tropical evergreen hardwood hammocks

In southernmost Florida, where killing frosts are rare and fire is excluded, tropical hammocks develop with palms, broad-leaved evergreens, and a high diversity of west Indian species such as gumbo limbo, poison wood, mahogany, and royal palms. Remnants of this forest type are found in south Miami and in oval tree islands in the southernmost Everglades.

19.5 Forest conservation

19.5.1 Role of diversity in protecting forests

Different species of insects can feed on and digest various kinds of plants. Because each kind of plant is chemically different, any consumer has to be especially adapted to digest that kind of plant. No one consumer can be adapted to eat very many kinds. If any plant species increases to excess, its consumer insects or disease organisms will increase until the species is again brought into stable balance with the other species. As with the prey-predator models in Chapter 11, the consumers increase with the plants and then decrease again as the plant species are consumed and become less abundant. In a diverse forest, even if one kind of tree succumbs to disease or an insect infestation, the ecosystem and its many species are preserved. Collectively, the abundant tiny insects regulate anything in excess. Larger birds and animals, by eating many species of insects, regulate them further.

If stable, diversified forests are sprayed with insecticide, the protective network is destroyed, and the system can become unstable, with weedy growths and an outburst of resistant insects, including undesirable species such as mosquitoes.

19.5.2 Pine forests and epidemics

When there are dense stands of one species, epidemics of insects and diseases can occur. Many insects are adapted to eat parts of pine trees, and some, such as the pine bark beetle, may become epidemic, destroying many trees. The beetle drills through the bark of a pine tree and deposits its eggs. When the larvae hatch, they begin eating the layer of wood just below the bark and can kill the tree if they eat completely around. The beetles tend to attack trees weakened from drought or damage.

Most natural pine forests, however, do not develop epidemics, apparently protected by the web of species diversity and the birds. Bark beetles may consume scattered trees, but as a part of the normal cycle of growth and replacement. Controversy arises when those raising commercial pine forests, in order to protect their investments, want to destroy all the pine trees in the natural forests. The natural forests are generally not threatened because of their complexity.

19.5.3 Endangered species

Species are endangered when their *habitat* is lost. Most forests in Florida are young, because most of the old ones were cut earlier in the century. The land area in natural forests is diminishing rapidly. Animal species requiring old forests are now endangered. For instance, the red cockaded woodpecker nests in trees more than 75 years old. As old growth forests have become scarce, so too has the red cockaded woodpecker. The gopher tortoise is endangered because most of the drier pine areas are being developed by humans. The scrub jay is found only in the sand pine ecosystem.

19.5.4 Sustainable management

In Florida forests, most of the nutrients are in the plants and wood with very little stored in the sandy soils. After lumbering, an infertile site remains, particularly if the roots are also pulled out for industrial use. Succession then is slow and has to depend on the gradual accretion of nutrients from the rain, especially where there are few mineral particles to weather into the soil. Ordinary quartz beach sand provides little nutrition, but calcareous sands and marls of south Florida have considerable phosphorus and other nutrients. In Florida's history, when the original forests were cut, the second growth was poor. Scrub cattle could not get enough nutrition for reasonable growth.

Questions and activities for chapter nineteen

1. Define: (a) deciduous, (b) canopy, (c) understory, (d) hibernate, (e) humus, (f) strata, (g) temperate, (i) subtropical evergreen hardwood forest, (j) fire climax, (k) fire tolerant, (l) annual rings, (m) epiphyte.
2. Compare the climates of the temperate deciduous forest, the pine forest, and the subtropical evergreen forest.
3. Name some plants in the temperate deciduous forest.
4. Obtain a cross-section of pine and hardwood trees and interpret the history of growth conditions from the width of the rings.
5. Explain how fire influences the vegetation in the pine forest.
6. Label the plants and animal species in the picture of the evergreen hardwood hammock in Figure 19.5.
7. Explain how human populations have affected the forests in Florida.
8. Explain what kind of forest fire in Florida may be beneficial.
9. If you have access to a computer and the BioQUEST module, Environmental Decision Making with EXTEND, run the fire-grassland simulation. For source information, refer to Appendix A.
10. Take a fire ecology field trip to a forest where it is possible to see areas that have recently burned, areas burned 10 or more years ago, and areas not burned in recorded history.
    (a) Look for evidences of fire and its effects; compare sites.
    (b) Brainstorm possible positive and negative consequences of forest and range fires for wildlife. Specify kinds of wildlife, and give examples.
    (c) Make and record observations, e.g., variety and quantity of vegetation, evidence of wildlife, actual sighting of wildlife, and soil layers.
    (d) Count 100 trees, plants, or animals, recording the number of different species.
    (e) Name the tree or plant that covers the most area and the most common animal.
    (f) Count the number of leaves between the ground and the sky (leaf area index). For trees, stand under the tree and look up; count the leaves straight above you.
    (g) Dig a soil pit and record the depth of the top layer, color of the top layer, and kind of soil material — sand, clay, peat shells.
    (h) Make a systems diagram to include the things and processes observed. Start by modifying one in this book.

chapter twenty

Landscape mosaic of ecosystems

Today, as in the time of the Indians, wherever lands and waters are left to their work, the processes of nature are busy using appropriately adapted species to fit ecosystems to the environmental conditions. The past work described in Chapters 15 to 19 was a mosaic (patchwork) of ecosystems that originally covered Florida. Although most of Florida has been converted into agricultural and urban use, remnants of the original ecosystems can be found in every county. This summary chapter provides simplified maps to overview the ecosystems remaining on the Florida landscape. It is hoped that we can interest schools, residents, and visitors in looking for the original ecosystems in each local area.

20.1 Regional maps of Florida ecosystems

Each of four regional maps in Figures 20.1 to 20.4 identifies ecosystem areas in relation to familiar geographical features. West Florida (Figure 20.1) has deciduous hardwood forests on loamy upland soils along the Georgia border and evergreen hardwoods nearer the coast. Some of the pines on dry sandy soils are in a fire climax state, although larger areas have successional pine forests and forestry plantations. Bottomland hardwoods line the river floodplains, and coastal marshes exist where the streams join the sea.

Northeast Florida (Figure 20.2) has headwater wetlands (Okefenokee Swamp and Osceola Forest), central lakes with a wide range of natural fertility, evergreen hardwood hammocks on the central ridges, sand pine forests (sand pine scrub) in the Ocala National Forest, wetlands and salty springs along the northward flowing St. Johns River, and coastal salt marshes.

Central Florida (Figure 20.3) has a lake district in the sands and limestones in the center, headwaters of the Withlacoochee and Hillsborough Rivers from elevated areas of cypress ponds (Green Swamp), pine flatwoods, freshwater marsh headwaters of the St. Johns River, large estuaries, and coastal mangrove wetlands along both coasts.

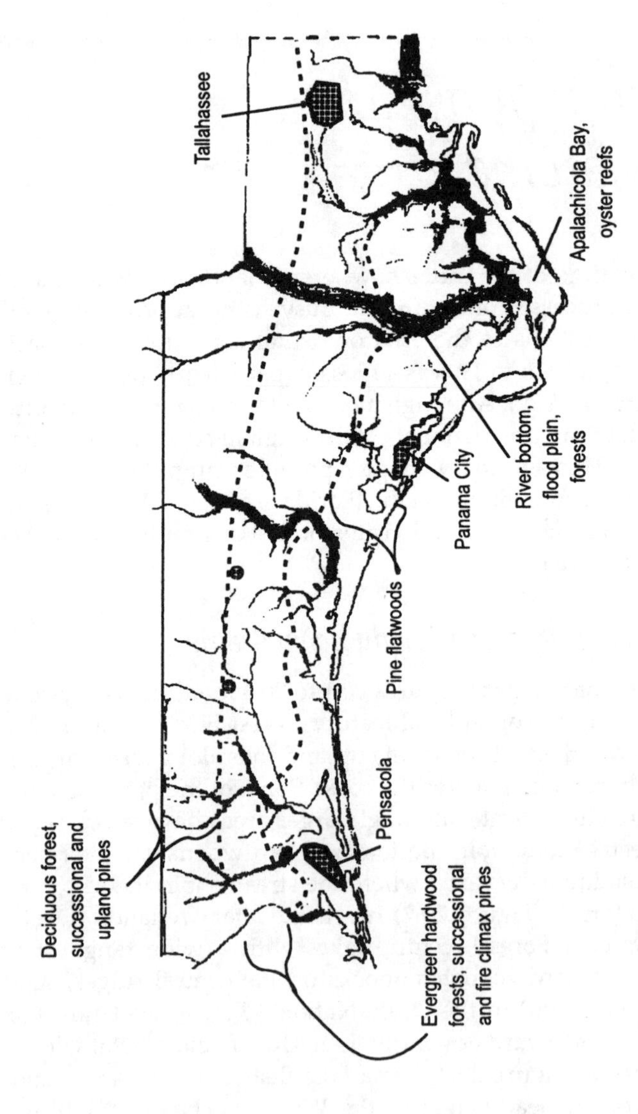

Figure 20.1 Ecosystems and land uses in the west Florida panhandle.

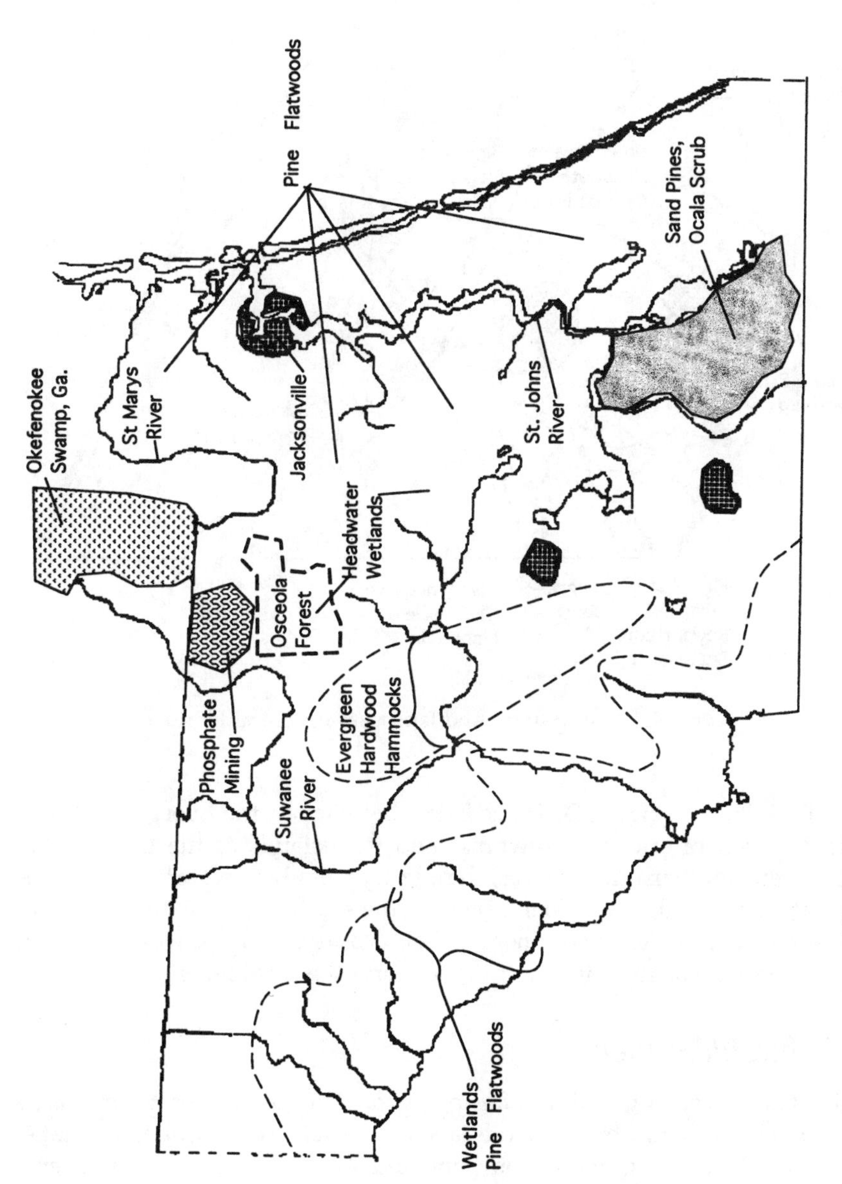

Figure 20.2 Ecosystems and land uses in northeast Florida.

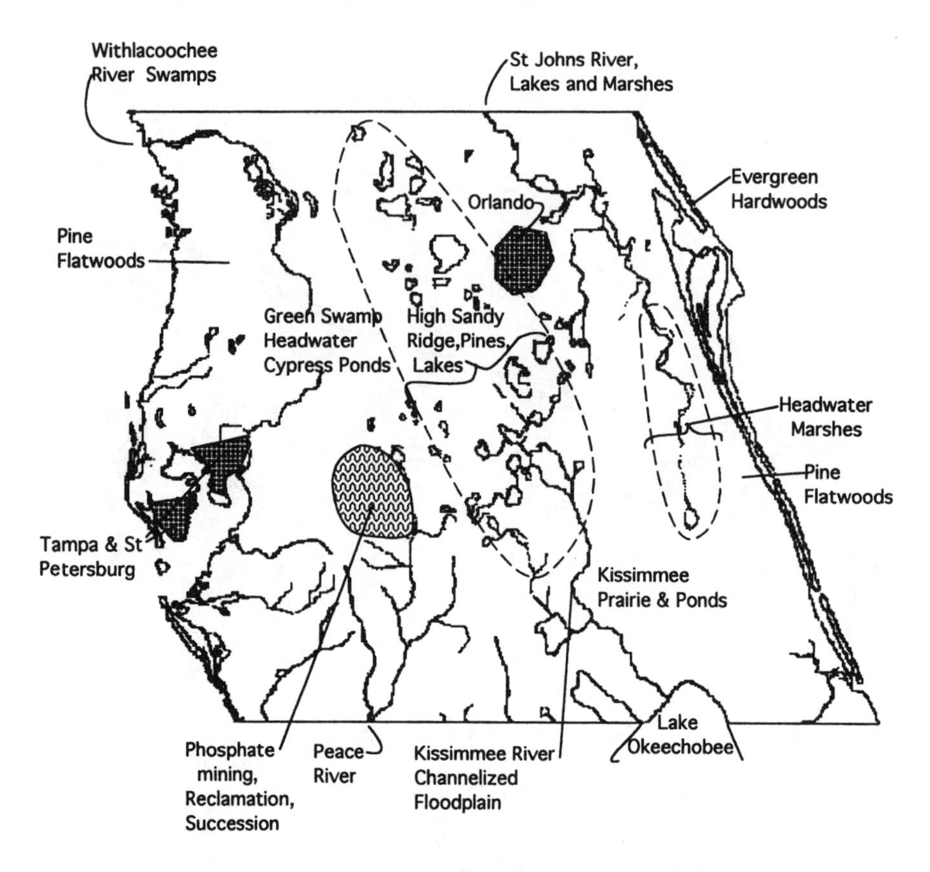

Figure 20.3 Ecosystems and land uses in central Florida.

South Florida (Figure 20.4) has Lake Okeechobee, the Everglades "river of grass", wet prairies, the sawgrass and tree islands of the Loxahatchee reserve, the southern mangroves, the Big Cypress headwater wetland, the Fakahatchee Strand, the rockland pines of the southeast coast, the grass flats of Florida Bay, the large continental shelf ecosystem of the west coast, the blue water ecosystem of the Gulf Stream, and the coral reefs.

20.2 Interdependence

While it is convenient, at times, for the purposes of discussion to separate a particular ecosystem from the landscape in which it is embedded, remember that it is linked with other surrounding and adjacent systems in an inseparable mosaic. Marine ecosystems are interconnected with the terrestrial environment, receiving runoff, silt, and organic matter. Wetlands are bound together with their surrounding uplands receiving runoff and sharing animals that often feed in both environments. Pine forests are often a mosaic of

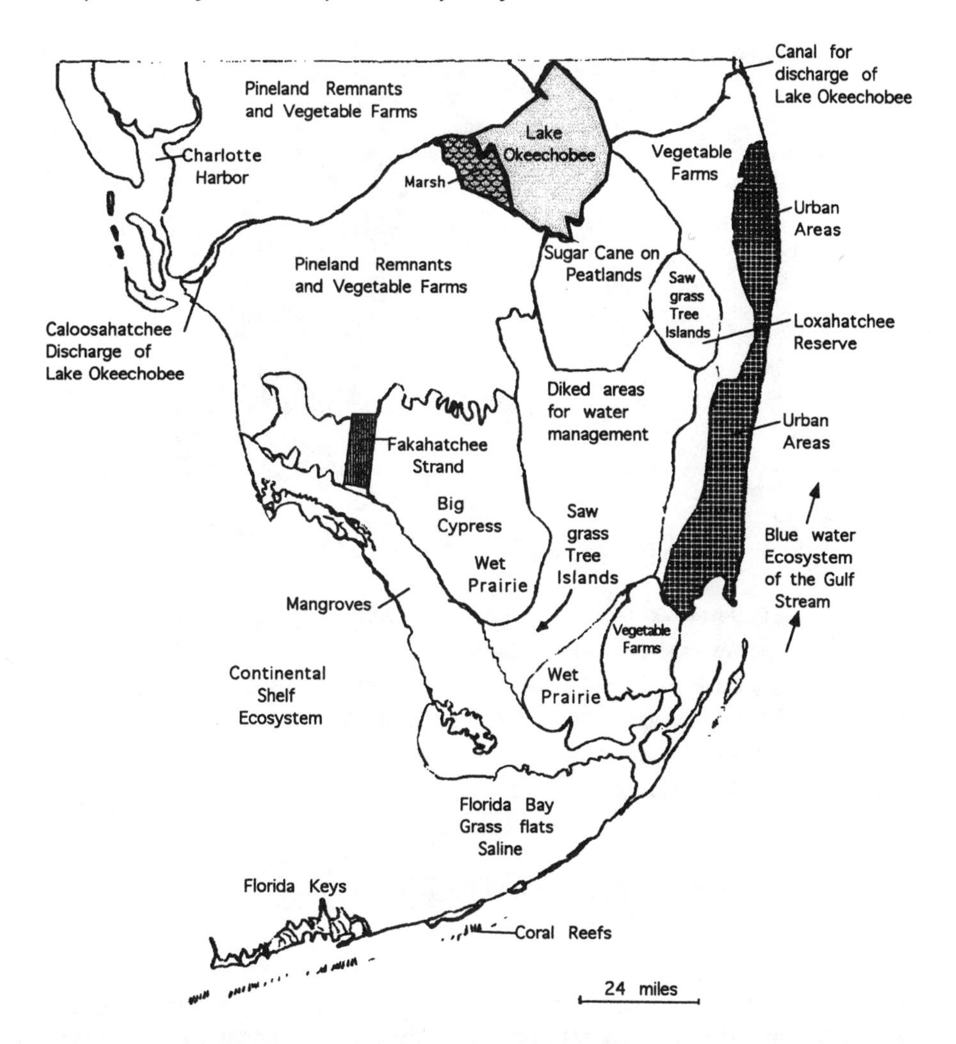

Figure 20.4 Ecosystems and land uses in south Florida. The sawgrass and tree islands area is called the Everglades.

wetland depressions and "islands" of hardwood forests that are interlaced so that the edges of one system blend into the other. At the edge of the land, where ocean waters lap the coastline, coastal wetlands exchange nutrients, organic matter, and animals with nearshore estuaries.

20.3 Soils of Florida

The underground part of an ecosystem is called the soil. As we already discussed in Chapter 19, the growth and succession of a terrestrial ecosystem eventually develop into a soil profile with layers like those in Figure 19.2 (O,

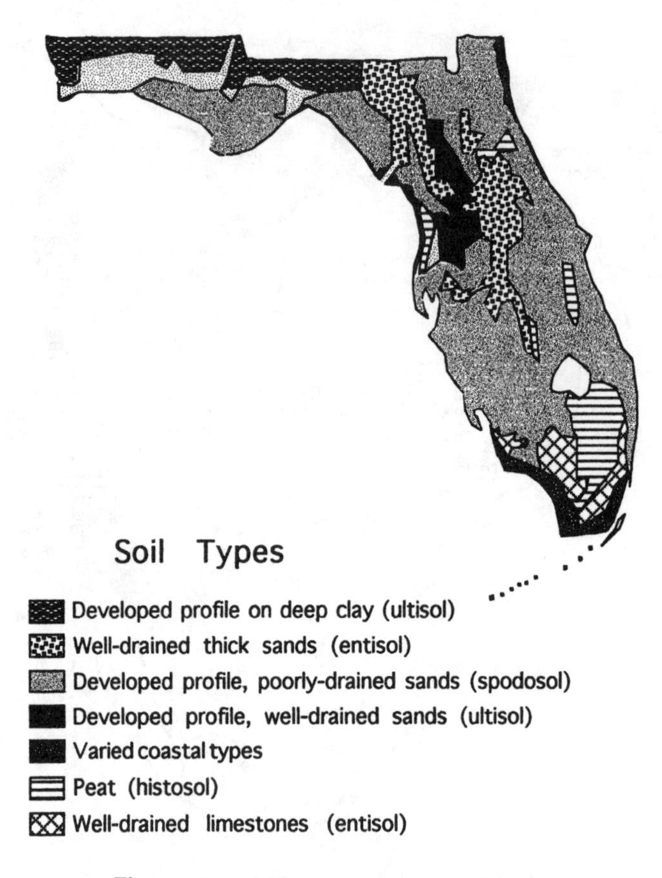

Figure 20.5 Map of soil types in Florida.

A, B, C, R). Although the type of ecosystem and soil formation depend partly on the climate, the groundwater and the kind of geological substrate that supply the earthen material are major factors also. Figure 20.5 shows where the primary kinds of soils are located in Florida. Because soils are by-products of ecosystems to some extent, soil maps are often similar to vegetation maps (Figures 18.1 and 20.5 and 20.6).

Some soils are uniform without much vertical differentiation into layers (technical name: *entisols*). This can happen if the original geological process or human work brought uniform materials to the surface without much ecosystem development. Examples are the well-drained, thick sands of former beaches in the north center of the peninsula. In the Miami area are slightly raised lands of little differentiated substrates of limestone and marl. Thick beds of peat (technical name: *histosols*) are found just south of Lake Okeechobee, deposited by custard apple wetlands and now used for sugar and vegetable farming (Figures 18.9 and 20.4).

Figure 20.6 Map of upland forest types in Florida.

Other soil types in Figure 20.5 have better developed soil profiles (more like Figure 19.2). Along the northern border there are soils developed on well-drained, deep beds of red clay by the deciduous forests in the colder climate there (technical name: *ultisols*). Farther south in the center of the state are soils developed by the evergreen forests on well-drained, sandy substrates. The soil type with the largest area in Florida is that developed by pine flatwood ecosystems in poorly drained, more acid conditions (technical name: *spodosols*).

After the above-ground part of a land ecosystem is cleared for agriculture, housing, or other use, the soil retains its characteristic for some time. In the long run, however, the land has to be allowed to develop multi-species ecosystems to restore the properties of soil texture and fertility.

20.4 Ecosystem overview

If we look at all of Florida from a greater distance as you do when you look at the *Florida Atlas* (Fernald and Purdum, 1992), a simpler overview of Florida results (Figure 20.6). Note the belt of deciduous hardwood forest across the north where winter temperatures are low. Evergreen hardwood hammocks are on the central ridge, and pine forests cover other uplands. Although economic development has replaced most forests with cities, roads, cattle ranches, and agriculture, a quarter of Florida is still in pine forests, although many of these areas are commercial forestry plantings (Chapter 22).

Questions and activities for chapter twenty

1. Define: (a) mosaic, (b) landscape, (c) soil, (d) topography.
2. On a field trip to a natural area, look for boundaries between ecosystems. See if some ecosystem types blend gradually into the next. Discuss what things cross the zone between any two ecosystems that you find.
3. Dig a soil pit in a forest and one where soil was turned over during construction a few years ago (2 ft across by 3 ft deep). Compare the layers and explain the differences. Are the layers in Figure 19.2 recognizable? (Be sure to fill up the hole to restore soil function and to prevent accidents.)
4. Collect a top soil layer from a nearby ecosystem. All you need is an area about a foot across and 6 inches deep (30 cm wide by 15 cm deep). Shake the soil loose from the small roots. Spread the soil layer in a seed flat exposed to sunlight and keep it moist but not water saturated. Observe the plants that germinate from the "seed-bank". Identify as many as you can. Are they weeds, pioneer trees, or climax species? Compare results from other areas.
5. Sterilize some soil. (This can be done by heating it in an oven for several hours to a temperature of 350°F — be careful!). Place soil in a seed flat outside and keep moist. Observe the number and species of germinating plants over a period of time. Where did they come from? Compare the number and species of plants appearing in this flat with those from in Question 4.
6. Describe the differences between ecosystems in north and south Florida.
7. Discuss management of a park in a central Florida forest area. How would you ensure a diversity of ecosystems for the public to see?
8. Describe the interactions between ecosystems: pine forest and hardwood, marine and shore, wetlands and uplands.
9. Discuss five ways in which humans have changed Florida's ecosystems.
10. Which four natural factors are most important in the geographical location of Florida's forests? For one type of forest, describe the specific combination of these four factors which has led to its location.
11. As you and a friend drive along a road in a rural area, write down the names of the ecosystems (natural and human-managed) you find and the way they are related to topography.
12. Locate your area on the maps of environmental characteristics and ecosystems in the *Florida Atlas* (Fernald and Purdum, 1992). Explain how they are related.

part three

Systems that use the environment

Part III is about the systems of the economy that use the environment and ways to keep them productive and sustainable. Chapters 21 to 28 consider the interface between environmental systems discussed in Parts I and II and the human economy discussed in Part IV.

Some of the systems yield products for human use and the economic market. The *yield systems* of Florida include forestry (Chapter 22), agriculture (Chapter 23), fisheries (Chapter 24), and mining (Chapter 25).

Development converts environmental systems into housing for retirees and immigrants. Natural ecosystems are displaced by *urban ecosystems* with new associations of plants and animals. Cities utilize clean waters, green belts, and suburban lands to buffer the concentrations of noise and air pollution. Many industries utilize the environmental lands, winds, and waters for raw materials or waste dispersal.

Areas set aside for parks and wilderness are used by millions of people. These systems usually undergo less change from the "human footprint" than from uses that displace ecosystems or harvest them. Parks derive aesthetic values from the environment for recreation and tourism, a major part of the economy of Florida.

The work of humans in society is coordinated by the circulation of money. Each person who has a job receives money as salary or wages and uses that money to buy what he needs from other people. But the circulation of money depends on real wealth from environmental work that includes the production of food, clothing, housing, minerals, fuels, and aesthetic experiences. Chapter 21 explains how the environment is connected to the economy.

chapter twenty-one

Economic use and development

The environment supplies raw materials and energy that humans use to produce goods and services essential for their lives and economy. At the interface between the environment and society, economic activity generates goods and services to sell. The circulation of money distributes these products among people. This chapter explains the basic plan for making economic use of the real wealth of the environment and what is required to make the development of human society and ecosystems prosperous and sustainable. A look at the history of rapid economic development of Florida suggests ways to make the interface more symbiotic (mutually helpful).

The study of the production and distribution of goods and services and their relation to the circulation of money is called *economics*. The economics of environmental inputs is the subject of *microeconomics* (small-scale economics). Measuring the environmental contributions and the intensity of human use on a common basis is an objective of *ecological economics*, for which we use emergy (spelled with an "m"; see Chapter 7). The circulation of money on the larger scale of the whole state system is *macroeconomics*, considered in Part IV.

21.1 Typical economic use interface

Wherever there is a farm, a business, or other uses of the environment by humans, there is an *economic use* system (see the overall plan in Figure 21.1). As shown on the left, the environmental system makes resources available for free (for example, forest wood). Human economic use harvests and sells the products (on the right). Money from the sale of the environmental products (for example, wood) is used to purchase the energy, materials, and services necessary to perform the work (above on the right). Money is shown with dashed lines going in the opposite direction from the items purchased.

The simplified summary shown in Figure 21.1 shows the way in which part of the real wealth comes from nature's work (from the left) and part is

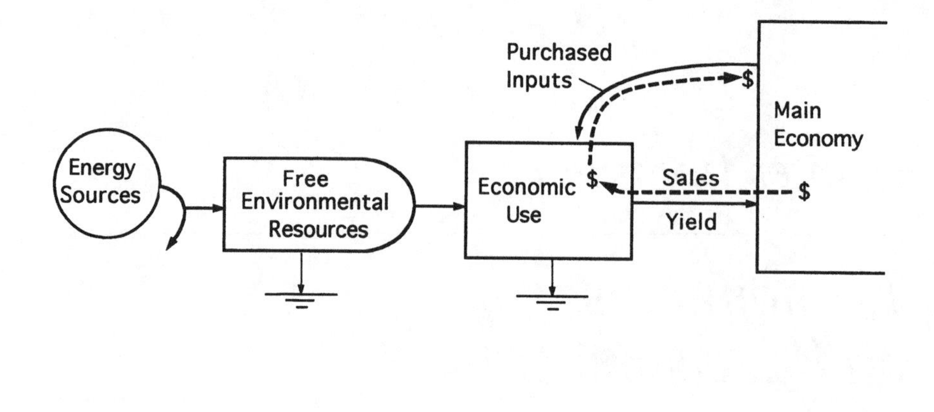

$$\text{EMERGY Investment Ratio} = \frac{\text{EMERGY of Purchased Inputs}}{\text{EMERGY of Free Environmental Resources}}$$

Figure 21.1 An economic use system with local free inputs from the left, purchased inputs from the upper right, and money exchanges in dashed lines.

contributed by purchased inputs (from the right). The systems of resource use in Chapters 22 to 28 fit this common plan.

Figure 21.2 shows more details of the relationships of money circulation (dashed lines) to the processing and use of resources (solid line). The economic users have to pay labor and buy fuels, electricity, goods, and services to gather, transport, and process the products. After selling products to markets in the economy, the earned money goes into the short-term storage (Figure 21.2). Out of this, the user must pay rent and taxes. If the environmental-use business borrows money, it has to pay interest and repay the loan. Profits can be kept in a bank or invested elsewhere

21.2 Money and real wealth

As everyone knows, money circulates among people who control the flow of goods and services. People receive money as salary and wages. They use the money to buy what they need, such as food, clothing, housing, and recreation. Money is exchanged, going in the opposite direction from what is bought (Figures 3.9 and 3.10). Money is not real wealth but is the means to buy it. The circulation of money helps make the work of society efficient.

In the typical environmental-use system (Figure 21.2), money (dashed line) is paid only to people and never to nature. For example, the money paid by farmers for irrigation water (dashed line in Figure 21.2) goes to the people for the emergy of their labor and services. Even where the purchased products come from elsewhere, such as fuel and goods, the money only goes for

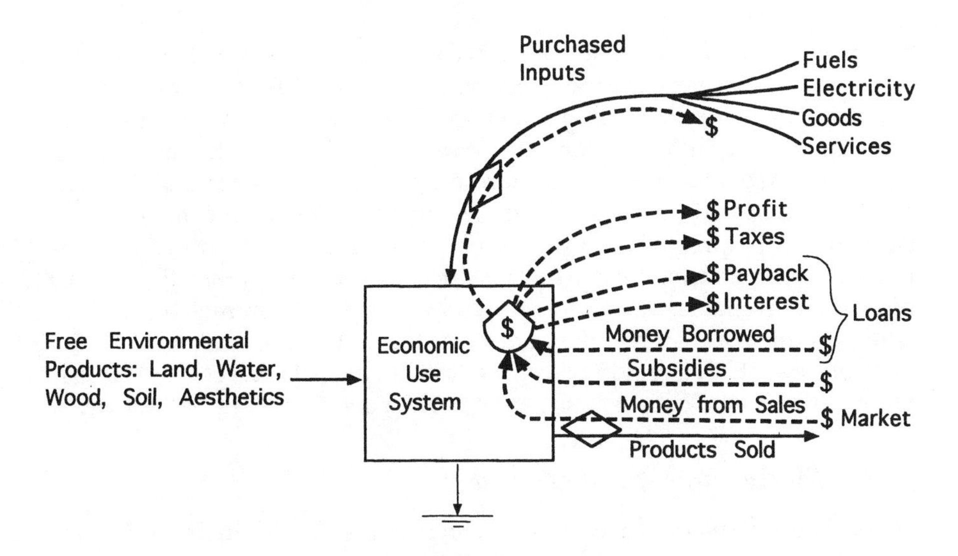

Figure 21.2 Flows of money (dashed lines) in an economic use system.

the human part of processing the products, not for the environmental contribution to the product.

21.2.1 Emergy evaluation

As already explained in Chapter 7, the work of people and nature is evaluated on a common basis using emergy. Emergy measures the previous work done in developing a product or service. Products and values from forests, farms, mines, and parks contain the emergy from nature and the economy (Figure 21.2).

Products delivered to people have more emergy than what is paid for with money, since only the human contribution to the work is reflected in the purchase price. When irrigation water is purchased, most of the emergy in the water comes from the unpaid work of nature.

Emergy is a common measure of the many kinds of real wealth (clothing, houses, information, fuels, electric power, foods, clothes, schools, cars, music, art, and environmental aesthetics). The annual emergy use per person is a measure of standard of living. This quantity is different from the dollars spent per person because it includes the free values of nature. In Chapter 29, emdollars are defined as a currency based on emergy.

21.3 Costs, prices, and scarcity

When resources are abundant and close at hand, little effort is required to obtain, concentrate, or transport products to market for human use. Little money has to be paid to people for services. Costs are low. When the supply of products is larger than the need, people do not regard the products as valuable. They are not willing to pay much money for them. For example, pioneers in a virgin forest paid little for wood, which was all around them.

When resources are scarce, more human effort is required to obtain, concentrate, or transport them to market. In other words, more money has to be paid for the processing. Costs are high when products are scarce. The number of people who want to buy products may be greater than the number of products available. In addition, prices are increased by human behavior — when products are scarce, they are regarded as being more valuable. For example, people in a region where the forests have all been cut have to pay more to those who collect and transport the wood from forests far away.

21.3.1 Market supply and demand

People buy and sell products in stores and markets. The prices depend on the quantity of products available and the number of people who want to buy the products. The quantity of products available is called the *supply*. The number of products that people are willing to buy is called the *demand*. When the demand is greater than the supply, prices are high; when the supply is greater than the demand, the prices are low. In other words, prices are proportional to demand and inverse to supply. The response of producers and consumers to prices helps keep products from being in shortage or excess for very long.

21.3.2 Real wealth and market price

When the resources from nature such as fish, water, and wood are abundant, the free market gives them a low price. Yet, this is when everyone can have plenty. Standards of living are high, but the economic system rates the products as low in value, and they have a low price. For example, when redfish were plentiful, many people fished and caught a large quantity of them. People could refrigerate the fish and eat them often. The living was good, but on the market the price for the fish was low.

Later, if a resource is scarce, the market gives it a high price. Yet, this is at a time when the products are contributing to the economy little more than what is required for processing. Using the redfish example, when the fish are scarce people do not catch many when they go fishing. Their meals are not so rich, but on the market the price is high. In other words, market values do not measure environmental contributions, and are often inverse. Prices are low when nature's contributions of real wealth are greatest.

To summarize, when real wealth is abundant, the prices are low. When there is little real wealth, the prices are high. Market values (prices) tend to be inverse to real wealth. Market values cannot be used to evaluate environmental resources correctly.

21.4 Economic development

The word *development* means to make resources usable. Economic development generates the most real wealth by retaining the environment's production while matching it with additional real wealth bought from outside. For example, a fish camp uses the environmental production of fish while bringing in fuels, fishing equipment, and fishermen. The fish camp operation that does not pollute, overload, or overfish the waters can yield more fishing with less cost. This attracts more business and maximizes its economic contribution. Good development fits the economic-use activity to the environmental processes so that they are symbiotic.

Sometimes governments try to encourage an economy to grow by making new money available at low interest rates for new developments. The idea is that circulating more money will draw in more resources without much inflation. This only works if there are new resources available to be developed or more efficient ways are found to process and use resources. There is controversy today concerning how much development is appropriate, where it should occur, and what kind of development is most suitable.

21.4.1 Predicting economic success with the emergy investment ratio

The *emergy investment ratio* is a useful index of the relative importance of emergy from the economy compared to that from the environment. Shown in Figure 21.1, the investment ratio is defined as the purchased emergy divided by the emergy supplied free by the environment. In undeveloped areas, the ratio may be close to zero. A typical average value for use of environmental resources in Florida is about 7, but uses of this intensity do not leave very many ecosystems intact. Parks that provide wilderness values have ratios less than one.

When the investment ratio is large, there is more concentrated activity of people and machines and more wastes going into the environment. There is more impact on the environment. Intensive developments of agriculture or industry may have values of 25 or more. The ratio is high in cities.

When the investment ratio is small, the free contributions of the environment are large. Products are produced with less human service and less cost. Enterprises with low investment ratios tend to succeed economically because they provide products at low cost. When some new use of environmental resources is proposed, the emergy investment ratio can be estimated to see if the ratio is too high, and thus too hard on the environment and requiring too many purchased inputs to be economical. Projects should be planned that have the same or lower ratios than other environmental uses in the same region.

An example of the use of the investment ratio is a proposal to build a condominium on the southwest coast of Florida. If you build one condominium overlooking the beach surrounded by woods, all the units look out on nature. The investment ratio is low. Units sell for high prices. However, as more units are developed, parking lots and noise increase, and nature is displaced. There is less nature per unit; the investment ratio is high. It is not so good a bargain.

21.4.2 Wildlife habitat

Florida's once continuous cover of green vegetation has given way to urban development and agricultural lands. The few "wildlands" that remain are now fragments of forests and wetlands separated by vast areas of housing developments and highways. The loss of habitat means loss of carrying capacity for wildlife. A discussion of wildlife corridors and greenways can be found in Chapter 26.

21.4.3 Water and Florida's economic development

When development fits the Florida economy to the system of water and wetlands symbiotically, it increases the potential for prosperity. This is because the original system of water flows was already doing what the economy requires: namely, to maximize the quantity and quality of water available.

The earliest settlers in Florida were farmers and ranchers. During these early years, the pattern of development was primarily clearing and preparation of land for agricultural uses. Cities were small, and most people lived in rural, agricultural settings. The best lands were settled first. Much of the remaining flat sandy landscape was not well suited for agriculture or cattle because of too much water at times. The groundwater table was just below the surface of the ground (Figure 19.4), and during rainy periods it rose to the surface. Lowland soils became waterlogged, and wetlands filled to capacity. Lands with high water tables were unsuitable for crops, cattle, or homes, so ditches were dug to make the lands drier. Water tables in southwest Florida were lowered 3 to 4 ft. In some areas, such as portions of the Everglades, wetlands were drained so that the rich organic peat could be used for agriculture.

As more and more people came to the state and the best lands were already occupied, development began to infringe on wetlands. With cheap fuels and better technology, these lands were drained for agriculture. Lands downstream from developed lands flooded more frequently and more severely. The response was to dig more canals and ditches and to straighten meandering rivers to get the water off the landscape more quickly. More waters were sent into the sea.

Florida was drained, and the landscape became drier (Chapter 13). While there was less flooding in the wet season, the dry season brought more serious droughts. The drier conditions made it easier for exotic plant species that were introduced to Florida as landscaping plants and ornamentals

(*Melaleuca* and Brazilian pepper, for example) to replace native species less adapted to this new hydrology. *Melaleuca* lives in wetlands by drying them out with high transpiration, whereas the native pond cypress with low transpiration prevails by keeping the area wet.

Because water may be the ultimate factor limiting the economy of Florida, as much as possible of the original water system needs to be restored, keeping more water from running off to the sea, recharging more groundwaters, letting nature work for us again with wetland filtration. Developments can be designed that do not require water tables to be lowered and that use wetlands for water storage and waste-water recycling (Chapter 34). Eliminating some channelization is already underway in the Kissimmee River and elsewhere.

Areas short of water have proposed taking water from less developed regions. For example, Tampa-St. Petersburg legislators want to divert Suwanee River waters from north Florida. When Los Angeles diverted water from other regions, it became overdeveloped and too dense for optimum living while other areas were left underdeveloped. Taking a larger scale view, a better economy results if each area optimizes its development.

Patterns and problems of growth are illustrated by the development in south Florida. The maps in Figure 21.3 show three stages. Beginning in 1900, when there was essentially no development, the environment of south Florida was altered over time to accommodate an increasing population's need for roads, buildings, and agricultural lands. Each year there were greater concentrations of cities on the coasts and more natural land converted to agricultural use. Miami became an international city with an economy based on trade with the Caribbean area and Central and South America (see Chapter 27); however, its economy still depends on its environmental basis — the lands, waters, and natural services of wetlands, estuaries, and beaches.

21.4.4 Planning and designing with nature

The early developers of Florida paid little attention to preserving the environment. Development was driven mostly by the idea that humans could make the environment of Florida a better place to live by making it conform to their ideas and wishes. Many of the newer residents did not recognize the environmental losses, as they were not here in earlier times when millions of birds inhabited the Everglades, coastal fisheries were plentiful, black bears were a common sight, and panthers roamed the entire state.

As interest in the environment and awareness of the importance of a healthy environmental system have increased, more attention has been given to controlling development to minimize negative impacts and maximize prosperity. The free services of the environment keep the costs of development and operation of cities down. The environmental systems surrounding developed areas cleanse air and water, build soils and stabilize them, store water and recharge drinking aquifers, provide aesthetic relief and recreation,

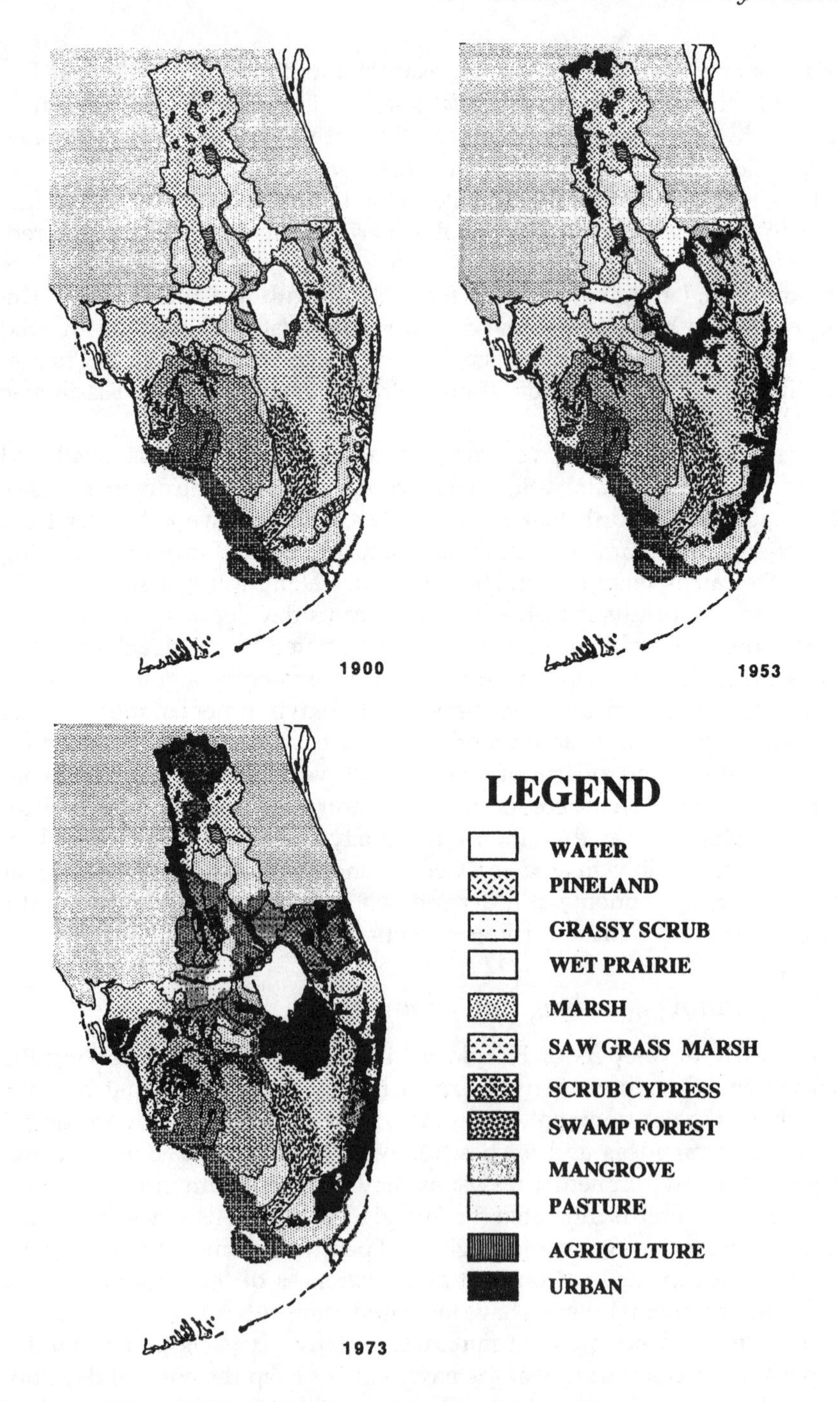

Figure 21.3 Maps showing stages in the development of south Florida beginning in 1900. Note the peatland farming area directly south of Lake Okeechobee and the expansion of urban Miami.

and assimilate and recycle waste by-products of all kinds. However, when environmental systems are not included in the design of development, these services are soon lost and must be replaced with costly technological solutions.

Planning and design of urban and agricultural areas can incorporate natural areas, such as forest and wetlands, as necessary parts of development (since they provide services, free of charge). Otherwise, the only leftover areas are around their fringes. Since ecosystems are good at self organizing to fit conditions, they can adapt to human development, within limits. Planning developments for which some of the landscape interface is left to the ecosystems to design is called *ecological engineering*. The idea that the environment is something that needs to be protected can be helped by the concept that the environment is a necessary part of humanity's prosperous existence.

Questions and activities for chapter twenty-one

1. Define: (a) macroeconomics, (b) microeconomics, (c) development, (d) groundwater, (e) water table, (f) emergy, (g) wealth, (h) market, (i) supply, (j) demand, (k) emergy investment ratio.
2. Explain how the circulation of money links the work of people.
3. Diagram the way circulation of money flows as a counter-current in opposite direction to the production, processing, and sale of goods and services.
4. Explain the difference between macroeconomics and microeconomics, using examples.
5. Find an article in a current newspaper about the interaction of resources or the environment with economic activity. Identify the pathways of that system using Figure 21.2.
6. If you buy fish in a market, do you receive more, less, or the same real wealth than the fisherman gets from your payment? Explain.
7. Explain the difference between money and "real" wealth, using examples.
8. Even though you always buy the same kind of fish, you find that prices are low when the fish is abundant and high when it is scarce. Explain this difference in price.
9. Explain how supply and demand work, using an example.
10. If you were buying a house to live in, how could you use the emergy investment ratio to help make decisions as to where to live? Where in Florida do environments contribute more?
11. Visit a new development in your area and list the ways that the environment has been changed by construction of buildings and roads. How could the development be changed to incorporate more of the free services of the environment?

12. Using a large sheet of paper and working in teams of three or four students, redesign your town or neighborhood to incorporate more environmental services and make it fit within the landscape better. Consider wastes and minimizing distances to essential urban services such as schools and shopping.
13. Watch the newspapers for a week or so and discuss the most important development issues in your community.
14. Discuss the balance between development and the environment. Why should development plans include the environment?
15. Draw a user system in which environmental resources are brought into the economy.
16. List the most important development issues and discuss their environmental impacts.
17. Locate on a map of Florida large-scale development projects that have caused environmental losses.

chapter twenty-two

Forestry

Operation of forests for commercial products is a major use of environment in Florida. In Florida, slash pine is the predominant species used in forest plantations. Slash pine is a native species found naturally in much of the north and central parts of state where forest land and water for paper mills are available together. In the past four decades, slash pine has been planted extensively, to the point where it now comprises 41% of pine plantations in the southeastern U.S. In Florida, slash pine occupies more than 5,000,000 acres (2,000,000 hectares). Other important pine species that are grown for the pulpwood industry include loblolly and short-leaf pines. Australian eucalyptus is also being tried.

Pine plantation harvests provide lumber and pulpwood for paper, packaging, paper-diaper filler, and even the thickening ingredient in milkshakes! Two thirds of the tree trunk is *cellulose*, a strong, fibrous starch-like compound used in industry. Trees are an important renewable resource and are likely to become even more so as fossil fuel shortages make plastics more expensive to produce.

22.1 Pine plantation system

Figure 22.1 depicts life in a pine plantation in north Florida. Whereas the upper canopy layer was planted, an understory and ground plants have developed that are more typical of the natural pine forests. To maximize the production, plantations of trees are grown with the same principles used in agriculture. The ground is prepared, fertilizer may be added, seedlings are planted, and often weeding is done until the trees get above the competing plants. Sometimes control burns are used to eliminate the underbrush and stimulate more pine growth.

To maximize height, trees are initially planted close together so the tops will shade out the lower limbs. Later, they may be thinned to give faster growth per tree. The trees are cut when they are large enough to be used for paper making, poles, or lumber. Plots of several acres are cut at one time. This practice reduces cost.

Figure 22.1 Life in a slash pine plantation. (Illustration by Elizabeth A. McMahan.)

Under prime conditions, plantation trees can be harvested in about 20 years, at which time many acres can be clear-cut. Ideally, the roots, branches, and needles are left behind to allow their nutrients to be recycled for the next cycle of growth.

The systems diagram in Figure 22.2 summarizes the main features of a pine plantation. Notice that the work is shared between the environmental inputs and the feedbacks from the economy in planting, cutting, and transport.

22.2 The money part of the business

The forest plantation (Figure 22.2) is a good example of economic use of the environment (see Figure 21.2). The bank account of the forestry company is shown as a storage of money which has come into the business from sales of logs and goes out to pay for labor, goods, services, and fuels. If loans from the bank are needed to pay for expenses before the trees are cut and sold, money has to go back to the banks to pay interest and to pay back the loan. The diagram shows the input of a government subsidy sometimes provided for forest acres set aside from real estate development. Money also goes out of the account to pay for federal, state, and local taxes. Profit is the difference between money inflow and outflow. To sustain operations, there must be a profit.

22.3 Comparison of plantations and mature forest

Pine plantations and mature forest are compared in Figure 22.3. In the plantation (Figure 22.3a) there are few consumers, and much of the photosynthesis of the trees goes into wood which is later harvested. In contrast, the mature forest (Figure 22.3b) puts its energy into a complex diversity of consumers that gives the forest efficiency and stability. The natural, self-maintaining forest has more gross photosynthesis than the plantation but uses its organic matter to maintain a complex structure of live and dead biomass and plant and animal diversity. The plantation may require fertilizer to replace nutrients lost in the harvested wood. The mature forest gets its nutrients from recycle after microbial action on the litter.

Plantations usually require planting of seedlings. Another way is to leave seed trees at regular intervals. In the mature forest, planting is done by natural reseeding with the help of animals. The mature forest can be sustained in its climax stage if only a few trees are cut during "selective logging", with care taken not to damage other components. However, if selective logging takes the best trees, inferior genetic stock may remain for seed production and regeneration. Except for protection, natural forest management costs are small; however, if tracts are clear-cut, long periods are required for regrowth after cutting.

Pine plantations are denser than natural pine forests and more of a *monoculture*, which is a uniform population of a single species which is normally susceptible to epidemics of insects and disease. However, slash

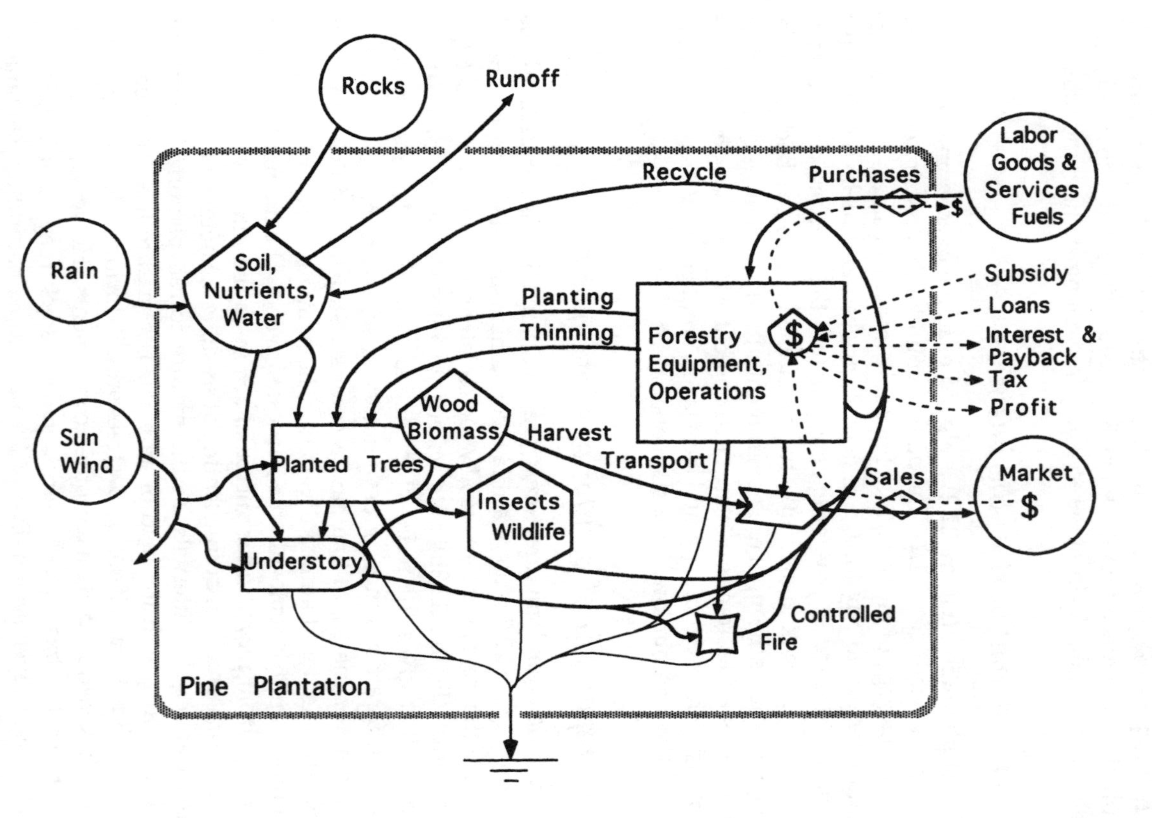

Figure 22.2 A Florida pine plantation system.

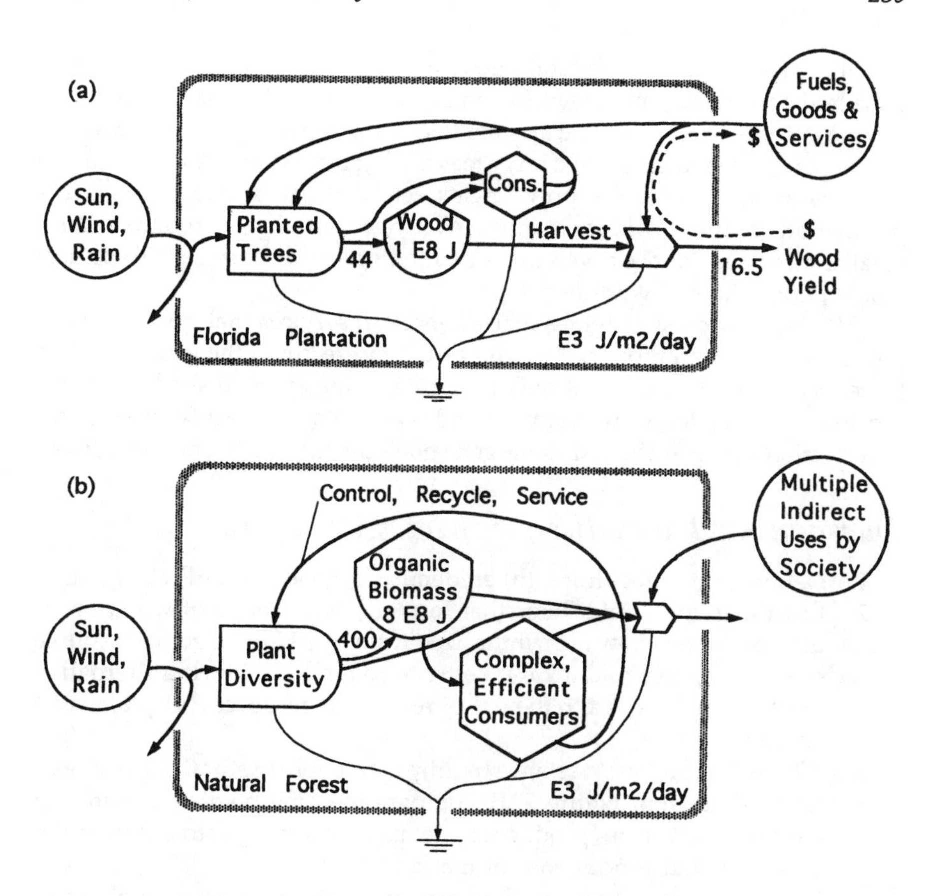

Figure 22.3 Diagrams comparing forest managements in Florida. (a) Plantation forest managed for harvest; (b) mature forest managed for watershed protection and multiple use.

pine plantations in Florida are fairly resistant, perhaps because slash pine is a native species and part of a system already adapted to most insects and diseases. The diversified understory of smaller trees, shrubs, and diversified animals may be a protection. Pines and many other trees require fungi (called mycorrhizae) which are symbiotic with roots to process nutrients efficiently.

22.4 Natural tracts and diversity

Areas of unmanaged forests are needed in plantation regions as seed sources and *gene pools*. The idea of a gene pool is to keep enough different plants, animals, and microorganisms so that their genes are mixed through reproduction to maintain biological variety and vitality. Sometimes the legitimate needs of simplifying plantations and keeping complex gene pool reservoirs are in conflict.

Numerous state and federal agencies own large tracts of land within Florida that are being managed for timber yield. Under these management regimes, single species of pine are planted, plantations are even-aged, clear-cuts are large in size, and few if any areas are left in "old growth" trees. Since these lands are almost the only large tracts that can support reasonable populations of some wildlife, their management with longer rotation times, smaller clear-cuts, and a diversity of planted species is necessary to protect viable populations of wildlife.

The mature forest provides many beneficial services that do not always get recognized even though they contribute indirectly to the economy. The forest rebuilds soils, concentrates nutrients, protects watersheds, reduces erosion, offers aesthetic diversity, provides habitats for wildlife, contributes to recreation, cleanses air, and stores gene pools for succession and future uses.

Questions and activities for chapter twenty-two

1. Define: (a) monoculture, (b) epidemic, (c) gene pool, (d) clear-cut.
2. List three household items that formerly were made of wood products, but which now are synthesized from fossil fuels. From a resource standpoint, do you think it is better to continue to develop alternative products or to revert to the use of renewable resources? Explain your answer.
3. Why is it important to retain a healthy gene pool in natural populations?
4. Using data from Figure 22.3, compare the pine plantation with the natural forest: gross production, net production, investment from the economy, and production for market.
5. How do monocultures, such as pine plantations, affect animal populations?
6. Obtain permission from the landowners to gain access to two areas of forest land: one should be a slash pine plantation and the other a natural stand of slash pine. Perform the following sampling techniques in each forest type.
   (a) Run a 10-m line transect and count the number of plant stems that intersect each line. Also identify each stem and develop a measure of the species diversity by dividing the number of species by the number of individuals counted. Repeat this technique two more times and then average your values.
   (b) Randomly select 10 slash pine trees from each site and measure the diameter (in cm) at breast height (dbh). Determine the average.
   (c) If available, use an increment borer to sample the 10 trees (from part b) from each site and count the annual rings. If a borer is not available, ask the landowners the year when the forests were last cleared and the trees planted.
   (d) Measure the leaf litter in each site and describe any differences in litter and soil.

(e) Use a light meter to record and compare the light intensity (at ground level) in open sunlight, the pine plantation, and natural pine woods.

(f) Make some general observations about the diversity and ecological quality of these two sites. Make notes about any animals, insects, etc. that you see.

(g) Compare species diversity in the plantation and the natural woods. Which ecosystem would you expect to be most resistant to epidemics of disease and insects? Which system will support the largest community of plants and animals?

(h) Record the range and mean for dbh in each ecosystem. Relate tree size to age for each ecosystem by making a graph. Which ecosystem generates the fastest growth for slash pine?

(i) Which system has the lowest light intensity on the forest floor? Explain the importance of light intensity in terms of competition, productivity, and planting pattern.

7. Investigate other planted forests, such as pecans or eucalyptus. How do they compare to pine?
8. The pulpwood industry is developing new technology that will allow them to utilize entire trees (roots, trunks, foliage). Do you think this is a good idea? Explain in terms of nutrient recycling.
9. Economically, which woods are considered to be most valuable, pines or hardwoods? Explain why one is more costly.
10. What proportion of Florida land do you think should be used for forest plantation, natural forest, and agriculture? On what basis should this be decided?
11. If you have access to a computer and the BioQUEST programs with EXTEND, run the logging system simulation (Figure 22.4). For sources, refer to Appendix A.

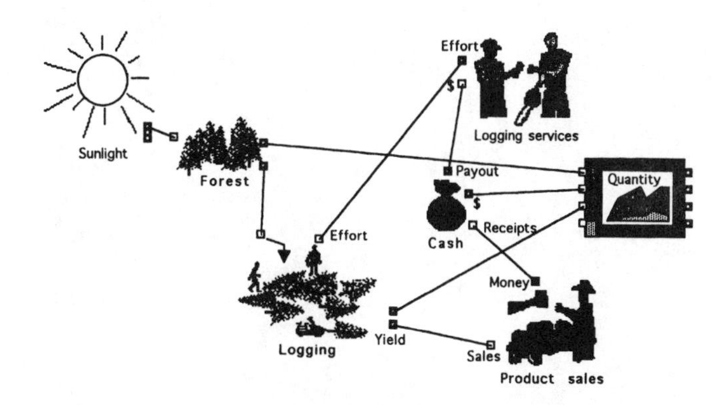

Figure 22.4 Computer screen of the simulation of a logging model from *Environmental Decision Making,* a BioQUEST teaching exercise. The CD-ROM can be obtained from the address provided in Appendix A.

chapter twenty-three

Agricultural systems

Agricultural systems, sometimes called *agroecosystems*, are the world's principal means of providing food and fiber for its populations. Agricultural ecosystems fit the general plan for economic use of the environment (see Figure 21.2). Components include a farm area with soils formed by previous geological and ecological processes, a producing crop plant, and farm equipment for planting, weeding, and harvesting. A market is necessary to buy the produce and to provide the money required for the fuels, fertilizers, labor, goods, and services needed to grow, harvest, and transport the crop. This chapter examines agroecosystems with examples from Florida.

In Florida, agriculture is a main part of the economy, with different crops in different soils and climatic zones. Because of the easily leached sandy soils in much of north and central Florida, tree crops and pastures for cattle are more effective than row crops in holding soils. After killing freezes in recent years, the production of citrus fruit moved to central and south Florida. Beef cattle pastures and ranches are found throughout the state. Winter vegetable crops are found in frost-free southern areas, and sugar cane is grown on the peatlands south of Lake Okeechobee. At present most of the agricultural endeavors in Florida are highly industrialized. They receive high inputs of fertilizer, chemicals, machinery, and technological processing.

An agroecosystem is a domesticated ecosystem in which humans operate as managers and consumers, replacing the animals that consume and manage in the wild ecosystems. The two are always in competition. Wild organisms constantly scatter seeds and invade the territory of the agroecosystems. If farmers did not control these by pesticides, weeding, plowing, and other methods, the wild ecosystem would re-establish itself.

Farms are able to prosper where the previous work done by the wild, high diversity ecosystems developed fertile soils. Most farming gradually depletes the soil even when it is fertilized. Letting the wild organisms have the land back temporarily to start natural succession restores the soil. This is called *rotation*.

Two centuries ago, most farms were highly self-sufficient family operations. A farmer produced according to the needs of his own family and sold only a few items for cash. Now most people in urban countries derive their foods from highly diversified markets. These markets obtain food from many different intensive farms, each of which specializes and mass produces only a few cash products for sale.

Primitive, low-energy agriculture used the labor of humans and farm animals without fuel or electrically run machinery. Modern *intensive* agriculture involves large inputs of fuel and electricity using machinery. It takes a lot of energy to produce all the goods and services used in modern agriculture as well as to process and transport the products. Intensive agriculture displaces the more primitive agriculture if there is a constant, cheap supply of energy.

The emergy investment ratio (ratio of purchased emergy to local free environmental emergy; see Chapter 21) is a good measure of the farming intensity. Intensive agriculture with high investment ratios (5 to 25) gets more yield but has to buy more inputs to do it. Two of the important questions of our time are when and how the intensive agriculture will change because of increasing energy costs (see Chapters 33 and 37).

There is already interest in less intensive farming methods used earlier. If this trend continues, use of purchased inputs (fertilizer, services, etc.) will decrease. With lower intensity, farmers rotate land between alternate uses, including time in natural succession, so that it can replenish its soil fertility.

23.1 Agroecosystems of Florida

With a systems viewpoint, we consider next the main kinds of agriculture in Florida and how global trends may affect these in the future.

23.1.1 Citrus groves

In central Florida, south of the worst killing frosts and centered among the many lakes and sandy hillsides, are the extensive orange and grapefruit groves that produce Florida's major export crop (Figure 23.1). The fruit is sold directly or processed into frozen concentrated juice, and the pulpy by-products are processed into cattle feed. Several years are required to produce yielding trees, after trees are started from cuttings and grafts. Inputs include fertilizer, pesticide treatments, and much labor used in picking. The fruits ripen in December. The citrus grove agroecosystem is summarized in Figure 23.2.

On very cold nights when there is danger of freezing, microclimate temperatures are raised by spraying water, burning tires or kerosene, and stirring air with motorized fans. Studies show the water method is the most cost effective. Recently, killing freezes have occurred several years in a row, causing the loss of thousands of acres of citrus in central Florida. To escape the killing freezes, many citrus growers have moved farther south. Thousands of acres of low land in south Florida have been converted to orange

Figure 23.1 Picture of a grove of citrus trees.

groves. To maintain lower water tables, the lands in the western parts of Indian River, St. Lucie, and Martin counties have been criss-crossed with drainage ditches.

Occasionally, diseases such as citrus canker and insect infestations such as fruit flies have appeared in nurseries and established groves, requiring a ban on shipment of the citrus and destruction of the infected groves. Increased imports of foreign sources of citrus, primarily from Brazil, have lowered prices, making it more difficult for Florida citrus farmers to compete.

23.1.2 "Truck farming" of winter vegetables

In south Florida, starting in the early 1900s, much of the uplands on the east and west coast were cleared of vegetation, and vegetables were planted. Because there were no killing frosts in the south, on many of these farms two, and sometimes even three, crops could be grown per year. The farms are called "truck farms" because of the heavy dependence on truck transportation during all phases of crop processing and selling. Many of the farmers live in the city, and rent the farm land on a yearly basis.

Insects that eat foliage and nematodes that eat the roots reduce yields where there is repeated farming of the same crop. As a consequence, in the

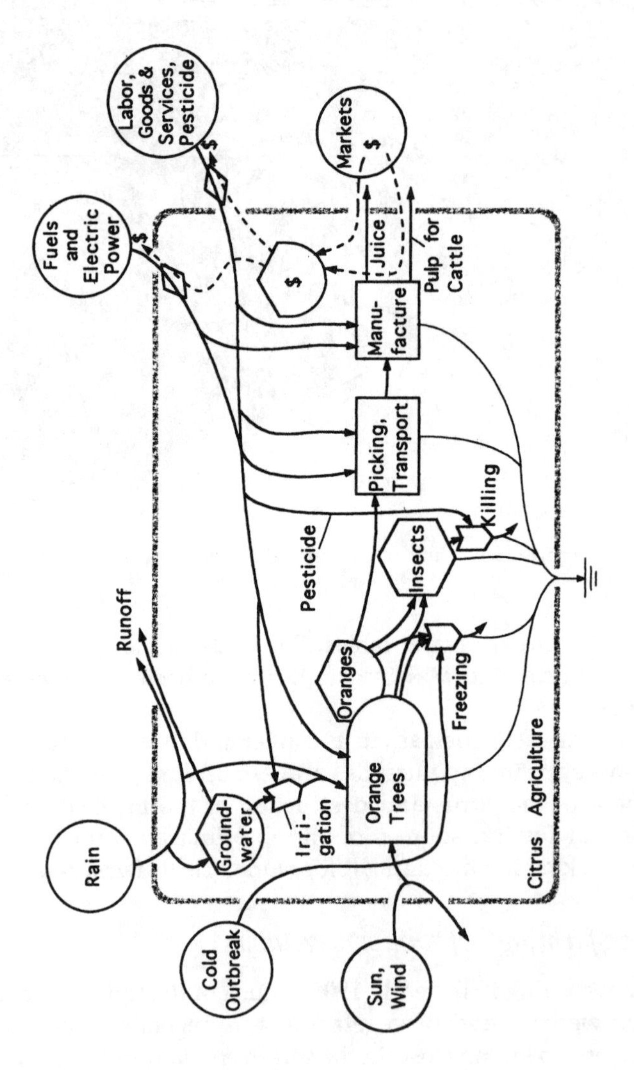

Figure 23.2 Systems diagram of a citrus agricultural system.

early years of farming in south Florida, lands were utilized for only one or two years then abandoned. Sometimes old lands were farmed again after they had been left for natural restoration for a number of years, but in many cases they were converted to pastures for cattle.

To maintain a vegetable farm in one place year after year in south Florida now requires that insect pests and nematodes be controlled through the use of pesticides. Without killing frosts that reduce populations of pests, the amount of pesticides needed is much greater in south Florida than elsewhere in the U.S.

Truck farming requires fertilizer, farm machinery, irrigation equipment, and fuels, as well as human labor, for planting and picking operations. Irrigation has become increasingly important as water tables have been lowered throughout south Florida; however, less groundwater is available for crops.

Analysis of the emergy requirements for growing vegetable crops in south Florida shows that vegetables grown in the winter for export to the north require about 40 times more resources than the same vegetables grown in the northern parts of the U.S. As a consequence, prices necessary for a profit are high.

With "free trade" agreements (trade without tariffs, taxes on imports), there is more international competition for sale of agricultural products on world markets. For example, in 1994 the North American Free Trade Agreement among the U.S., Mexico, and Canada was negotiated. This trade without tariffs means that Florida growers must compete with growers in Mexico, who have lower costs for labor and environmental protection and can sell at lower prices.

With urban development, costs of water, rent, and taxes rise. Many farms are being converted into residential housing. Some farms have "you-pick" operations, where local residents in nearby urban centers travel the relatively short distances to the agricultural countryside to pick fresh vegetables for their own consumption. The future of intensive vegetable farming in Florida to produce vegetables for export or for regional use is uncertain.

23.1.3 Sugarcane

Sugarcane is grown along the south shore of Lake Okeechobee on wet peatlands which were drained (see Figure 21.3). Sugarcane became a major crop in Florida after the embargo imposed on Cuban sugar following the 1959 Cuban Revolution. Because Florida's sugarcane is grown on soft organic soils where machines cannot operate, it is harvested by contract laborers, mostly Caribbean migrants. Although it costs more to grow sugar in the U.S. than in many other countries, federal subsidies (9 cents per pound in 1996) keep the price of U.S. sugar competitive. When used for farming, peatland is rapidly lost by oxidation (see Figure 18.9). Peatland farming in Florida is like a mining operation and is temporary.

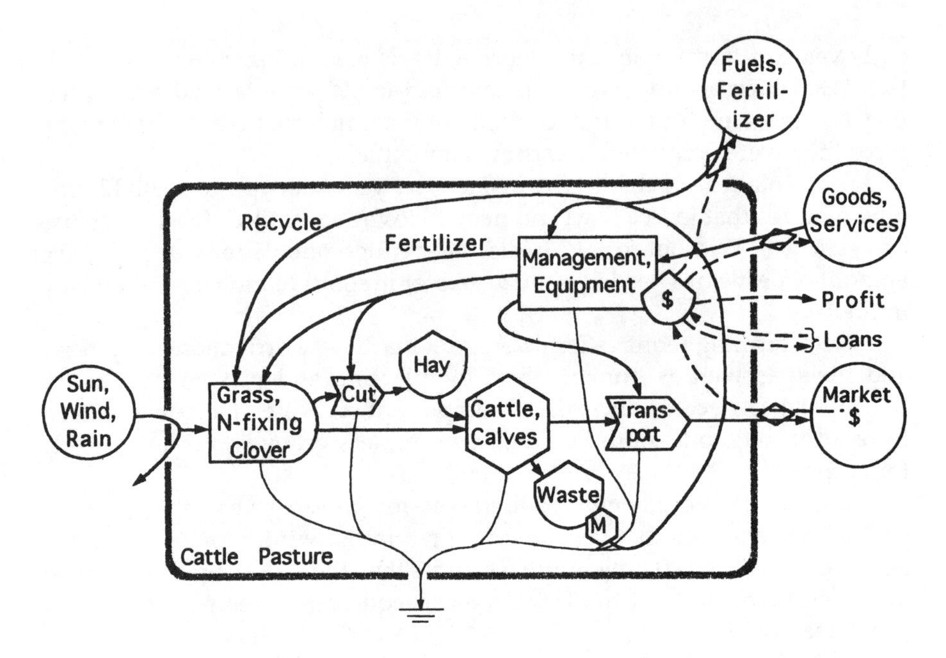

Figure 23.3 Diagram of cattle pasture system.

Excess nutrients from fertilizer and from oxidation of the peat runs off into drainage ditches to the Everglades and sometimes into Lake Okeechobee. To control these nutrients, programs involving the government and the sugar companies are setting aside some of these lands to become a wetland filter again (see Chapter 34).

23.1.4 *Beef cattle system*

Cattle ranching has been important for a hundred years in Florida, which is now second only to Texas in the number of ranches. Earlier, tough, self-sufficient cattle with long horns were allowed to roam freely and were managed in the cowboy style that one often associates with Texas. Later, higher yielding beef cattle systems were developed with fertilized, managed, and fenced pastures and with better producing varieties that require more care and costs to manage their cycles of breeding, calving, and nurture.

Figure 23.3 summarizes the cattle pasture system. On the left, the grass and clover, which fixes nitrogen from the air, use renewable energy sources of sunlight, wind, and rain. At the right of the diagram are the flows to and from the economy. Inputs of equipment, labor, fertilizer, pesticides, salt blocks, etc. are paid for by the money obtained from sales of cattle. You can see the money from the sales going into the farmer's money storage and from

Figure 23.4 A cattle pasture with small wetland used for shade, water, and waste absorption. (From Browder, J. et al., *South Florida: Seeking a Balance of Man and Nature*, Bureau of Comprehensive Planning, Division of State Planning, Tallahassee, 1977.)

there to buy inputs and to pay back loans and taxes. Herds in Figure 23.4 use small wetland ponds for shade and water, while the vegetation and peat absorb excess nutrients.

Because grass production is limited in the winter by low temperatures and less daylight, farmers harvest and dry surplus grass in summer and autumn as hay and feed it to their stock in the winter. As the systems diagram shows, pasture grass is kept fertile by the recycle of nutrients from the cattle wastes, but fertilizer has to be added to compensate for the nutrients that are lost when the cattle are carried to market. Adding fertilizers and harvesting hay require fuels and machinery. In some areas, irrigation is needed in dry periods.

Ranch cattle are often moved to feed lots for fattening on grain to generate high-priced luxury meat cuts. Now, concerns about unhealthy fat and high cost are changing the diets of Americans. The cattle agroecosystem may shift towards lower intensity and lower costs with less special feeding.

23.2 Energy and food requirements

With rising populations and declining resources, there will be more and more concern with providing enough inexpensive food for everyone in years ahead.

23.2.1 Migrant labor

An unsolved problem is how to obtain enough labor to harvest crops with costs that make the food available to everyone. Harvesting is often done by migrant labor, people that move from one area to another as crops mature. Migrant labor is used on vegetable and sugarcane farms in many parts of Florida. Many are new immigrants, both legal and illegal, from the Caribbean or Central and South America, who speak and understand only limited English. Wages are low; housing is often substandard. Children of migrant laborers rarely receive services available to the rest of the population. School attendance is intermittent and education limited. The system is not making good use of human resources.

23.2.2 Energy and vegetarian diet

Frequently asked is the question, "How much energy could be saved by leaving out the meat in the human food chain and having people eat only plant products?" It is an appealing proposal, since as much as 100 units of energy of plant production are used to make about one unit of beef protein.

A small amount of high-quality protein is necessary for good human nutrition. It can be obtained by eating a small amount of meat regularly or by eating a diet with the right balance of vegetable proteins. Cattle can produce meat by digesting grass which humans cannot do. Cattle do it with the help of symbiotic one-celled animals (protozoa) in their stomachs. Growing and harvesting the cereals, vegetables, and nuts necessary for a healthy diet require a great deal of emergy. The most efficient diet may be largely vegetarian with a small regular meat contribution. Many humans in developed countries eat much more meat than they need.

23.2.3 Self-sufficient agriculture

Another question is, "How self-sufficient can people be, living on the land?" Figure 23.5 summarizes an energy analysis of the agroecosystem of a diversified family living in our times. The farm shown was awarded a prize by "Mother Earth News" for having a high degree of self-sufficiency, comparable to farms of pioneer settlers.

The Taylor family farm was very diverse, with many different kinds of crops and livestock. The only inputs from the main economy were equipment and hogfeed. Manure from livestock was used for crop fertilizer. Oxen instead of tractors were used for cultivation and plowing. The farm produced

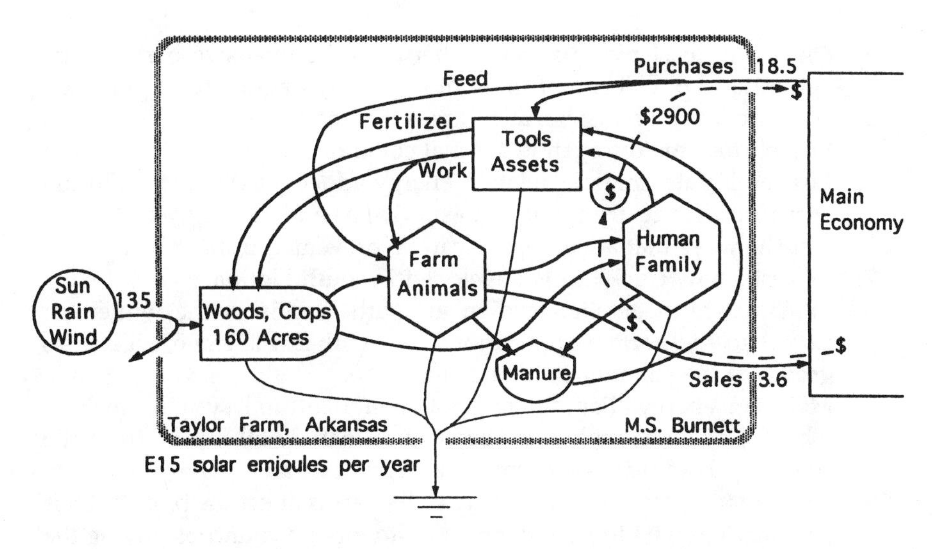

Figure 23.5 Diagram of Taylor farm, a relatively self-sufficient agroecosystem. (From Burnett, M.S., Energy Analysis of Intermediate Technology Agricultural Systems, M.S. thesis, Environmental Engineering Sciences, University of Florida, Gainesville, 1978.)

almost all the food needed by the family of four living on the farm. They exported enough food to feed almost 16 other people. They used the money from exports to purchase some outside goods and to pay the mortgage and taxes on their land.

As indicated by its lower emergy investment ratio (0.13 compared to the 7.0 typical of Florida), the Taylor farm required less inputs than the cattle pasture system (Figure 23.3) but could not operate without some inputs from the main economy. Economic survival in our present-day society probably requires that some inputs be bought so as to gain the benefit of cheap fuel directly or indirectly.

In the past, industrial farms increased in size and energy input but used fewer workers. If world resources become more scarce and costly in the future, the way of life on the Taylor farm may become more common. The farms of the future may be more self-sufficient, more diversified, and use more people.

The eight inhabitants of Biosphere 2 supported themselves on their own gardens for 2 years in the 3-acre glass mesocosm in Arizona, but they had unlimited electric power.

Questions and activities for chapter twenty-three

1. Define: (a) agroecosystem, (b) low-energy agriculture, (c) intensive or high-energy agriculture, (d) diversified, (e) purchased input, (f) land rotation, (g) yield.

2. Discuss at least two differences between the intensive high-energy agricultural practices and those of low-energy agricultural practices.
3. Where does the phosphorus in the soil come from?
4. Where does the nitrogen in the soil come from?
5. List the inputs (materials, labor, energy added, etc.) that the Florida agricultural industry requires to survive as a high-energy system.
6. List the ways that a fruit grove can be prevented from freezing.
7. Describe how truck farms developed in south Florida.
8. Explain why vegetables grown in south Florida in the winter for export to the north require more energy than the same vegetables grown in the summer in the northern U.S.
9. Make an energy diagram of another agricultural system, such as tobacco, peanuts, or horses. Be sure to include all inputs from the economy. Find real figures for the energy flows and money if you can.
10. How would government subsidy to the farms affect the price of their products charged to poor people and to foreign countries buying the products? Discuss.
11. Dig up, dry, and weigh one square foot (0.09 m^2) of grass. Count the number of cattle on a measured area, either by using an aerial photograph of cattle pasture or by taking a trip into the country, perhaps estimating the area by the odometer on the car. One cow weighs about 1500 pounds. How many kilograms (1 kg = 2.2 lb) of grass are required to support 1 kg of cow?
12. If you were a cattle farmer, how could you cut down on your fuel energy inputs? What do you think will be the future of farming in the lower energy world? Explain.
13. Discuss the positive and negative reasons for being a vegetarian.
14. If you have access to a computer and the BioQUEST Environmental Decision Making program, run the grassland simulation. The CD may be obtained from the address given in Appendix A.

chapter twenty-four

Fisheries

Many kinds of fish are found in Florida's extensive seas, estuaries, lakes, and streams. Many of the larger species are regularly harvested by commercial fishermen and sold for the restaurants and home dinner tables of Florida. In the historical development of Florida, there were many fishing villages around the long coastline, and fishing was a principal way of life (Figure 24.1). As the population of citizens, tourists, and retirees has increased, more people are catching fish for recreation and home consumption. For many people, good fishing is one of the main reasons for visiting and living in Florida. The ecosystems that grow fish and the economic system that uses them are called *fisheries*. Figure 24.2 is a typical fishery system. With coastal developments, many of the areas that produce fish have been impacted, reducing stocks.

24.1 Yield and overfishing

Managing fisheries is complicated because there are natural oscillations (Chapter 11). The economics of fishing also cause oscillations. When fish are abundant, banks finance more boats and tourist businesses encourage more fishing. As the fish begin to be scarce, the prices go up, which encourages commercial fishing beyond what is healthy for the ecosystem. The stocks become depleted, bankruptcies occur, fishermen move to other areas, and eventually the stocks may come back, causing the cycle to repeat.

Unfortunately, many of the fisheries of the world have been overfished. *Overfishing* occurs when so many of a species of fish are caught that not enough remain for reproduction. As a result, the stocks of fish are far less than could be supported by the aquatic ecosystems. With few fish, the contributions to the economy are small. Regulatory bodies then restrict fishing, but the stocks do not always come back. After harvest of the desirable species reduces their reproduction, the species less desirable to fishermen may increase their populations, taking over the food chains (Figure 24.2).

In 1996, after stocks of larger fish were becoming scarce, commercial fishing with nets was banned in Florida's coastal waters, a great experiment to determine if overfishing was a major factor. By the next season, the stocks

Figure 24.1 Coastal scene in Florida with shrimp trawlers.

of mullet, redfish, and sea trout increased. These fishes have a lifetime of about 3 years.

It is sometimes suggested that if people harvested the sea more efficiently it would produce much more food. This is exaggerated. Most of the open oceans have sparse food webs and very few nutrients. Most of Florida's commercial fishery products — fish, crabs, lobsters, and shellfish — are caught on the continental shelves. These areas are already heavily fished. As explained in Chapter 16, the estuaries are nurseries for the young fish, but harvesting the small fish there is bad policy as it prevents a much larger harvest when the fish move out of the passes a few months later.

The tonnage of marine fish caught around the world and in Florida (Figure 24.3) showed a sharp rise in harvest from 1900 to 1970, after which yields changed little or declined. Mechanical trawling devices and fish factory ships bring up so many fish that the numbers of some species available to reproduce have become seriously depleted. Harvesting may have gone beyond the *maximum sustainable yield*.

Every system needs controlling and reinforcing feedbacks if it is to survive. As diagrammed in Figure 24.4, the human economy has been taking yield from the oceans, but has not been putting much back to keep the system productive.

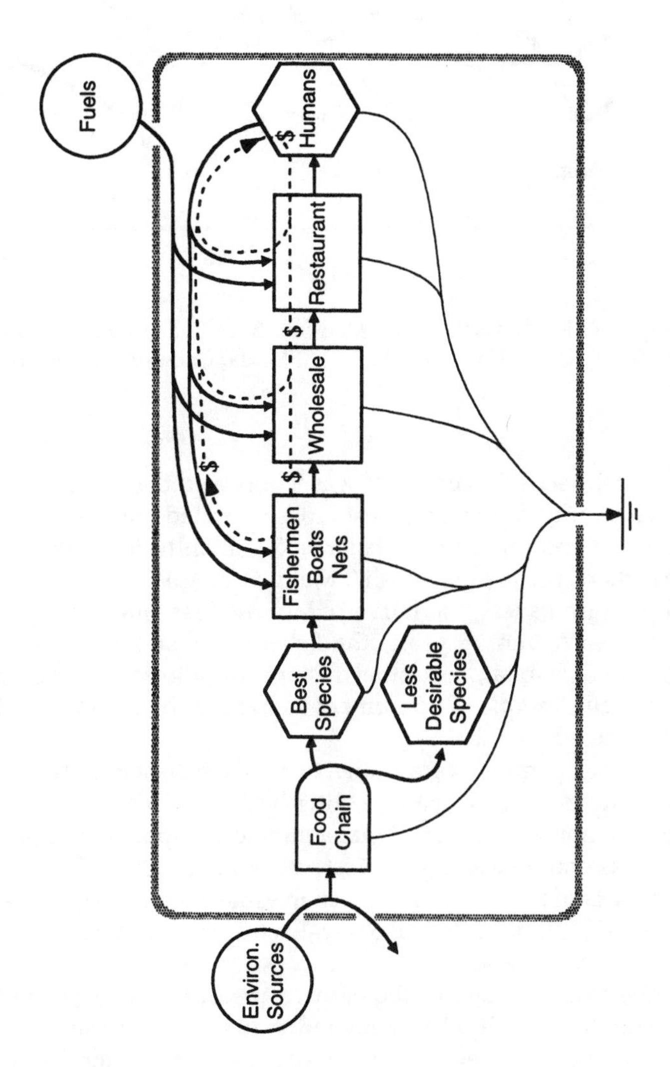

Figure 24.2 System of world fish production and economic use.

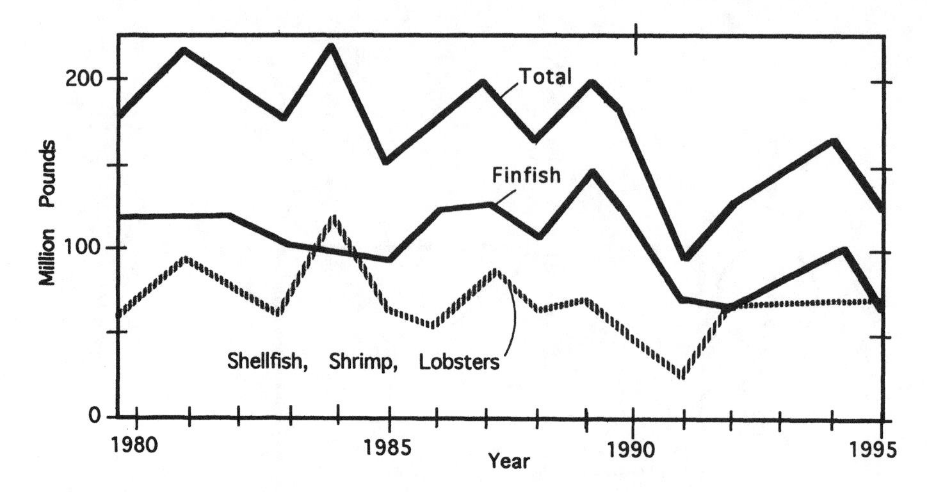

Figure 24.3 Graph of Florida's fish catch. (Modified from Fernald, E. A. and E. D. Purdum, Eds., *Atlas of Florida,* University Press of Florida, Gainesville, 1992, p. 190. With permission.)

Not until after the second world war was it observed that the small bait-sized commercial shrimp caught in the estuaries migrated out to the shelf to become jumbo-sized and reproduce offshore a few months later. When this was discovered, there was a "gold rush". Many 50-ft (20-m) trawlers were built, and as this virgin fishery was harvested for the first time, some people became millionaires. Within 15 years, the fishery was so reduced due to heavy fishing, loss of habitat, and other factors that the industry became depressed. This is an example of consumers taking a yield without feeding back to reinforce the energy chain.

For many years attempts were made to manage fisheries in natural waters as if each species was a crop growing by itself. This was a failure, because the ecosystem is a food web of many interacting species, and predictions of what stocks could be safely harvested were not correct. Figure 24.5 shows the main parts of the fishery food web in estuaries and coastal waters on the west coast of Florida at Crystal River. Each species population by its variation can affect the others.

As the numbers in the diagram show, there are more fish at the bottom of the food web (on the left). It takes many individuals at the lower levels to support a few at the top. In general, the higher levels in the food web are larger, more attractive to sport fishermen, better tasting, and priced higher in markets. Overfishing the valuable species causes the lower food chain species to become very abundant. As the preferred fish decrease, people begin to consume more of the lower food chain fishes, such as mullet. If fish at the bottom of the chain are heavily fished, the more desirable species may still be further reduced because their food supply has been decreased.

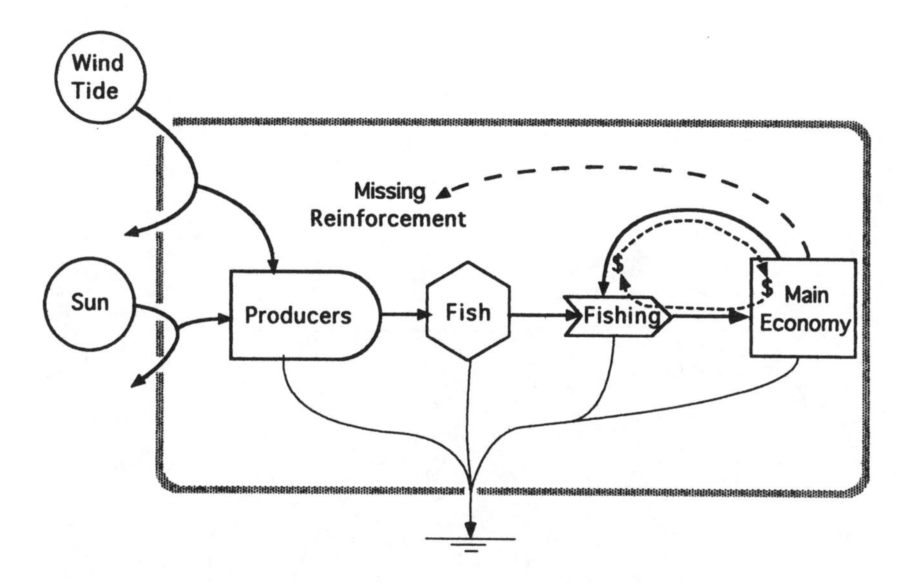

Figure 24.4 Diagram showing how fisheries can draw yield from public waters without the reinforcing feedback needed to sustain the food chain.

Important commercial fisheries in Florida include the species in Figure 24.5 that are also important to sports fishing such as sea trout and redfish, and especially shrimp, spiny lobsters, stone crabs, mullet, and blue crabs. In offshore waters, party boats obtain species related to tuna and favorite game fish such as tarpon, sailfish, and marlin. Oysters are harvested from managed reefs, especially at the Apalachicola River estuary (see Chapter 16).

Waters that are low in nutrients tend to have a high diversity of species that are good for sports fishing, but not very many of each. As Florida lakes and estuaries become more and more eutrophic, different species develop in larger quantities, but these are less prized for sport and food. Heavy growths of plants that thrive in eutrophic lakes also interfere with boats and fishing.

24.2 Competition between sports and commercial fishermen

Figure 24.6 shows the way sports and commercial fishing are related and compete to use products of the same food chain. Sports fishing takes fish at the top of the food chain, whereas commercial fishing also includes the larger populations in the middle, such as mullet. Both systems attract investments, money circulation, and jobs.

In the 1950s, there was conflict between sports and commercial fishing regarding large-scale seining for freshwater fish then allowed in the larger lakes of Florida. The game fish (bass, bream, and crappies) were thrown back.

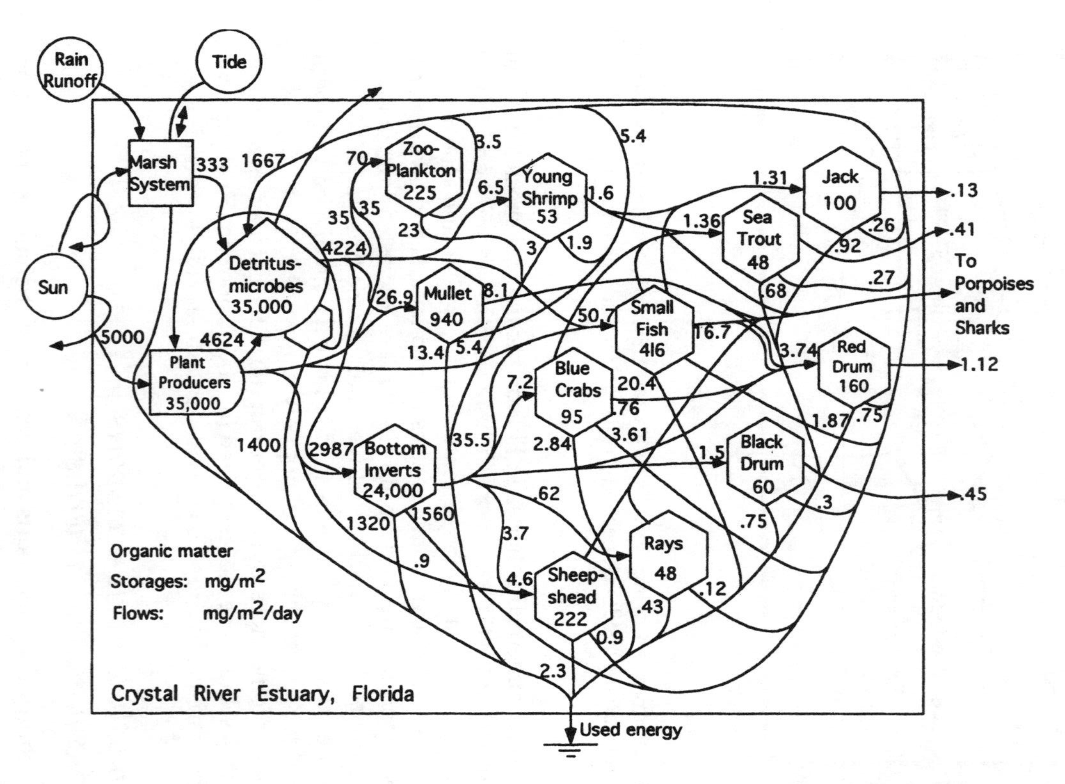

Figure 24.5 Diagram of the food web in the estuary at Crystal River, showing the daily consumption of each species (mg = milligram = 0.001 gram). (Calculations were made by M. Kemp using weight measurements by M. Homer). (From Odum, H.T., *Environmental Accounting*, John Wiley & Sons, New York, 1996. With permission.)

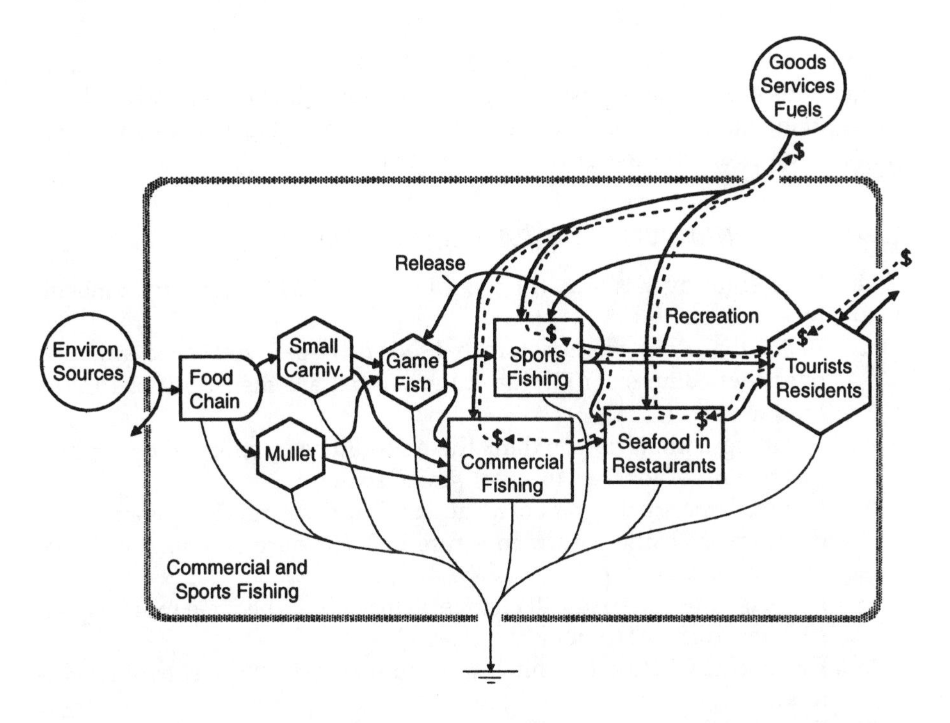

Figure 24.6 A systems view of the commercial and sports fishing system showing the competition for stocks and the way both contribute to the lives of tourists and residents.

Large carnivores (gar and bowfin) were destroyed, and plankton-eating fish, mullet, catfish, and others were harvested for food, animal food, and fertilizer. Later, freshwater netting was stopped.

As the number of saltwater fishermen increased and the marine fish stocks decreased, competition between commercial and sports fishermen became intense, with each group trying to restrict the other by influencing the political processes and government bureaus responsible for regulation. The net ban eliminated many commercial fishermen, reducing the local fish available in markets and restaurants. This raised the costs of seafood to residents and tourists. As already considered in Figure 21.1, harvested fish contribute much more emergy to people who eat them than is represented in the money paid to those who catch them.

24.3 Aquaculture

Increasingly, fish are being produced in ponds using the agricultural practice of simplifying the system. For example, restaurant catfish are produced from ponds in southern Georgia in this way. In aquaculture, only a simple food chain is allowed to develop. Just one or two species of fish are introduced.

In some aquaculture ponds, food for fish comes from fertilized plankton. Such ponds are like cattle pastures. In other ponds, pellet food may be added; these ponds are like cattle feed lots. After a crop of fish is harvested, everything is removed and the growth is restarted.

Questions and activities for chapter twenty-four

1. Define: (a) overfishing, (b) maximum sustainable yield, (c) aquaculture.
2. Visit a market or grocery store that sells fish; see if you can identify the ones that come from Florida waters. What are their positions in the food chain?
3. What regulations do you think there should be on sports and commercial fishing of lobsters or other scarce seafood?
4. Draw an energy diagram of an aquaculture system. Compare it to the diagram of a cattle pasture in Figure 23.3. How are they similar? What are the differences?
5. Discuss how fisheries will change as fossil fuels become more expensive and difficult to obtain.
6. Discuss ways of maintaining maximum sustainable yields of all seafood.
7. Discuss the new trend to increase the consumption of fish for health reasons. What effect might this have on Florida?
8. In the diagram of sports and commercial fishing (Figure 24.6), what is the effect on sports fishing if commercial fishing reduces the mullet population? What happens to the mullet population if sports fishing reduces populations of the larger carnivorous fishes (game fish)?

chapter twenty-five

Mining

Many of the materials used in developing our civilization are mined from deposits in the Earth where they were accumulated by the slow geological processes of the past. Since we use these up faster than they are being accumulated, they are sometimes called *nonrenewable resources*. After the deposits are gone, society will have to get its materials from recycling.

Except for a few oil and natural gas wells, most mining in Florida is done by *strip mining* with giant electric drag lines. The surface soils, vegetation, and upper layers are stripped off and put in piles, the minerals underneath are removed, and the original materials are used to fill up the holes again. Then the surface is smoothed out, water supplied, and restoration of vegetation and soils started. Some holes become lakes. Mining stops after the best mineral deposits are used up. Public controversy develops over the land disturbances, heavy truck traffic, and air pollution when mining conflicts with the needs and daily work of human settlements.

Major mining in Florida (Figure 25.1) includes the phosphate deposits, mineral sands containing titanium, peat deposits, and quarries for limestone and sand. We can think of soil as being mined, also, wherever agricultural practices use it up faster than it is being formed.

25.1 Phosphate mining

The most valuable mineral resource in Florida is phosphate, located about 20 ft (6 m) under many parts of the state (Figure 25.1). It is now being mined with open-pit strip mines in the northern region near White Springs and in the "Bone Valley" region near Bartow, in and around the Peace River drainage basin.

The process by which phosphorus is concentrated into deposits that can be mined commercially is driven by both geologic and biologic processes. In ancient Florida, limestone rocks formed from coral reefs and marine shells that contain up to 1% phosphate. Through geologic processes they are continually being uplifted to replace rocks that have weathered and eroded. These rocks are constantly being exposed to acidic waters that result from

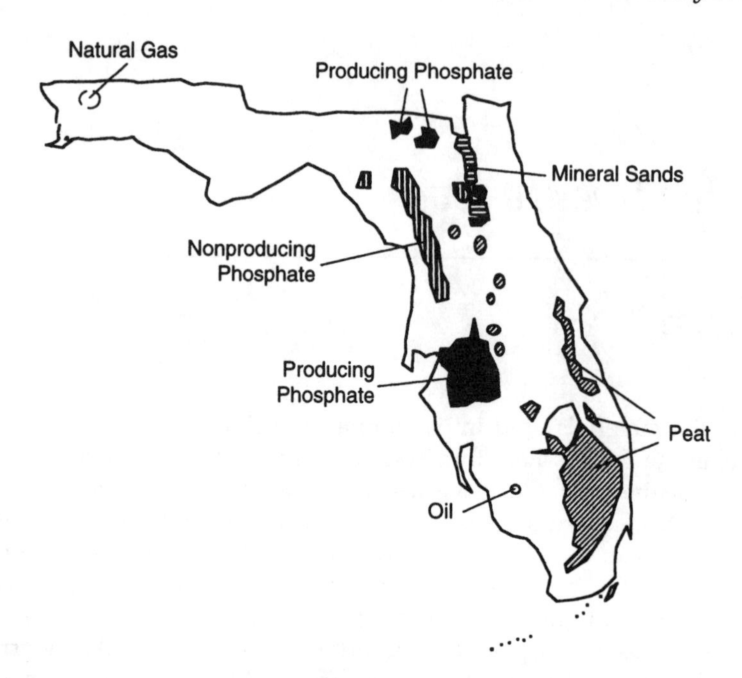

Figure 25.1 Map of the major mineral and material resources of Florida. (Modified from Fernald, E. A. and E. D. Purdum, Eds., *Atlas of Florida,* University Press of Florida, Gainesville, 1992, p. 189. With permission.)

decomposing organic matter in swamps. As the water percolates downward from the swamps above, some of the limestone is dissolved and is washed out as hard groundwater. Phosphate remains behind because it is less soluble, and it becomes more concentrated. In addition, phosphate percolating downward from the leached topsoil is caught in the limestone as calcium phosphate. After 25 million years of the combined accumulation of calcium phosphate and the increased concentration of phosphate in the eroded limestone, deposits of 5 to 20% phosphate are formed. Chemically, it is calcium phosphate, and its mineral crystals are called *apatite.*

These phosphate deposits are now being commercially mined (Figure 25.2). Drag lines remove the soil, upper sediments, and rocks (called overburden), putting them in temporary piles. The phosphate rocks and sediments are lifted out and the associated clays washed off and stored in giant slime ponds. The slimes settle gradually as wetland vegetation invades. Eventually they dry out, becoming suitable for farms, forests, or settlements. Most of the overburden can go back into the mined cavities, the ground smoothed over, and vegetation replaced. The calcium phosphate may be used directly for fertilizer or processed to make *superphosphate* (H_3PO_4). Calcium phosphate is

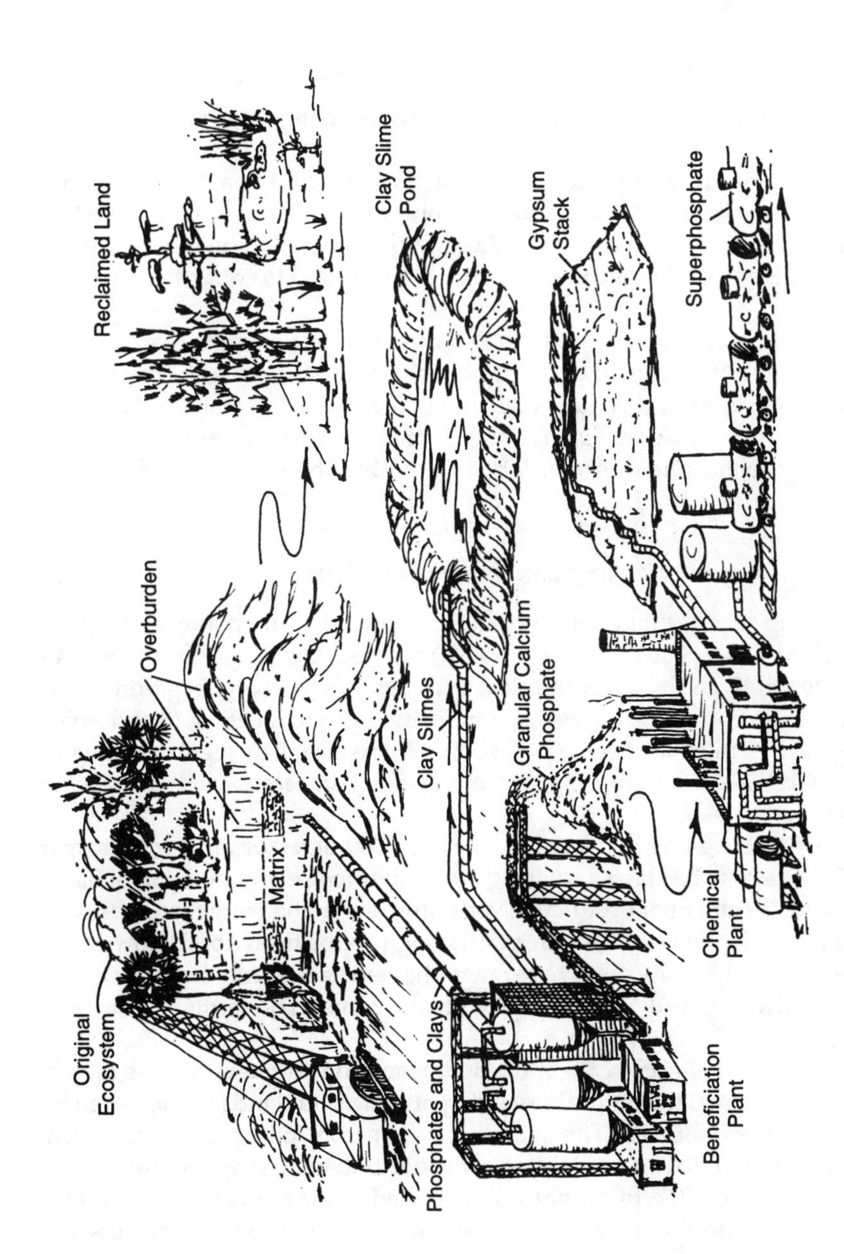

Figure 25.2 Stages in the mining and processing of phosphate in Florida.

converted to superphosphate using sulfuric acid brought in from out of state. The sulfuric acid is combined with the apatite to yield superphosphate and gypsum. The chemical reaction is as follows:

$$3H_2SO_4 + Ca_3(PO_4)_2 \rightarrow 2H_3PO_4 + 3CaSO_4$$

$$\text{(sulfuric acid)} + \text{(apatite)} \rightarrow \text{(superphosphate)} + \text{(gypsum)}$$

The main by-product, gypsum, is stored in huge mounds near the conversion plants. Since gypsum contains some strong acid and traces of radioactive uranium, it is difficult to find a good use for it. Vegetation on the mounds is sparse. Leaching under one mound caused the ground to collapse into a new sink hole.

Superphosphate is very soluble and costs less to transport than calcium phosphate, but the original calcium phosphate (apatite) lasts longer (is less soluble), is cheaper, and contains calcium that is needed in many soils. It may be that, as processing costs rise, more phosphate will be used as apatite rather than as superphosphate. About 90% of the phosphate mined in Florida is exported.

25.1.1 Environmental impacts and restoration

Strip mining requires that the overlying structure of living plants and topsoil be stripped away. Once the resource is removed a void is left, the size of which depends on the amount of material removed. Reorganization of the landscape after mining requires a combination of engineering (to fill voids and level spoil piles) and ecological science (to re-establish vegetative cover). Much water is used in mining, but increasingly in recent years it is being recycled for reuse.

Phosphate mining accounts for nearly 5000 acres of strip-mining per year. Florida now has laws requiring the *reclamation* (reorganization of lands and re-establishment of productive vegetative cover and commercial uses) of mined lands. However, there are conflicting propositions about how much mining should be done, how to arrange the new landscape, how to control the groundwater, and what to do with the large clay settling areas and gypsum stacks.

Because of the large scale of phosphate mining (some mines are tens of thousands of acres) the natural reseeding of mined-out lands is hampered by distance to seed sources. The soils that result from phosphate mining are high in clays and do not resemble the topsoils that were stripped away prior to mining. About 60% of all phosphate-mined lands are covered with clay storage areas, and the remaining 40% of the land is in former mine cuts and gypsum piles. To accelerate soil restoration, human actions are needed to help succession by planting and reseeding lands; otherwise, they may remain in a state of arrested succession.

25.2 Peat

Peat is partially decomposed plant matter that results from the incomplete decomposition process of the anaerobic environment of most marshes and swamps. There are many peat storages throughout Florida in wetlands; some are only several acres (acre = 0.4 hectares), while others are hundreds of acres. Like topsoil, peat is being used at rates faster than its production in swamps and marshes.

Peat is mined and used for soil additives and commercial potting soil. Its use as an energy source is discussed in Chapter 33. However, peatlands have alternative uses that are more valuable than mining. Many peat deposits are located in valuable wetlands that are stream headwaters, aesthetic conservation areas, or sites offering unique products and services. The most extensive peat lands are the areas just south of Lake Okeechobee and immediately north of Lake Apopka in central Florida, both of which are being used for agriculture (Figure 25.1; see Chapter 18).

As long as peat is kept wet it does not oxidize; however, when peat contacts dry air, it is consumed by microorganisms and chemical oxidation (see the record of peatland loss in the Everglades depicted in Figure 18.9). Peats are being lost to oxidation all over Florida wherever drainage allows air to react with the organic matter. Although some of the drainage is intentional so that peatland can be used for agriculture, much of the drainage is the result of lowering water tables in adjacent lands to accommodate development.

25.3 Sand

Sand is a large part of Florida's land at the beach and under the topsoil. Silica sand, made of quartz grains, was brought down from rivers farther north by the wave-driven currents along the shores. In south Florida, calcareous sands develop from the break up of shells and coral reefs. Over millions of years, deposits of sands have been left in old dunes that are now more or less in the center of the state. Sands are mined throughout the state for concrete and construction. A small amount of silica sand is used in the production of glass. Sand production is greatest in Polk, Putnam, Marion, Hendry, Broward, and Lake counties.

25.3.1 Titanium mineral sands

Scattered in beach sand are crystals of minerals containing titanium that are heavier than quartz crystals. Titanium is used in industry for paint pigments and heat-resistant alloys. In Duval and Clay counties in north Florida (Figure 25.1), there are commercial deposits where the heavy sand grains were previously concentrated by waves and winds.

25.4 Limestone

Underlying most of Florida is limestone (calcium carbonate, $CaCO_3$) from ancient coral reefs and marine shells from organisms that were living when the peninsula was submerged. The limestone is part of the underlying formations that make up the deep aquifer from which most potable (pure enough for drinking) water is withdrawn for drinking and agricultural uses (Chapter 13).

Where the limestone formations are close to the surface, they are mined, primarily for road-bed construction and for aggregate in concrete. Lime obtained from limestone is one of the major raw materials in cement. Limestone is mined from open pits. The space left from lime-rock mining often fills with water to form ponds. The highest concentrations of limestone quarries are in Dade, Broward, and Marion counties.

25.5 Fossil fuels

Fossil fuels (oil and natural gas) are extracted using deep wells. Crude oil is often pumped to the surface, while natural gas, usually under considerable pressure, does not require pumping. Florida does not have coal or lignite deposits.

Natural gas is produced from gas wells near the town of Jay in the panhandle (Figure 25.1). This site contributes about 16% of the natural gas used in the state. Gas from the Jay field joins the major gas pipeline that starts in Louisiana, passing east across the Florida panhandle and then southeastward diagonally across Florida down to the populated east coast of south Florida.

Small groups of oil wells are located in Sunniland wells in southwest Florida around the Big Cypress swamp (Figure 25.1), but most of the wells have exhausted the supply. All of the crude oil produced in the state is shipped out of state for refining.

Oil companies spent billions of dollars in past decades drilling for oil on the continental shelves off the Florida coast, but to no avail. Because of the possibilities of major oil spills, there is concern whether there are net benefits of offshore oil processing where the coastline is heavily used for tourism and recreation.

25.6 Net benefits of mining

In the states where mineral resources are concentrated, far more is usually extracted than can be used by the state, and the majority is exported to other regions. Some states are questioning the environmental degradation they must suffer locally as a result of mining, when the resources are shipped out of state, mostly benefiting other economies. The values of the mined product need to be compared with the loss of productivity of the years of interrupted land productivity during mining and the 50 years sometimes required for soil restoration. Taxes are placed on exported resources to help restore the landscapes. For example, the Florida Institute of Phosphate Research at Bartow uses tax funds to sponsor the research necessary to utilize mining wastes better and restore land sooner.

Mined and exported mineral and material resources are important sources of money and jobs. Yet, as measured with emergy, much more real wealth is exported with the minerals than can be purchased with the money that is received from the sales. A more equitable basis for trade is needed. Florida could generate more wealth by using its phosphates to boost its own soil fertility than by selling it for cash. For mining of fuels, see Chapter 33.

Questions and activities for chapter twenty-five

1. Define: (a) strip mine, (b) mineral sands, (c) peat, (d) limestone, (e) apatite, (f) superphosphate, (g) reclamation.
2. Draw a systems diagram of phosphate mining that includes reclamation.
3. Take a field trip to a nearby mine (almost every county in Florida has some kind of mining). After the trip, discuss the environmental impacts of the mining operation. What is being done to reverse these impacts? What kind of ecosystem is likely to result after the mining is finished? How might the mining operation be changed to speed up reclamation and revegetation?
4. Discuss the economic and environmental impacts from an oil spill in the Gulf of Mexico off the coast of Florida.
5. List and describe five resources mined in Florida.
6. Peat can be considered a renewable resource. How would you propose managing land to produce peat?
7. If you inherited a farm with worn-out soil, what would you do to restore it? What, if anything, should the government do to help?
8. The text suggests that phosphate should be used in Florida rather than exported. Explain.

chapter twenty-six

Parks, tourists, and biodiversity

Biodiversity is the variety of life. The great biodiversity in Florida is sometimes said to be our most precious natural resource. Many areas have been set aside as nature parks by local, county, state, and federal governments and managed with the dual purpose of preserving the species of the natural ecosystems and to provide places for people to enjoy nature — hike, camp, fish, and wonder at the way it was when all of Florida was undeveloped. According to one theory, the human species requires the aesthetic influence of natural environments for good mental health. Florida residents and tourists use the parks, beaches, lakes, and the biodiversity of natural areas. This chapter considers the way the economy is stimulated when people enjoy the ecosystems and sustain the biodiversity.

Figure 26.1 shows the system of humans experiencing nature. Nature produces the aesthetic attractions of natural areas and wildlife. Tourists are attracted by the environments, mild climate, and good facilities for vacations, as well as by commercial attractions such as Disney World and Marine Aquaria. Tourists bring money that they spend on hotels, restaurants, boats, and guides to show them around. This money then goes back out of state to buy the fuels, goods, and services that give Floridians their standard of living. In other words, the processes of nature make jobs by attracting people.

26.1 A measure of environmental use

In Chapter 21, we represented the connection between the economy and environment as a process combining nature's work and some inputs from the economy (see Figure 21.2). The emergy investment ratio was used to relate the inputs from the economy to the local and free inputs from nature (see Figure 21.1). We can use this same ratio to measure the loading that people put on the environment. In Figure 26.1, the emergy investment ratio is the ratio of the feedbacks from the economy, F, to the inputs from the

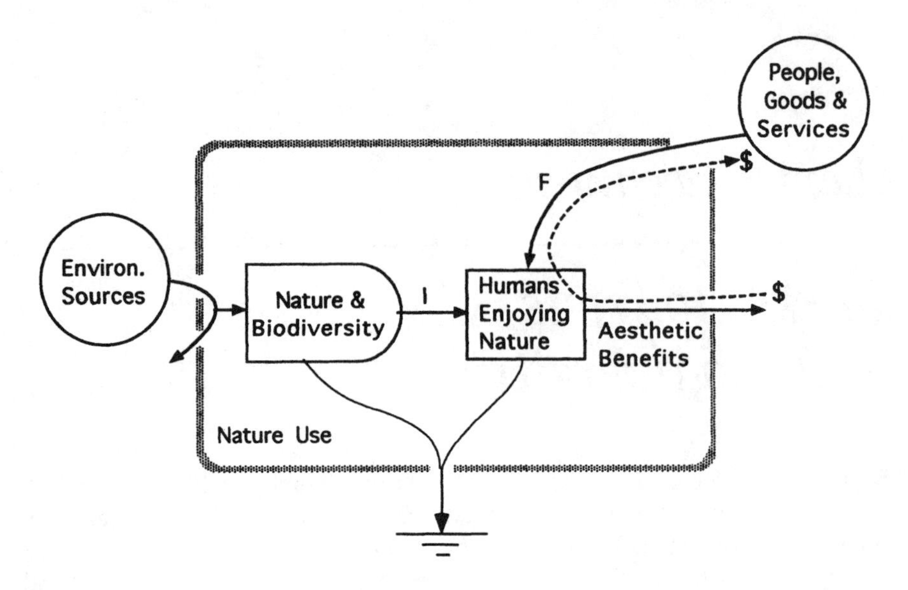

Figure 26.1 Overview of the way the economy is involved when people enjoy nature.

environment, I. The typical investment ratio for developed landscapes of Florida is about 7. When the ratio is high, as in a city, there is so much activity by people and machines that there is little space and opportunity for nature's work. In a wilderness area without many people, the ratio is much less than one.

26.2 Parks

Figure 26.2 shows some of the special parks and nature reserves in Florida. Most of the kinds of ecosystems can be found in these areas. Sometimes areas are set aside as *wilderness areas* and managed with little access by people. However, experience shows that such areas are not usually appreciated, and in places such as Florida that have dense growing populations, political pressure develops to make more economic use of them. To maintain natural areas requires that people use them enough to be appreciative of them but not so intensively that the biodiversity is lost. A low investment ratio is appropriate.

26.2.1 Everglades National Park

Figure 26.3 shows the economic interface between nature's work in the Everglades National Park and the uses of the park by managers, tourists, and businesses that profit from being next to the park. Here the emergy investment ratio is less than one. Everglades National Park is not used as much as

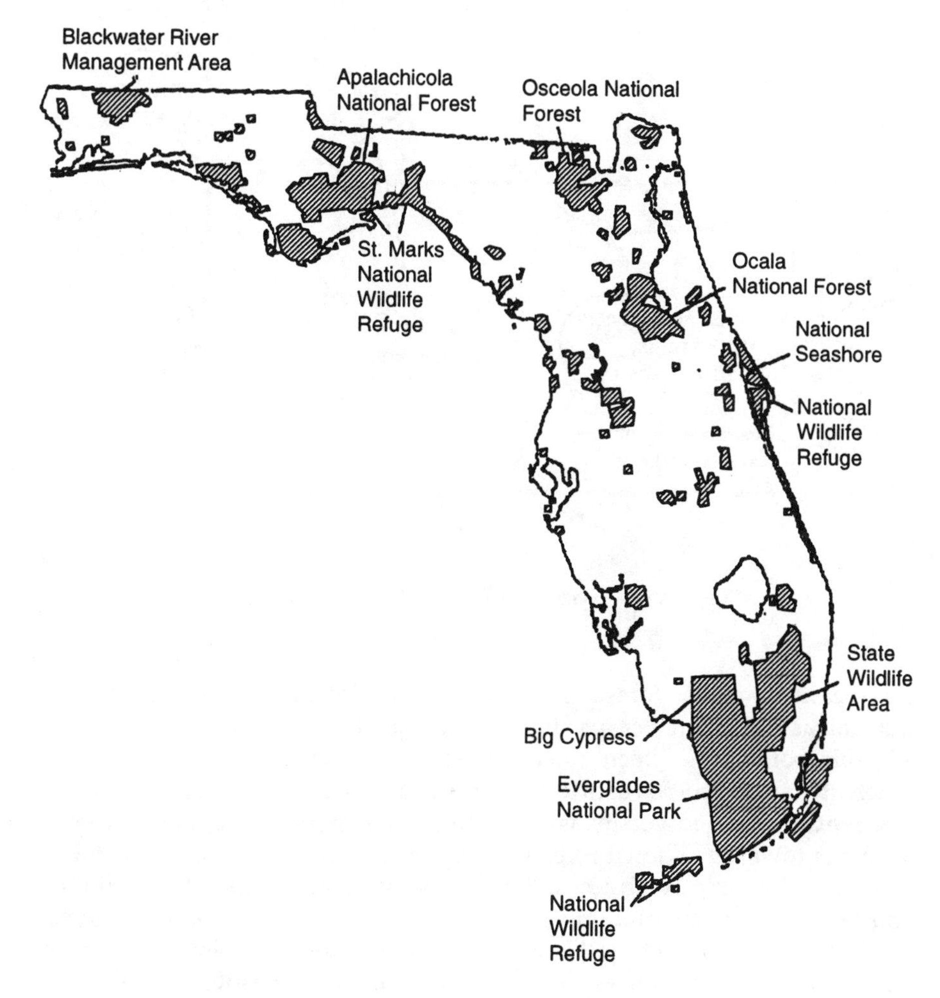

Figure 26.2 Map of some parks and reserves in Florida.

many other national parks. This may be partly due to the huge numbers of biting mosquitoes in the summer. Saltwater mosquitoes are worse than they should be because of the diversion of freshwaters that used to flow through the park. Large areas of dwarf mangroves have developed that have many temporary pools of saltwater where mosquitoes breed.

26.2.2 National forests

Unlike the national parks which were set up to protect and maintain trees and wildlife, national forests were originally set up to develop a sustainable system of growing and harvesting trees. Notice the large areas of national forest in north Florida (Figure 26.2). There are many controversies about whether plans

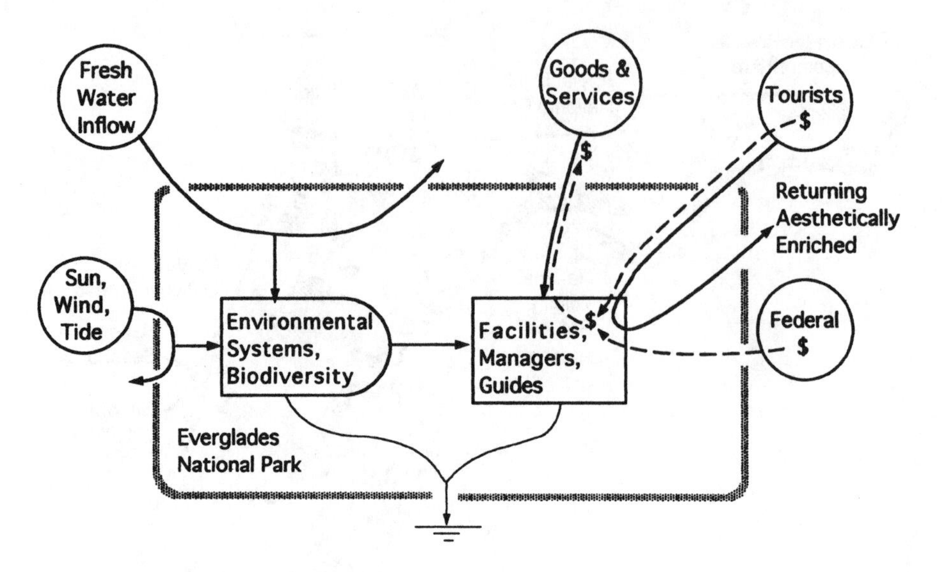

Figure 26.3 System of Everglades National Park.

for operating national forests should include mining the trees or cutting with a sustainable rotation. Because Florida pines grow to marketable size in only 25 years, rotation has been practiced more here than in many other states, where it may take a 100 years to replace a stock of trees. Where forests are cut frequently, the biodiversity is less. As human populations have grown, attitudes towards national forests have changed. Many people want them used more as parks, with less clear-cutting, more management for wildlife and biodiversity, less draining of the wetlands, and fewer economic uses.

For example, the Ocala National Forest contains the unique sand-pine scrub ecosystem that grows in deep sand. It has a beautiful pattern of interesting species, some endangered. To save some of this ecosystem will require new policies or the setting aside of some of the area as a national park.

Managing for natural conditions is complicated by questions of fire management (Chapter 19). The natural pattern for the sand-pine scrub was to grow for 20 years and then, when the vegetation was thick enough, to be burned by natural lightning during dry periods, causing succession to start again. All the plants and animals have adapted to survive fires. Fires can be set by managers when the ground is damp and winds light, so that the natural cycle can continue; however, as people build more homes in the region, it is becoming increasingly difficult to burn without danger.

26.2.3 Aquatic parks

Some parks have been created for the preservation and enjoyment of underwater ecosystems. They include many springs, lakes, streams, estuarine

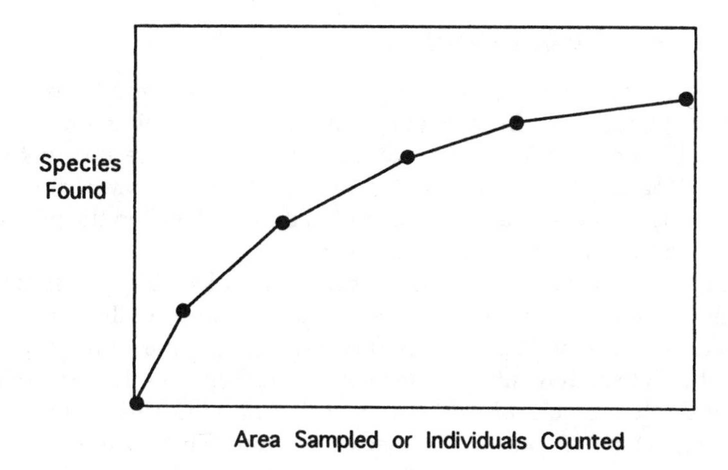

Figure 26.4 Graph of species found as more area is examined.

areas, and coral reefs under protection. As with land areas, high-energy activity from humans is not compatible with the ecosystems and wildlife.

High-energy boating areas should be in different waters from those used for wildlife, canoeing, and quiet enjoyment of nature. Similarly, spear fishing does not fit with visual enjoyment of reefs by snorkelers and scuba divers. Unprotected reefs soon lose their larger fish population. Then, urchins and other unregulated smaller animal species overpopulate, causing the reef's normal functions and biodiversity to be lost.

26.3 Biodiversity

The variety of life is a valuable resource that required millions of years to develop. We receive this wealth in the genetic information passed from generation to generation by each kind of life. To survive, a species has to have an ecosystem to which it is adapted. Environment suitable for a species is called its *habitat*. There must be enough of that habitat to support a large enough population of that species to reproduce with genetic variety. To sustain biodiversity, enough area must be set aside to keep the populations of the ecosystems healthy.

The larger the area and resources, the more species can be supported. Figure 26.4 shows a typical graph of the increasing number of species found when the area of a particular ecosystem is being increased. The graph curves to the right showing that it takes more and more area and resources to support additional species. More and more resources have to go into species interactions. Species interact in various ways — competing, cooperating, and avoiding each other by being active at different times of day and night. The more species there are, the more interactions there are among the different species.

26.3.1 Measuring biodiversity

The diversity of species is easily measured, which is a good class exercise. Just count plants or animals as you come to them, keeping a record of the total number of individuals counted. Set aside in a box each new kind that you find. You can count shells on the beach, leaves in a forest, animals that you pull up in a seine or in dipnets, or plankton observed in the microscope. If you have time, count until you have 1000.

Ordinary Florida environments may have 15 to 30 species per 1000 individuals counted. If there has been a disturbance, pollution, or recent destruction, you may find only one or two species per thousand. The ecosystems with highest diversity are the most tropical ones — the coral reefs and some of the hammocks of southernmost Florida where frost is infrequent. These might have a count of 50 species or more. There are many international discussions among governments now about preserving tropical biodiversity.

26.3.2 Speciation

A *species* is a population of interbreeding organisms. A new species can develop when a population becomes isolated for a long enough period in special environmental conditions so that the processes of genetic change make the population different. When an isolated population has changed enough in color and behavioral habits, it may no longer breed with members of the original stock. When interbreeding stops and genes no longer flow back and forth between the old and new population, the newly changed population is said to be a new species. Although data are few, estimates suggest some species required 10,000 years to evolve.

For example, deer spreading into the Florida keys became isolated for thousands of years. As in all healthy species, there is variation in the genes in the population. In the Florida Keys, those individuals that were born somewhat smaller apparently had an advantage, and more of the smaller ones survived and reproduced. When organisms are different but can still interbreed, they are said to be a variety or *subspecies*. Key deer are a subspecies of the white-tailed deer.

26.3.3 Endangered species and subspecies

With so much time required for developing species and subspecies, we need to appreciate what is required and how long it would take to replace a species that has become extinct. Florida has many endangered species, for which the number of individuals in the remaining population may be close to the minimum number necessary to reproduce the species. The endangered Florida panther (a subspecies of the mountain lion or cougar) is so important to the people that the school children voted it the state animal (Figure 26.5). The

Figure 26.5 The Florida panther. (Modified from Fernald, E. A. and E. D. Purdum, Eds., *Atlas of Florida,* University Press of Florida, Gainesville, 1992, p. 177. With permission.)

state and federal government together set up a panther refuge just west of the Everglades marshes to try to protect the 40 to 50 remaining panthers.

In an effort to increase the panther population, several plans have been developed. Some of the cubs are being captured and bred in captivity, a procedure that worked well in increasing the numbers of whooping cranes. Others are left to breed in the reserves. Wild panthers have been wearing radio collars so their movements can be tracked. Because it has been learned that the panthers do go under highway bridges, I-75 across the state was built with wildlife passageways under it. Hunting with dogs is not allowed in panther areas, as the dogs scare the panthers and they will leave an area that is hunted with dogs. Since panthers eat deer and wild pigs, preventing the hunting of these species in panther reserves allows an area to support more panthers. All these management methods are the subject of much discussion; some people think it is better to leave them alone.

26.3.4 Wildlife corridors and greenways

Larger wildlife species require large areas to support healthy populations, but many parks are small, isolated, and surrounded by housing and urban developments. When natural fluctuations cause populations to become extinct locally, there is no pathway for others to move in. Wildlife crossing from one habitat to another are killed on roads at night. Because the populations of wildlife are separated, inbreeding can cause loss of genetic vitality.

Now there are efforts by public and private agencies to join the fragmented natural areas with wildlife corridors (greenways) connecting these "patches" of suitable habitat. The greenways are also good for hikers. Old railroad tracks and powerline strips can be used. Some greenways include a narrow lane for bicycles. Where strips of wild vegetation are made continuous along the electric power lines, the highway rights of way, and the streams, wildlife can travel between suitable areas. Strips of wild vegetation help maintain viable gene pools as well as distribute populations more evenly. By designing continuous greenbelts, urban areas become an integral part of a green landscape instead of a development that disrupts, cuts, and fragments the natural landscape.

26.3.5 Ecosystem reserves for automatic reseeding for restoration

When forest patches and corridors are evenly distributed over the landscape, seeds and birds to spread them are available to restore forests and soils when land is abandoned from other uses. As population, agriculture, and development have increased, forests have been displaced. For the future, special efforts may be required to preserve enough forests for reseeding bare lands.

26.3.6 Land rotation

Land rotation is the plan that alternates forest growth with forest use (Chapter 22). It is the normal way of sustaining forests and soils in a landscape. Retaining strips of high biodiversity aids in land rotation, soil restoration, and recreational use of the environment.

26.3.7 Residential biodiversity

In the past, the custom in many housing areas has been to surround houses with smooth lawns and a few ornamental shrubs — a low-diversity environment. Maintaining simplicity and monocultures is costly, requiring fertilizer, irrigation, pesticides, mowing, and labor, with runoffs that damage the environments elsewhere. Now it is becoming respectable to have complex, high-biodiversity surroundings, often using native vegetation. Costs and efforts are much less with a more natural residential ecosystem.

26.3.8 Hunting

A controversial practice in environmental management is hunting. Hunting, sometimes with dogs, was a part of the rural culture and a practice prized by many residents to this day. As populations have increased, areas where hunting is safe are being reduced. Hunting reserves probably have to be separate from other recreational uses of environment. Many people want to see and photograph wildlife, not have it shot or driven away from view. Many of those who manage deer believe the human hunter is needed to replace the carnivores that once regulated deer populations. Coyotes are moving into Florida and may help control deer populations. Hunting the 200 or 300 bears left in Florida hardly seems appropriate; however, when the bears become friendly, they come into housing areas in search of garbage and make problems for authorities worried about injury to children and pets.

Questions and activities for chapter twenty-six

1. Visit the nearest bit of natural ecosystem and make a diversity count. Compare this with the count you get by counting plants on the lawn (see Question 10 in Chapter 15).
2. Count 100 plants in several lots in your neighborhood. (If you have the time and people to help, count 1000). Which have the highest biodiversity? Can you tell why from the way the vegetation is managed?
3. Using a city or county map, locate the parks and protected areas in your town. What ecosystems and prominent species are represented there?
4. Discuss which natural areas might be attached as a wildlife corridor. Connectors could be power lines, abandoned railroad tracks, or stream banks.
5. What are the natural attractions for tourists in your area? Discuss why they are attractive.
6. What is the status of fish populations in your area? Talk to wildlife managers to learn of problems.
7. Discuss the pros and cons of hunting.
8. Define biodiversity and explain why it is considered valuable.
9. Discuss the value of endangered species. Use a local one as an example. How much public time and money should be put into protecting it?
10. Of the $259 E9 gross state product, $60 E9 is from the tourist industry. Calculate its percentage. Discuss tourism and environment.

chapter twenty-seven

Cities

The largest system of the landscape is the city with its huge concentrations of people, cars, and industries. Its energy comes from fossil fuels and electric power. It requires water, foods, and materials from its surroundings and generates great volumes of waste in the air, waters, and solid-waste dumps that impact the environment. The city is spatially organized hierarchically from the rural edge to the concentrations of business and information at the center. This chapter relates the city to its resources.

In only a half a century, Florida has changed from a rural state with an economy based on raw commodities to an urban state with many large cities. Florida's population has grown at a tremendous rate since the first census. In 1830, the year of the first official census in the state, there were almost 35,000 people living in Florida. By 1997, the population had grown to almost 15,000,000 people (see Chapter 35). In the early years of growth, most of the incoming people settled in rural areas and small settlements. In the last few decades, most of Florida's new population has been concentrated in the large metropolitan areas of the southeast coast (Miami to Palm Beach), the southwest coast (Sarasota, Ft. Myers, and Naples), the Tampa Bay region (Tampa, St. Petersburg, and Clearwater), central Florida (Orlando), the northeast coast (Jacksonville), and the extreme northwest coastal area surrounding Pensacola. In 1992, 93% of the population of the state lived within metropolitan areas.

27.1 Urbanization with fossil fuels

Early cities were small and depended for their support on the surrounding agricultural lands (i.e., agrarian cities). Then, with fossil fuels and nuclear power came industry, cars, and power plants with many people moving to the city for jobs. Figure 27.1 shows the change.

27.1.1 The agrarian city

The simple diagram in Figure 27.1b shows the relationship of a town to the surrounding agriculture. Food and other rural products were taken to the

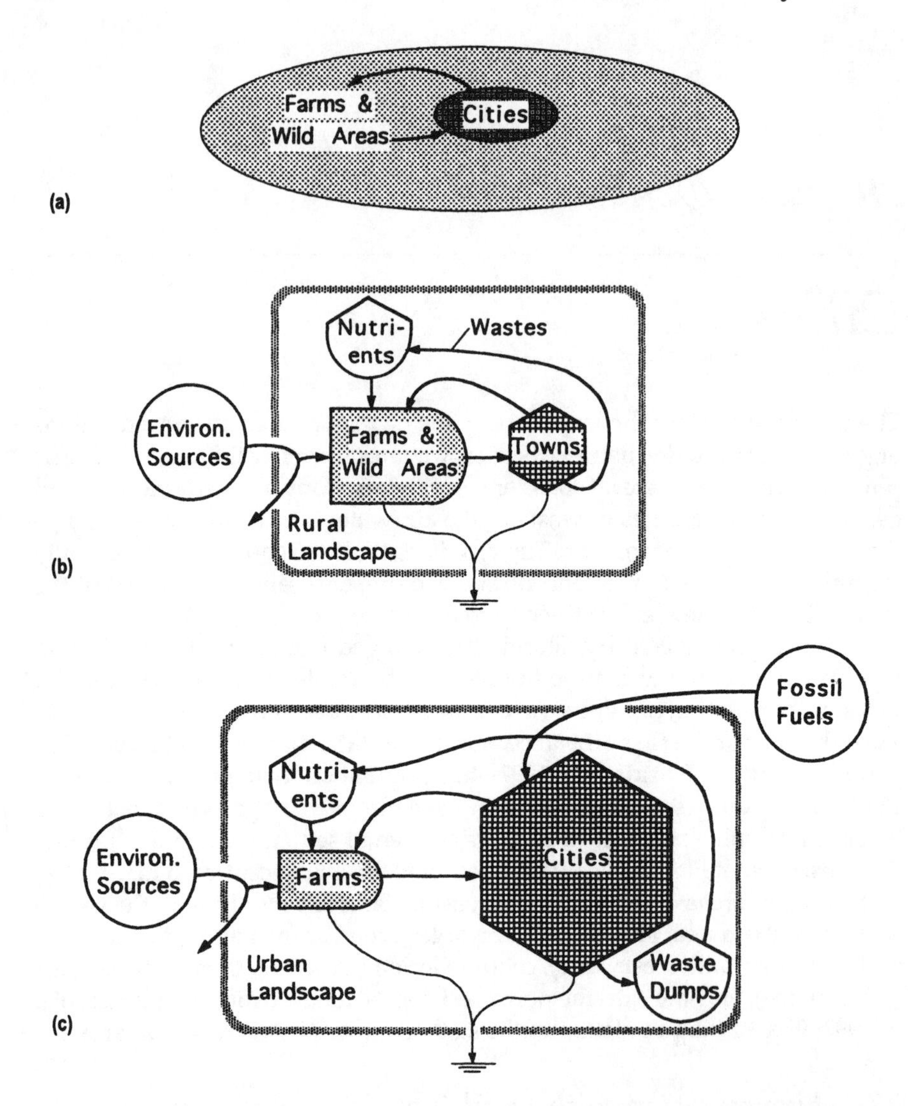

Figure 27.1 Relation of towns and cities to their supporting landscape and the recycle of wastes. (a) Spatial view showing the concentrated city and the large rural area required for its support; (b) early towns based on rural production only; (c) giant cities based on fossil fuels but also requiring rural resources.

city to support the people who lived there. The city in return provided simple tools, labor, and other products to the farms.

Notice the other feedback pathway in the diagram, the recycle of wastes from the city back to the agricultural lands. This pathway was very important during early times before fertilizers were available. Wastes of humans and horses were picked up at night by the farmers and returned to the farm to be

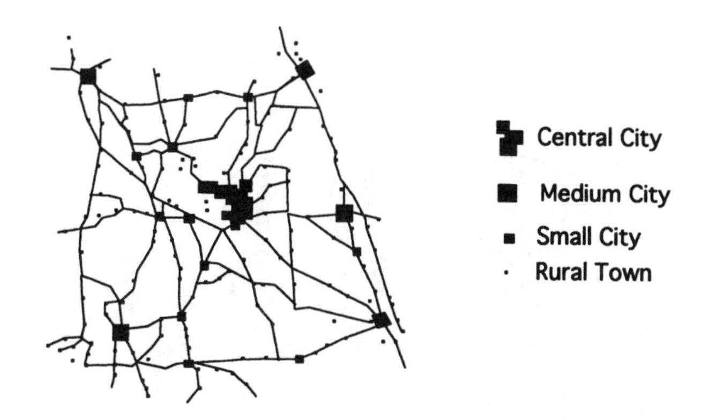

Figure 27.2 Map of a region showing the way roads connect the city to the rural surroundings. This is the region around Orlando before the Disney World era.

used as fertilizer. In this way, the farms and cities were linked with a closed-loop cycle of nutrients which dispersed waste and maintained the productivity of the farms. With the advent of fertilizers, this circle was broken. Fertilizers were obtained from mines and the wastes accumulated in dumps and polluted waters (see Chapter 34).

Human settlements are organized as a hierarchy, with rural people interacting with small towns, and towns connected by roads to the regional city. Early in the century, Florida towns and cities were centers of communication and exchange for a landscape engaged mostly in agriculture. Figure 27.2 shows Orlando before it became an international center of tourism.

27.1.2 The fuel-based city

Much of the growth of Florida's urban centers is the direct result of an increase in availability of fuels (fossil and nuclear). As populations and energy use increased, city transportation was reorganized around the automobile. Industries and jobs attracted people to the city. Housing districts developed around the city. Prime agricultural lands were covered by streets, parking lots, and buildings. With their population and money, cities controlled regional development. Wastes were released, often causing pollution of the air flow, rivers, streams, and lakes.

27.2 Overview of the city system

Some of the main zones of a city are related in the systems diagram of a city in Figure 27.3. The rural, green areas are on the left with the central business district on the right.

We all appreciate the vegetation and wildlife in the parks and lawns of residential areas, but perhaps we are not aware that the renewable energies

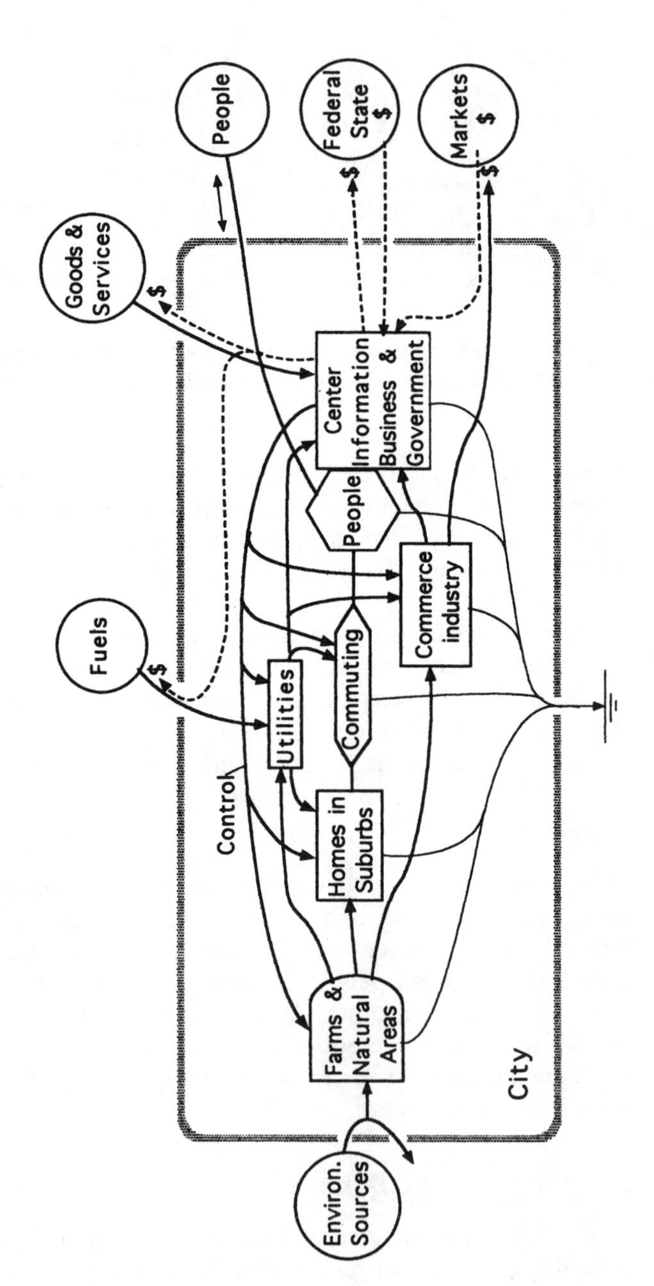

Figure 27.3 Systems diagram of a city with the center on the right.

do much more for the city. The sun, wind, and rain (and tides and waves, if the city is located on the coast) are important to industry as well as to the microclimates affecting people directly. The winds blow away automobile exhaust and industrial smoke. The water in streams, wetlands, and tides carries off solid and liquid wastes from industry and housing.

Housing was developed in suburbs for people (who can afford it) to live in a green environment while holding jobs in the middle of the city, but this requires commuting — spending time, energy, and money on going back and forth. In Florida, instead of developing mass transportation (commuter railroads and buses), private cars were used and much of the money available for public expenditure went into roads instead, filling the city with noise, congestion, and air pollution. Bikeways were not generally included the way they are in European cities. A major problem for society is how to make cities more efficient and livable. Part of the solution may come from reducing the number of cars. But, Americans love their cars and the freedom they give them, at least when there is no gridlock (so many cars on the road that traffic is almost stopped). Now shops and stores are spread out along highways (strip development) or clustered in distant malls. The way cities are organized now, without sufficient mass transportation, a person cannot function without a car.

Industrial production of a city is processed by the commercial sector — some products being sold to the people and agencies within the city, some sold to markets outside. The people supply the labor for industry, commerce, and governmental services. The various departments of government, such as health, education, and police, have a controlling influence on the other sectors of the city.

Money circulates through the local city's economy, flowing in from the sale of exported products and from state and federal sources, circulating around and around within the economy and eventually flowing back out again either to purchase goods, services, and fuels or to pay taxes. The closer to the center of the city you are, the more money circulates and the higher the costs of land, goods, and services.

Cities have connections to the state and federal governments. People pay federal income taxes and receive federal *transfer payments* (money from federal government for social security, welfare, postal service, urban renewal, and federal courts). The state government also levies taxes with money being returned to the city to pay for state police, courts, welfare, school programs, and community colleges.

Another inflow to the city that has a large effect is the in-migration of people from elsewhere. Many cities have growing populations that put pressure on all parts of the city. Government must provide more police protection, roads, libraries, and schools. Remaining areas of open land are often paved or built upon to meet the increased needs of housing or parking. To pay for the additional services required by the growing population, government must often raise taxes to keep up with the growing demand for services. When the city becomes too intense, people begin to look elsewhere and move out in search of lower taxes and a "higher quality of life".

Compared with cities of northeast U.S. or the older cities of Europe, Florida's cities are new, built during recent, very rapid growth. But, much of the construction is temporary, made of wood and plastic, which are not permanent in a subtropical climate. Each of the parts of the city contains buildings, machinery, and equipment that depreciate with use. Since these assets as well as the *infrastructure* (roads, water lines, sewer lines, powerlines, telephone lines, and television cable lines) were all built within a relatively short period of time, they will all begin to need repairs and replacement at about the same time — a very big problem ahead.

27.3 Spatial organization of the city

Each city is arranged in a spatial hierarchy, which we showed in the arrangement of city parts from left to right in Figure 27.3. The center of the city is most concentrated, having the largest buildings, highest density of people, and greatest flows of emergy and money. Surrounding the central city are rings of less and less concentrated activity going away from the center. Streets that radiate out into the less concentrated rings from the central cities get smaller and less traveled as they extend into the surrounding city. Commercial activity is concentrated on main roads, especially those that connect the city with other cities. Large cities have a beltway (highway that circles the city). Where the beltway intersects the roads between cities, there are intense areas of activity such as shopping malls and industrial parks. The activity is noticeable from the air at night. The lights of the city resemble the shape of a star with the central city as the apex and the brightly lit heavily traveled streets the arms.

Figure 27.4 shows a map of land uses in Jacksonville. The greatest concentration of activity is typically in the center of the city and is called the *central business district*. The biggest buildings and highest concentrations of business activity are found here. As you move away from the central business district, the level of activity is less concentrated and land uses are primarily residential with some concentrations of commercial activity at shopping centers along major roads. Industrial land uses are usually associated with transportation arteries such as railroads, rivers, major highways, and airports.

There is much controversy over the planning of cities and how the land uses should be organized within them. The tendency in the past has been to separate uses that were thought to be incompatible, such as residential neighborhoods and shopping centers or industry. So, many of Florida's cities have separate districts where industry, commerce, and residential uses are located. They are often on opposite sides of the city, separated by many miles and great traffic congestion. When shopping centers and offices are zoned to separate districts and removed from the residential districts, people must use

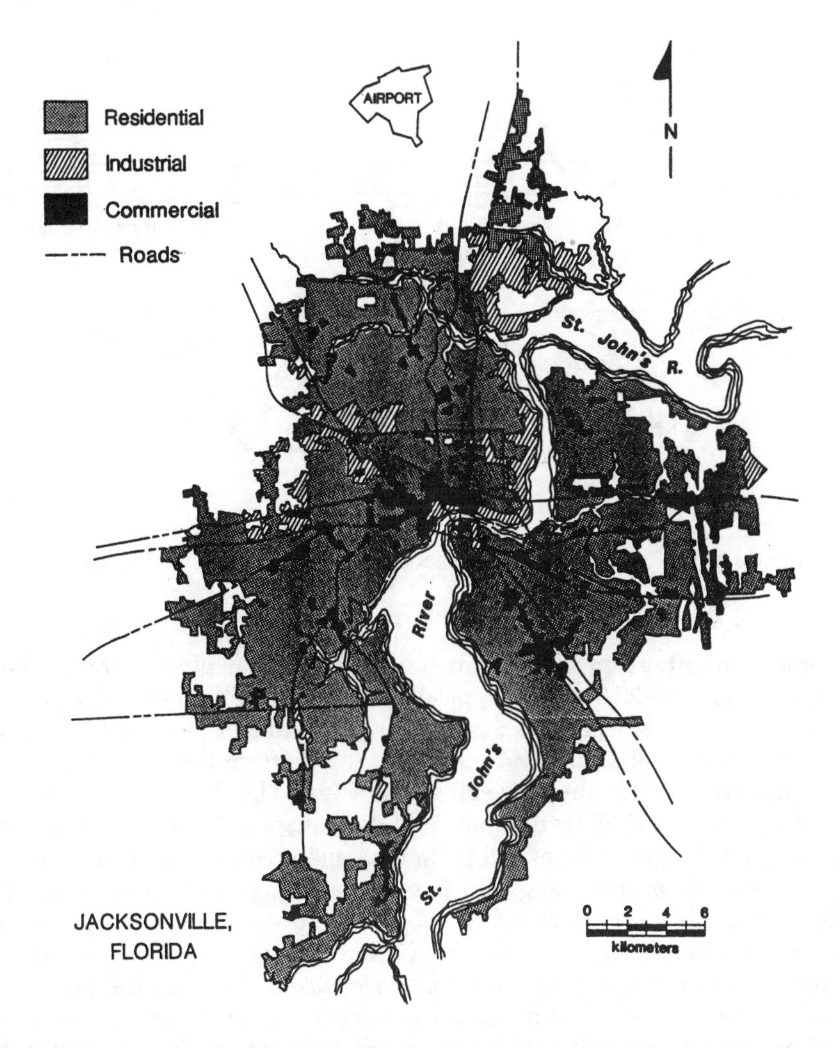

Figure 27.4 Aerial map of Jacksonville, Florida, showing the dense concentration of urban structure and activity in the central city, and less density in outlying areas. Commercial areas are black; industrial areas are cross-hatched, and the stippled areas are residential. (From Brown, M.T., Energy Basis for Hierarchies in Urban and Regional Landscapes (Ph.D. dissertation), Environmental Engineering Sciences, University of Florida, Gainesville, 1980.)

their cars for even the simplest of shopping trips. Much of the traffic congestion of cities is due in large part to the separation of land uses.

27.3.1 Residential neighborhood

Affluent cities of America have developed an outer zone of residential housing away from the congestion, noise, and pollution of the inner city. Many

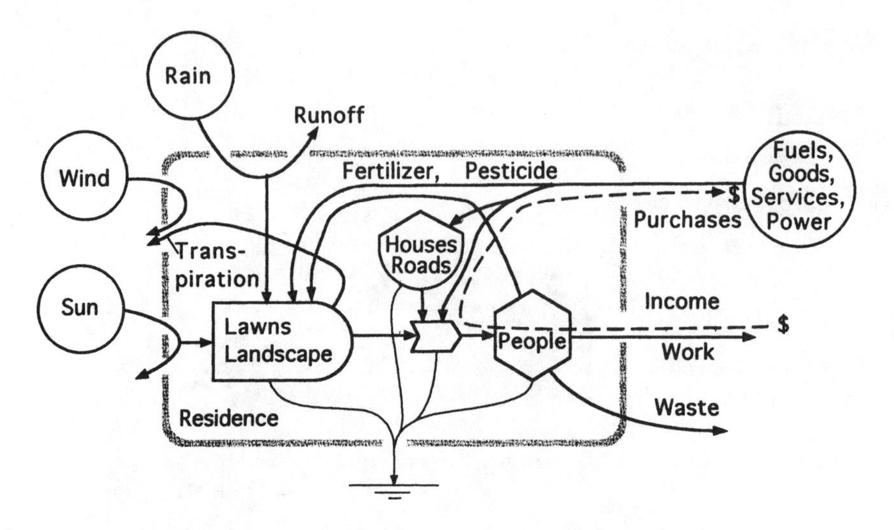

Figure 27.5 Residential neighborhood system with house and landscape.

people commute from this suburban area to jobs downtown. The systems diagram in Figure 27.5 shows a typical residential neighborhood. The pattern is like other human activities that use the environment — a combination of free environmental inputs and purchased inputs from the economy.

Sunlight and rain are utilized by the lawns and landscaping. Some of the solar heat is carried off by the cooling of the transpiration of grass and trees, creating a better microclimate than in the paved-over zones of the city.

Human time and energy go into managing, planting, weeding, and mowing the lawn. Fertilizers are used to supply nutrients for growth of lawns, bushes, and trees. As the lawn is cut and the cuttings removed, the nutrients lost must be replaced by more fertilizer. The chemical pesticides used to eliminate weeds and insects can impact the health of birds, squirrels, and other small animals. Not only is some of their food destroyed, but the poisons often affect them directly. The residential areas consume and pay for fuels, electricity, water, goods, and services, including disposal of wastes. Every day, at least one member or, in some instances, two or three members from each household work in industry, commerce, or government. The money earned is used to purchase electricity, water, goods, and services. So, the residential neighborhood provides the labor for other productive processes of the modern economic system.

27.4 City waste

Wherever there is energy and resource consumption, as in a city, there is an accompanying production of waste by-products. Sewage, urban runoff, garbage, and air pollution are by-products of Florida's urban systems. In the past, when cities were small, and the numbers of people were not so large, wastes were easily assimilated by the environment. Sewage was handled by individual septic tanks in back yards, and garbage was composted. Now municipal waste is first processed in sewage treatment plants, solid waste and garbage are deposited in landfills and dumps, and air wastes are processed through filters (see Chapter 34).

27.5 City malfunction

Partly because of the disruption of city organization by excess cars, parking lots, and highways and partly because a large part of Florida's population is without jobs, crime, drug use, and gangs have been increasing in Florida cities and suburbs. The city and its families function badly when many people are idle and disrupt useful work. The challenge is to get everyone fully involved in a job or in education for a job. Perhaps a combination of decentralizing cities into neighborhoods, better education, public works programs, and better maintenance of housing can reverse the trends.

Questions and activities for chapter twenty-seven

1. Define: (a) convergence, (b) hierarchy, (c) transfer payments, (d) microclimate, (e) circulation of money, (f) pollution, (g) toxic substances.
2. What factors usually determine the location of a settlement or city?
3. Describe the spatial hierarchy of cities within a landscape.
4. Describe what things are fed back from larger cities to smaller ones?
5. Why is it predicted that cities in the future will decentralize?
6. Why are residential areas generally considered consumers?
7. List Florida's major population centers.
8. List five probable reasons for the great influx of people into Florida since 1830.
9. List the environmental resources of the coastal zone that attract population to settle here.
10. As a group exercise, gather around a table with a large piece of paper in the center. Make a list of the major features of your town. Be sure to include energy sources, industries, public activity, and any unusual environmental factors. With one person doing the pencil work, combine ideas and diagram the listed features.
11. Clip one newspaper article each day for one week that deals with a problem in your city. The class can be divided into groups, with each given a city problem to investigate and present to the class for discussion. Are the problems related to growth of population and no growth of resources? Evaluate the problems and solutions using energy diagrams, being sure to consider the environmental energies of the area.

12. Count the number of cars passing a particular area in 10 minutes. Do this at five places, starting in the center of the city and again at each mile or two going out from the center until you are in a rural area. Plot the car counts on a graph as indicated. Why is the graph shaped the way it is? Where is the energy use the greatest? How does this relate to the dollar value of land? To taxes?
13. Discuss the possible future trends of cities. Consider center-city problems, possible decentralization, costs of services, political power, and how cities are controlled by the state.
14. Originally, county governments with larger territories were in charge of smaller towns in their area. Discuss ways to reorganize when a city has more budget than the rest of the county; when the city becomes larger than the county. Refer to the hierarchy concept in Chapter 7.

chapter twenty-eight

Industry, technology, and information

Industry combines the work of people with energy and material resources to make products for society at home and for sales out of state. A prosperous economy requires that the products compete well in world markets. The successful industries are often those that use the newest technology and information. The economy of Florida in recent years has been stimulated by the many high-technology industries that form the U.S. space program. This chapter considers industrial systems, the development of technology, and the role of information.

28.1 Industry

The typical industrial system connects fuel and electricity with raw materials, human labor, technological equipment, and information know-how to manufacture such products as airplanes, cars, boats, components for houses, clothing, etc. (Figure 28.1). Money received for the product sales has to pay for the many inputs. An industry cannot survive if it has higher costs than other competing industries. It requires low-cost fuels and electricity and reasonable wage rates. During the present time of rapid change in computer technology, an industry particularly needs to receive the latest information and to educate its employees, another large expense.

Complex industrial equipment and the information required to use it is called *technology*. To stay efficient and competitive, industries have to use the best technology. It takes money to buy new equipment; money to set up new assets is called *capital*. The industry gets the capital to expand or modernize by borrowing from banks or by selling stock. Stocks are paper certificates which promise to pay the owner a share of profits. People who buy the stock are investing in the industry and expect to receive money from the profits in exchange. The money paid to the stockholder is called a *dividend*. Paying dividends or paying interest on loans is part of the cost of doing business.

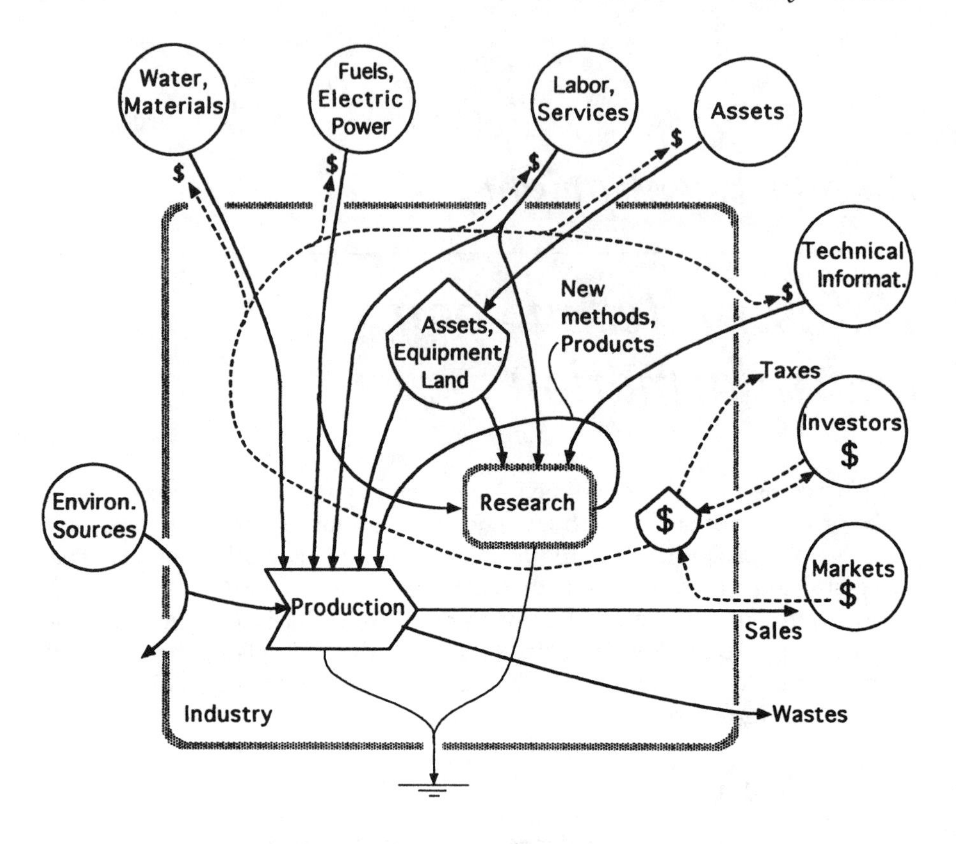

Figure 28.1 Diagram showing the system of an industry with its main inputs, processes, and circulating money.

Industries also have to pay taxes. Sometimes, local communities try to attract an industry to move to their area by making them exempt from tax. Responsible industries pay for waste processing that does not deplete public environmental values. At times it is hard for our industries to compete with some less responsible industries abroad that pay less taxes, dump their wastes, or pay wages too low for decent living. Some large companies have closed their production facilities in the U.S. and moved them to countries with lower production costs, resulting in the loss of many jobs.

28.2 Information

As used here, the word *information* refers to a set of parts and their connections. For example, the information in a cookbook on how to bake a cake has a set of parts: the ingredients and things the cook has to do. The recipe puts these parts together in a required order for the cooking to succeed. Another example of information is plans for assembly of a radio which show the parts

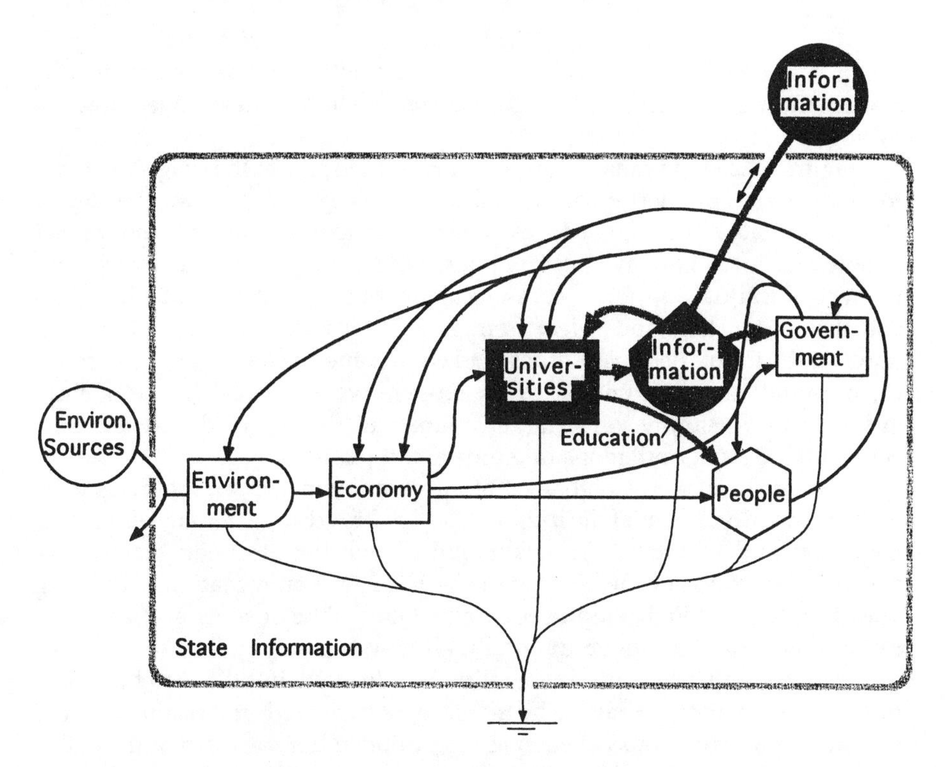

Figure 28.2 Role of information in the state system of Florida.

and how to connect them to make the radio work. Much of a student's life is concerned with learning information about how to be employed and to live in our complex society. Such information started with the basics of reading, writing, and arithmetic but now requires the ability to use computers and to learn new technologies.

28.2.1 Information system in the economy

Of all the sectors of the economy of humanity and nature, those concerned with information are at the top. Many resources are involved in the web of interactions that support institutions of education, high-technology industry, and government. Much energy goes through the web to develop and maintain information, and the result feeds back to organize the whole system (Figure 28.2).

The information that everyone shares together is sometimes called society's *culture*. By having many basic beliefs and ways of working together, all the separate individuals together constitute a powerful unity. For example, most of us are raised with common ideas about how to treat each other and what is right. For another example, consider the way millions of cars in a great city move about at great speed with only rare accidents. The restless motions of a city are possible because of the shared knowledge about driving.

Figure 28.2 shows shared information at the right of the diagram, at the top of the web of energy transformations. This example shows the role of state university education in developing new ideas, technologies, and trained people that spread out over society interacting, contributing, and reinforcing the web of mutual support. Representing the public, government employees use their education and select from new information to administer and regulate the industries, institutions, and environmental management to make the economy and environment work cooperatively. Ecosystems work similarly but more simply when larger animals at the top of the food chains control the other populations of animals and plants.

In order to sustain a high level of information in society, there has to be a continual enrichment of the information used to educate employees such as engineers and scientists. An evaluation of one high-technology industry showed that much of the emergy of the industry's information processing was supplied through employees trained in public schools and colleges, rather than from on-the-job training with new technology.

Not only must each new generation be educated, but the storehouses of information in libraries have to be regularly reorganized, and restored. Fungi and insects destroy books even in air-conditioned libraries so that the books may not last more than 50 years. Whereas copying information is relatively cheap, the continual copying necessary to sustain the great knowledge of our age in enough libraries to be secure is costly in energy and money. Our civilization depends on sustaining its information. During the decline of the Roman civilization, much of the accumulated information was lost and had to be redeveloped again in our times.

28.2.2 Global information processing

As information systems, space satellites have been highly successful in improving worldwide communication, developing information about our lands and seas, and feeding back information on defense to maintain peace. In Figure 28.3 is a nighttime view of Florida taken from a satellite in 1989. Notice the lights marking the towns and cities, and especially the urban areas around Miami, Tampa, Orlando, and Jacksonville. The lights show the spatial pattern of hierarchy in which the brightly lit areas are the centers of human activity, especially information and economic processing. Of course, the lights are running on electricity, which is the main intermediate between our basic fuels that operate power plants and the higher level human energy used for information processing.

Television has become very important in the lives of people, occupying most Americans several hours a day. In this highly varied spread of information, there is great amplification when many people share the same

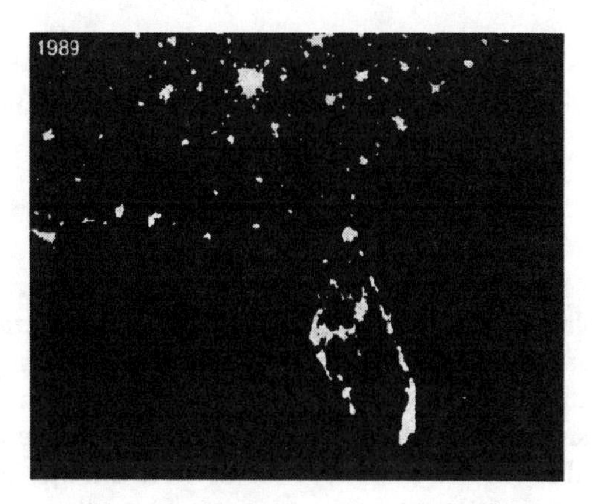

Figure 28.3 Image of Florida from satellite at night. (Modified from Fernald, E. A. and E. D. Purdum, Eds., *Atlas of Florida,* University Press of Florida, Gainesville, 1992, p. 131. With permission.)

programs. Perhaps unimportant information is being duplicated more than the important information that people need most to work together. In a sense, we are impacted by storms of information which our society has yet to learn how to use.

Sending messages by computer as "e-mail" is so easy that it is becoming the standard way to communicate in business or with friends. The latest miracle is the world Internet, by which anyone with a computer can get information formerly accessible to a few. There is so much information available that it is difficult to find what you want unless you have an Internet address for it.

Because information is so readily transmitted, information does not follow markets very well. Ideas and computer programs may spread worldwide without much buying and selling. Those in universities are dedicated to teaching and spreading information without limit, whereas those trying to make industries profitable use patents and copyrights so that they can make more money from the information. New information comes from research, and enough money must be made available to pay for continued research.

As people share common objectives through the sharing of information, old mechanisms of the world order such as wars and power politics can be replaced by a sharing of common information, trust, and ideals about world peace.

Because cities around the world are using the same fuels, cars, and electric power, their patterns of urban life are becoming more similar. But, perhaps it is unfortunate that similar ways of economic development worldwide are making people less diverse in their culture and life styles. People with pride and sense of security in their national traditions are concerned lest these traditions be weakened. Perhaps we can share globally what is necessary for good environmental and international relationships, while retaining what is special about life in our own particular countries and cultures.

28.2.3 Space futures in Florida

Florida has more than its share of high-technology industries and applications due to the National Aeronautical and Space Administration's program in and around Cape Canaveral. Space operations require very large emergy inputs from resources, energy, and engineering. Recently, some costs have been saved by using smaller, lower cost rockets to replace communication and weather satellites. More than 1000 times more emergy is required for a person in a space station than for the same person on the ground. Each year Congress questions the usefulness of putting people into space.

In considering the future of space programs, there is a conflict between those attracted to the science-fiction ideas of space ships, flying saucers, and Star Treks, and those possessing the scientific knowledge that there is not any place in space that is close enough to Earth and has the resources to support the people of Earth.

Questions and activities for chapter twenty-eight

1. Define: (a) culture, (b) information, (c) technology.
2. Visit the Kennedy Space Center. Discuss the pros and cons of the space program.
3. Discuss how you think the culture of our society is best transferred to new generations. For example, should everyone be encouraged to attend public schools to share a common culture?
4. Discuss how to finance education.
5. List the four most important ideas you have learned in school. Discuss how you learned each.
6. Discuss the idea that shared information can take the place of wars to solve international problems.
7. Draw an energy diagram of your school. Be sure to include the information flows.
8. In which of your classes do you learn the most? Discuss the best ways to organize teaching and learning.

part four

The Florida state system

Part IV considers the whole system of the state of Florida, its economy and people, starting with the resource basis (Chapter 29). The renewable energies of sun, rain, wind, waves, and tides are important resources that have attracted economic development. Development started with native American Indians building an environmental system of hunting, gathering, and simple agrarian techniques (Chapter 30).

The present Florida economy is examined in Chapter 31. It is now largely based on the inflows of bought fuels, goods, and services, paid for with the money earned from the sales of agricultural products, phosphate, timber, and income from tourists. Chapter 32 shows the way the state system is organized with networks of infrastructure. The future depends on present and alternative fuel sources and electric power, considered in Chapter 33. Chapter 34 analyzes the increasing efficiency of waste use and recycle.

Population and its support are discussed in Chapter 35. Directions for computer simulation of simple mini-models of future trends are in Chapter 36; the results clarify ideas about what scenarios are possible ahead. Chapter 37 looks at a world of decreasing energy and how to make that regime prosperous.

chapter twenty-nine

Resource basis for the Florida economy

Sometimes called the "carrying capacity", the amount of environmental and economic development in Florida depends on the resources available within the state plus those that can be brought in. The kinds of development depend on the kinds of resources available. Self-organizing systems develop unique work processes that correspond to the types of resources. In this chapter, we examine Florida's renewable sources, its stored reserves and the resources purchased from outside, and relate these to the circulation of money.

Every area of the earth has a different combination of energy flows. The main energy sources coming into Florida are shown in Figure 29.1, including those brought in by purchase with money. These sources are arranged from left to right in order of their quality; that is, in order of their transformities starting with sunlight on the left. The ones on the right require more solar emergy to generate them, so they have higher transformities. The joules flowing in are greatest on the left and least on the right. Florida's energies, their transformities and emergies, have been calculated in Appendix B, Table B1.

29.1 Renewable resources of Florida

Consider the renewable resources in the order they appear in Figure 29.1. There is much *direct sunlight* in Florida because of its low latitude and its relatively few clouds in winter and spring. Direct sunlight varies from winter to summer with the highest sunlight striking Florida in July (see Figure 12.3). The sunlight absorbed by the ocean evaporates water and generates the *winds* (see Chapter 12) that bring rains to the land.

Rain (see Chapter 13) has two main kinds of energy: its *potential energy* of height and its *chemical purity*. Most people do not think of water as an energy source except when it is elevated so as to drive water wheels and turbines. The potential energy of elevated water running downhill is minor in Florida because there are no high mountains to catch and hold water. Water also has purity, which is chemical potential energy relative to a salty condition. So,

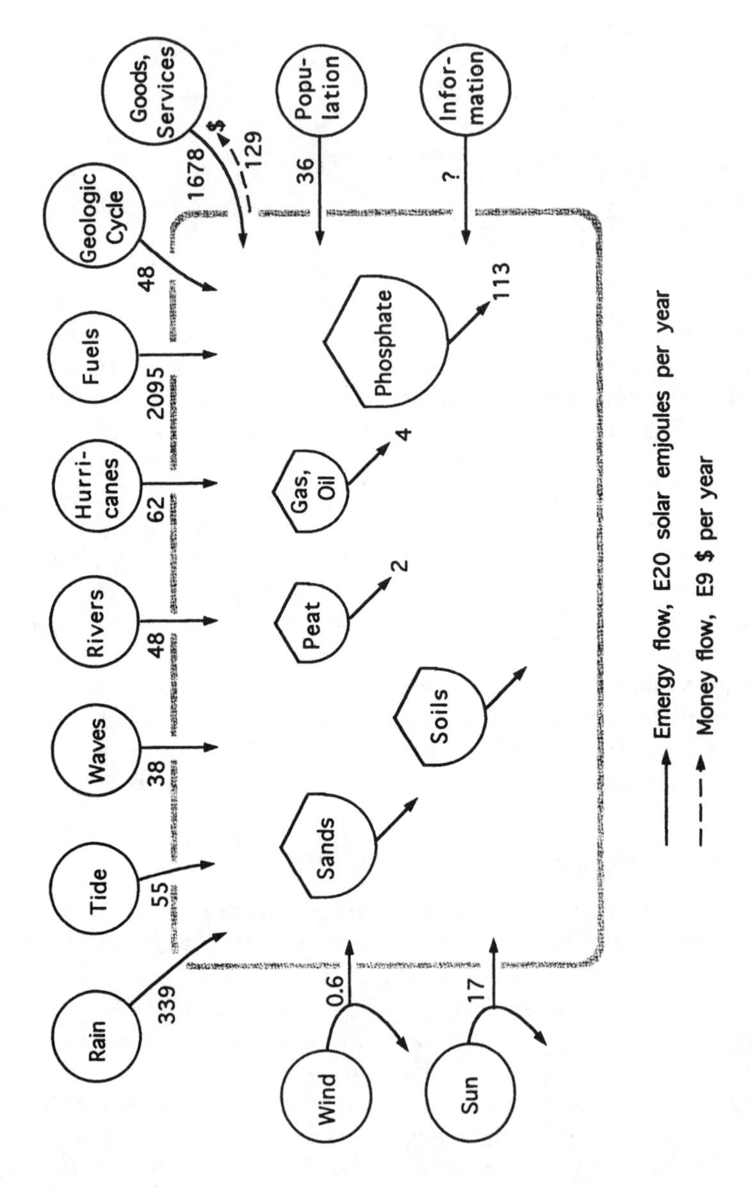

Figure 29.1 Annual use of resources in Florida expressed in emergy units, E20 solar emjoules (sej) per year. (See Appendix B, Table B1.)

freshwater can be used by vegetation and animals or to wash clothes, or it can be used for industrial processes. Freshwater reaching the estuaries drives currents (Chapter 14). Freshwater becomes more salty as it loses its chemical potential energy.

Florida's many coastal counties receive extensive marine energies of waves and tides. Wind-driven *waves* (see Chapter 14) are formed over the oceans. The beaches depend on the energy of waves breaking to maintain the beach form and to move sand back and forth. Beaches also depend on the winds, which blow sand up where dune vegetation can catch it. *Tides* (see Chapter 14) in Florida are not large, about 3 ft (0.9 m). As water rises and falls along the shore twice a day, the energy of the elevated water drives currents in and out of estuaries, helping to mix the systems of the rivers and the ocean. Tides help maintain productivity, fisheries, and geological processes, and they carry wastewaters out to sea.

A few *rivers* cross the boundaries from the north. There is a small hydroelectric plant on the Apalachicola River at the Jim Woodruff dam.

Although they do not come every year, *hurricanes* are intense concentrations of wind energy that can destroy cities (Figure 12.9). For vegetation that has adapted, there may be beneficial effects, such as the recycling of old trees and restoration of water reservoirs.

We take the land we live on for granted, but its existence requires continual *geologic uplift* (Chapter 13) to replace its gradual solution. As rains percolate through the land, they wash matter into streams and groundwaters, and then into the sea. Figure 29.1 includes the small geological emergy inputs associated with elevating the land. In Florida, this uplift is very slow, measured in millimeters per thousand years. The cycle of land uplift and erosion and solution in Florida is small compared to that in the Rocky Mountains.

29.1.1 Purchased fuels

Since the state has almost no fuels, they have primarily been brought in from outside. Natural gas used for heating and in power plants for making electricity comes by pipelines. Several power plants use coal brought in by rail and by marine barges pushed by tugboats. Other power plants use a heavy grade of oil brought in by tankers, mostly from Venezuela and Texas. Florida has five nuclear power plants: two at Turkey Point near Miami, two at St. Lucie, and one at Crystal River on the Gulf coast. They use nuclear fuel mined and concentrated outside the state. The generation of electricity accounts for the use of about half of the energy imported to the state. Electricity is easy to move about and use for many purposes (Chapter 33).

29.1.2 Goods and services

Many goods and services are purchased from sources outside the state, including cars and most things consumers buy in stores. Few consumer goods are produced within the state. In Figure 29.1 is an estimate of the emergy in these purchased goods and services.

29.1.3 Population and immigration

About 1000 people immigrate to Florida each day including retirees, young people from other states, and foreign immigrants. Some people are leaving. The net immigration is about 300 per day. Immigrants and tourists bring their emergy into the state in the form of education, money, activities, and possessions.

29.1.4 Information

Another emergy source, important to all parts of the state system, is information. We think of information as existing in the minds of people, in books, and on television, but plans of buildings and genes found in plants and animals are information, too. Information is in everything in Florida and comes into the state by many routes, such as in books, in the trained minds of immigrants and retirees, in newspapers and television, and in teachers brought in to teach new areas of knowledge. The actual energy in information is tiny, but the emergy (that required to generate, maintain, and duplicate the information) is large. Information flow also includes control actions from outside, such as market prices, federal government regulations, and new technology.

29.2 Stored resources reserves in Florida

The economy of humanity and nature runs on nonrenewable resources such as gas and oil as long as they last, as well as on the renewable sources that keep flowing in. The main resource reserves within Florida are shown as storage tanks with the rates of use in Figure 29.1 and are evaluated in Appendix B, Table B2. There are only small amounts of fuel reserves (peat, oil, gas) and mineral sands. Phosphate mining is a major contributor (see Chapter 25). Soil is nonrenewable; we use it up faster than it is being formed.

29.3 Resources and the circulation of money

The economy of Florida gets its real wealth from the various sources shown in Figure 29.1 and summarized in emergy units as 4369 E20 solar emjoules used per year (Figure 29.2). This wealth is distributed among people by the circulation of money. One measure of the amount of money circulating is the *gross economic product of the state,* which is roughly equal to the annual state income. The ratio of emergy to money circulation for 1992 is calculated in Figure 29.2. It indicates the real wealth (in emergy units) that a dollar can purchase.

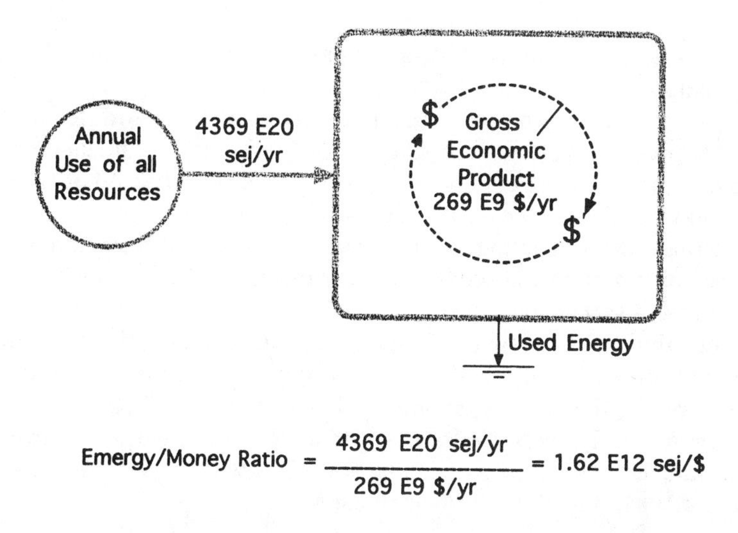

Figure 29.2 Summary of annual emergy use and gross economic product of Florida used to calculate the emergy/money ratio.

29.3.1 Emdollars

The emergy/money ratio can be used in a reverse way to estimate the dollars of gross economic product for which emergy is the basis. For example, if you have an emergy value for an environmental product such as wood, divide it by the emergy/money ratio to get the dollar equivalents. This dollar equivalent of emergy is called an *emdollar.*

Questions and activities for chapter twenty-nine

1. List five renewable energies that flow into Florida. For each one, give an example of its value to the economy.
2. List four nonrenewable resources found in Florida. For each one, give an example of its value to the economy.
3. Explain why the energy sources in Figure 29.1 that are farthest to the right have more emergy per unit energy than those on the left.
4. Use whatever light meter or solar energy measuring instrument that may be available to measure sunlight outside your building, inside, on the shaded forest floor, and during passing clouds. Estimate percentages of full daylight found in various familiar places.
5. Collect pictures from magazines that illustrate the processing or use of gas, oil, and coal, preferably in Florida. Explain the significance of each picture to the total system of this energy source.
6. Take a field trip to economic activities in your area that involve environmental inputs: agriculture, forestry, fisheries, mining.

7. Draw energy diagrams of the systems of environmental use visited in Question 6.
8. Assemble samples of peat, coal, phosphate rock, gypsum from phosphate processing, limestone, calcareous sand, quartz sand, and fossils. Relate these to the geology and geography of Florida. What environmental problems come with each?
9. Calculate the percentage of Florida's total emergy that comes from renewable sources. Discuss what this means to the economy as nonrenewable sources get used up.
10. Determine the emdollar equivalent of one of the environmental resource inputs in Figure 29.1. (Divide the annual emergy use of that resource by the emergy/money ratio from Figure 29.2).
11. Appendix B, Table B2, contains estimates of some resource reserves in Florida. Use the emergy/money ratio to estimate the emdollars of stored wealth in the unmined phosphates.

chapter thirty

Early human society in Florida

The first human societies to develop using the environmental resources of Florida were the American Indians immigrating before the end of the last ice age, more than 12,000 years ago. They hunted wildlife, harvested shellfish from the estuaries, and later cultivated gardens. But, starting in 1500, Indian societies were displaced by the European colonial system. This chapter explains the environmental basis of the early society and how it was replaced.

Information about the early Indian system of Florida comes mainly from archeological studies. *Archeology* is the study of remains of past societies. By studying shell mounds, grave sites, remnants of hunting implements in springs, and the written accounts of early Spanish explorers, archeologists have outlined many aspects of the history of Indians in Florida. Figure 30.1 summarizes the sequence of societies inhabiting Florida.

30.1 The Indian landscape

Many of the resources of the Florida landscape in Figure 29.1 supported the Indian society. Indian cultures reinforced the landscape pattern with fire, hunting practices, and early ways of agriculture. The native Indian practices may have helped develop diversity with clearing, patch-burning, spreading of seeds, and shifting agriculture.

30.1.1 Hunting and gathering society

During Pleistocene times (1 million to 9000 years ago) Florida was colder and, like the rest of North America, was the habitat for many large animals now extinct in America. Human remains, along with those of extinct animals of the ice age (about 10,000 years ago) have been found at the Cutler fossil site near Miami. They show that the early human inhabitants lived in close contact with mammoths, bison, lions, giant sloths, short-faced bears, forest

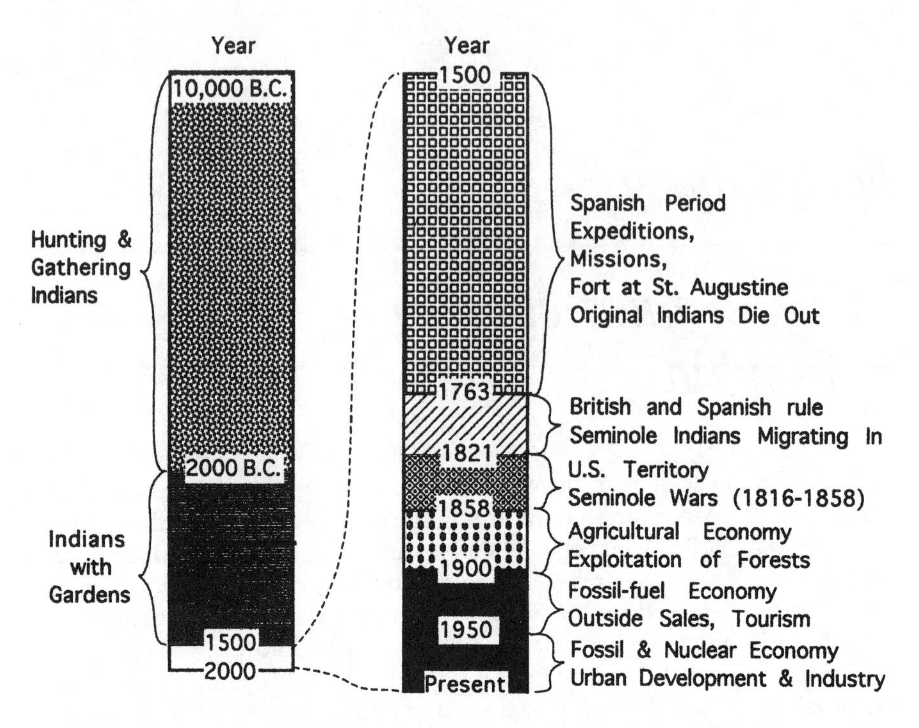

Figure 30.1 Chronological record of human societies in Florida.

peccaries, wolves as big as Shetland ponies, and about 50 other animal species now extinct.

The native Indian cultures at this time were hunters and gatherers and left no evidence of permanent settlements. Food, clothing, and tools were obtained directly from the environment. They hunted animals, consumed the meat, and utilized other parts such as bones and fur for tools and clothing. Plant materials such as roots, fruits, and nuts were gathered from the surrounding landscape and added to their diet. The economy of these early cultures was based solely on the renewable energy flows that were concentrated in wild plants and animals.

The climate and vegetation changed. For example, pollen records in lake sediments show cypress moving into Florida about 4000 years ago. By 2000 B.C. (B.C. means "before Christ"), the large game animals were gone. The native Indian system that developed was more complex. Some tribes had long-standing villages at the margins of the west coast estuaries where there were reliable supplies of shellfish, shrimp, and fish supported by the marshes and grassflats. The sites of these villages are marked, even now, by *shell middens.* Middens are shell piles accumulated over a long period. In a way, they could be called primitive garbage dumps.

30.1.2 Agricultural society

Sometime after the last ice age, agriculture started with gardens of primitive corn (maize) and other products that supplemented the hunting and gathering. The domestication of grain and other foodstuffs led to a more permanent settlement pattern. Indian populations grew as a result of the ability to cultivate crops and increase the productivity of the landscape. Also important was the transmission of information through migration and the trading of tools and arrowheads. Rare shells and stones were used in trading as a simple form of money.

Figure 30.2a shows the simple relationship of the pre-colonial Indians to their environment. The economy was based on energy from the sun, growing plants, and the food chain of animals. These early Indians occupied the top of a food chain based on solar energy. Just as we have seen with the energy transformation in natural systems (Chapter 6), the amount of top consumers that can be supported by solar-based food chains is limited by the productivity of the landscape. A large population could not be supported in any one place.

At the time of the Spanish explorer, Ponce de Leon (1513), and Spanish colonization of St. Augustine, the Indian population of Florida was estimated as 100,000. Four tribal groups have been recognized, each with some differences in the relics they left from their ways of life. Figure 30.3 shows the Apalachee in the panhandle, Timucuan in northern peninsular Florida, and Tequestas in southeast Florida. Calusa Indians in southwest Florida had affinities with Cuba and traveled at sea in marine canoes.

The Spanish conquistadors and other European settlers brought diseases to which the Indians had not developed immunity, and their populations were reduced by epidemics, in addition to slavery and war. By the time the Spanish gave up Florida to the British in 1763, most of the original Indian people were gone or had been assimilated by invaders.

A new wave of Indians, however, moved in from the north. The expansion of the American colonies with settlers and protective soldiers moving west displaced the Creek Indians from Georgia, Alabama, and the Carolinas. They began to move into Florida a hundred years before the American colonists, with a population of about 5000. Called *Seminole Indians,* they occupied Florida for several generations prior to the American colonists. This was long enough to develop characteristic ways of life adapted to the Florida landscape. There were also settlements of runaway slaves. The early Seminoles occupied north and central Florida, but were pushed south into the Everglades during the Seminole wars (1821 to 1845) by military campaigns of the U.S. government, who sought to control them. Although many Seminoles were moved to reservations in Oklahoma, a substantial population (about 400 people) remained in the Everglades without being conquered, becoming a proud part of Florida's varied peoples.

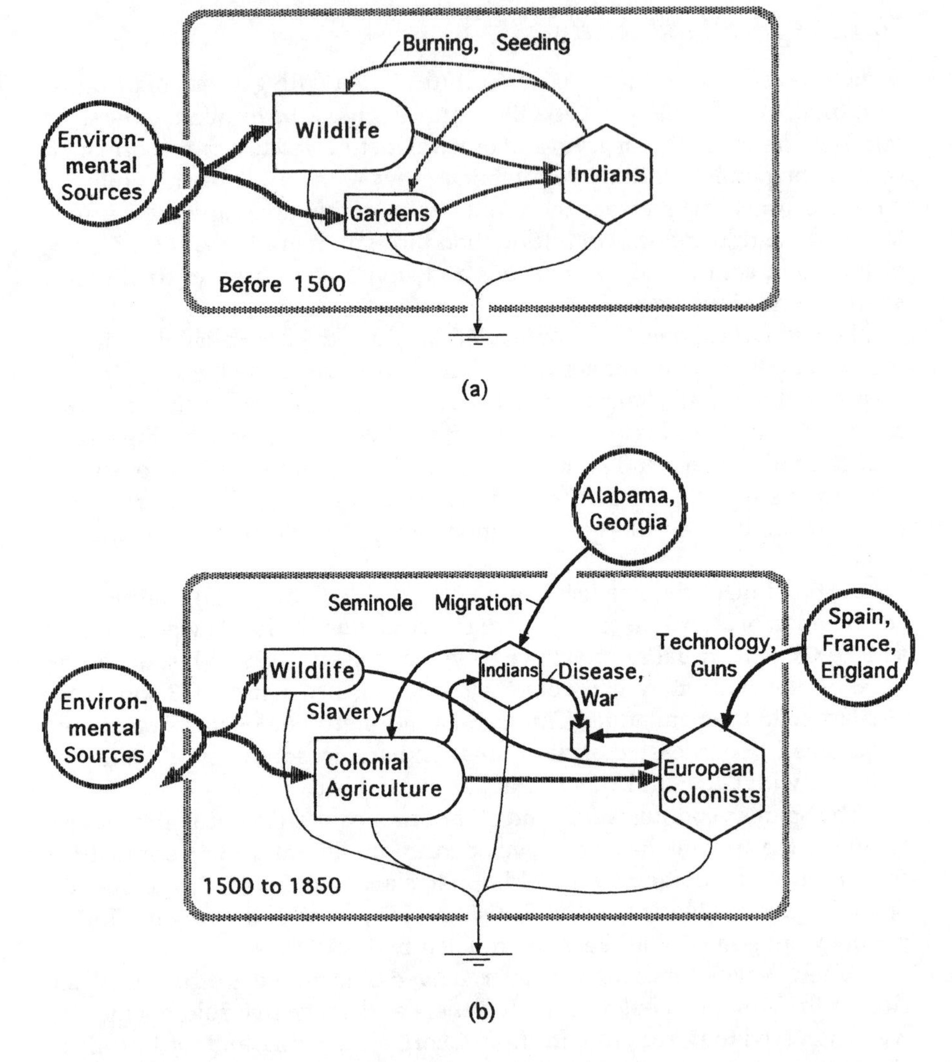

Figure 30.2 Diagram of the early Indian system in Florida and the displacement by European colonists. (a) Pre-colonial Indian system; (b) displacement of the Indian system by the colonial economy.

30.2 Colonial economy of Florida

Following the Spanish colonization in the 1500s, there was a competitive struggle for the resources of the landscape. With guns, the colonists could control the Indians. The Spanish colonists developed missions that used the Indians as workers on agricultural land tracts. Indians saw the new inhabitants reducing the game that was once more plentiful, cutting

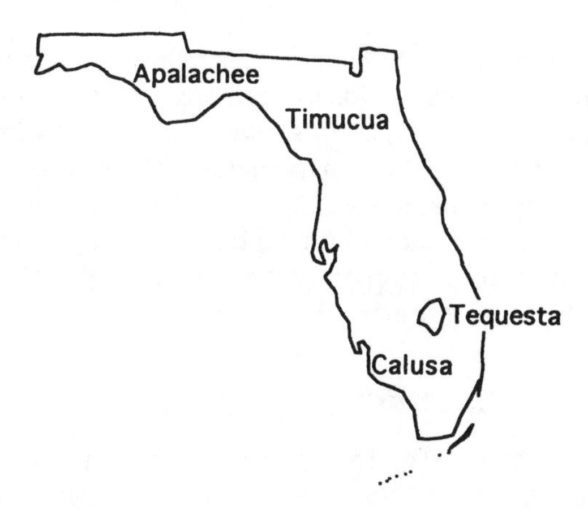

Figure 30.3 Indian tribal territories in 1500.

the forests to make homes, and clearing the land to cultivate crops. With technologies such as axes and plows brought from the "old world", the new culture's larger populations controlled the lands and displaced the Indian way of life.

30.2.1 Systems view of cultural competition

The diagram in Figure 30.2b summarizes the relationships between the Spanish and Indian cultures in conflict and competition for the same resources. While the Indian cultures were based primarily on the flow of local resources, the invading culture also received emergy, technology, and information from more highly developed economies of Spain, France, England, and later from the American colonies to the north.

The American Indian system lost out to Spanish conquistadors and later colonists because of their futile war against guns, disease, slavery, and better organization and because the wildlife system was depleted by, and lands replaced with, colonial agriculture. The more energy-intensive culture — that is, the one using more emergy — displaced the lower-energy culture.

30.2.2 Nonrenewable economy

The new growth economy of Florida was exploitive, not concerned with keeping resources sustainable. Environmental products were used up and then people shifted to use new resources. For example, the pencil industry at Cedar Keys stripped the region of the cedar trees without provision for replacement, so the industry eventually closed. The pattern of seeking short-run gain from environmental resources is still evident in Florida's growth economy.

We can learn from past cultures and their relationship to the landscape, perhaps developing a better symbiosis of humanity and nature. In the quest for a more sustainable economy, people often think of early Indian cultures as a good example of the way to manage nature. The midden records show long periods of shellfish consumption. Whether tribal stability in relation to environment existed in Florida remains to be learned from archeology and paleoecology (study of past ecology). What we know of the history is that it had invasions, population pulses, and change.

Questions and activities for chapter thirty

1. Define: (a) hunting and gathering, (b) carrying capacity, (c) old world, (d) archeology, (e) shell midden, (f) nonrenewable resource, (g) paleoecology.
2. Divide the class into groups, each studying and making a diagram of the details in the life of pre-industrial society including the Florida Indians and other examples: the Mayans in Mexico, the Incas in South America, England in the 1700s, early U.S. settlements, or early African communities.
3. Discuss three major differences between early native American Indian society and modern society.
4. What were some of the consequences of early native Americans' learning to cultivate land and grow crops?
5. How did the European colonies out-compete the native tribes for energy and resources?
6. Point out three ways that early societies organized the landscape.
7. Discuss the importance of imports and immigration to the European colonies. How did these inflows of energy change Florida?
8. Look at a map of your local area and search for names with an Indian origin. Try to find the early meaning for the name. (For instance, Okeechobee means "big water".) Search for other signs of early Indian influence in your area.
9. Research and discuss the native American population and communities in Florida now. What are some of their problems and possible solutions?
10. Sometimes early Indian societies maintained their territories with local intimidation and battles. How does this compare with the way countries have maintained their territories in the twentieth century?
11. With the Indian-colonist competition in mind (Figure 30.2b), and using the methods of model building and simulation in Chapter 9 and the diagram in Figure 30.4, write a computer program of the two competing systems. (In Figure 30.4, A is the natural energy inflow; B, the inflow from the old world; Q, the Indian system; and P, the colonial economy.) Run the simulation with different starting conditions. Vary input B starting at zero.

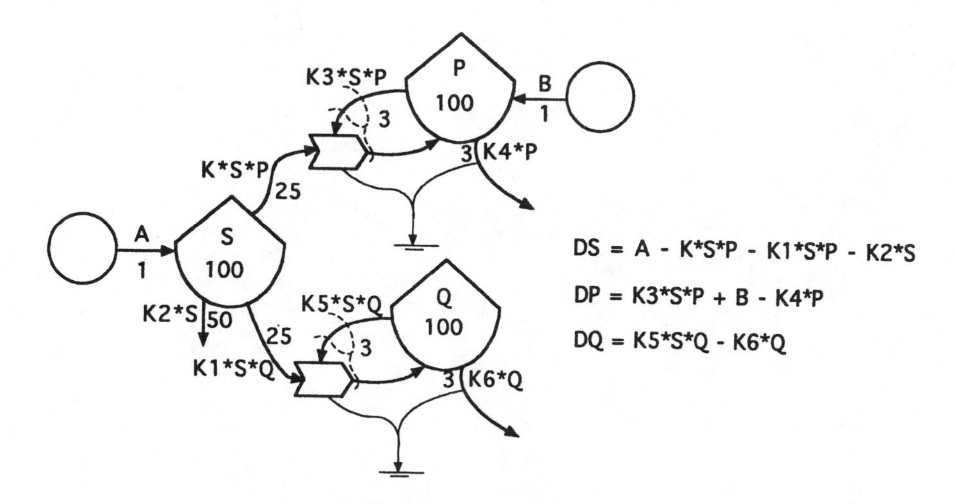

Figure 30.4 Suggestion for Question 11: a simple model for simulation of competition where one component is aided from outside.

chapter thirty-one

Overview of the Florida economy

The combination of resources available to Florida first produced a pattern of nature and Indian culture directly and indirectly based on renewable solar energy (Figure 30.2a). Now, however, 500 years later, a state system has evolved, dominated by a diversified urban-oriented economy that is more dependent on inputs from the outside world than upon its local resources. In this chapter, we present an overview of the Florida state system and the way it works.

31.1 Environmental-economic system of Florida

The system of the state of Florida contains the human economy and its connections to the environmental systems that we have studied in pieces in earlier chapters. An overview based on an emergy evaluation of the main parts and processes in Florida (Table 31.1) is shown in Figure 31.1. The diagram summarizes main storages and the inputs and outputs crossing the state line.

On the far left, the renewable energies — sun wind, rain, tide, and waves — flow into the state, supporting forests, lakes, estuaries, agriculture, and human settlements. Within the boundary (on the left) are the stored reserves of natural resources: soils, peat, limestone, phosphate, and sands. Because we currently use these faster than they are reassembled by nature, they are called nonrenewable (they really should be called "slowly renewable"). Some resources, mainly phosphate, are mined but not used much in Florida. Most are sold and exported out of state, carrying their wealth to others.

Purchased inputs are shown coming in from outside economies from the upper right. Especially important are the fuels, goods such as automobiles, and many kinds of services that are part of what goes into the items we buy from out of state.

Table 31.1 Summary of the Resources Supporting the Florida Economy (Data from 1992–1995[a])

Resources	Annual Solar Emergy Use (E20 sej/yr)
Sources within Florida	
Renewable sources[b]	441
Nonrenewable sources	118
Total	559
Sources from outside Florida	
Goods and services	1678
Fuels	2095
Population in-migration	36
Total	3809
Total resources used by the Florida system	4369
Exports	
Products and services	3460
Resource exported without use (phosphate)	1018
Total	4478

[a] For details, see Appendix B, Table B1.

[b] Renewable sources are the sum of the emergy of rain, tide, and river inflow.

The storage tank of "Florida Assets" represents all the wealth of nature *and* human developments. Maintaining those assets requires continual use of resources in production processes to keep up with the unavoidable depreciation.

Money is shown with dashed lines. Money to buy items from out of state comes into the state from several sources: from investors, from transfer of money from the federal government (money for military bases, federal projects, social security, and retirement), and especially from sales and services delivered to tourists and to people out of state.

The money is shown circulating within the state, as well as coming in and going out. The total money circulating each year through the production part of the economy is called the *gross state product*. The money that is paid to productive industries for their products circulates back to pay people for their labor.

Most of the dollars circulating as gross state product are at the same time the income of all the human workers who are buying and consuming the products. Notice that the circulation of money is only from human producers to human consumers. Money is not paid to the natural energy sources and ecosystems.

Figure 31.1 shows what makes wealth and prosperity in Florida. Anything that increases the amounts of resources being used productively increases

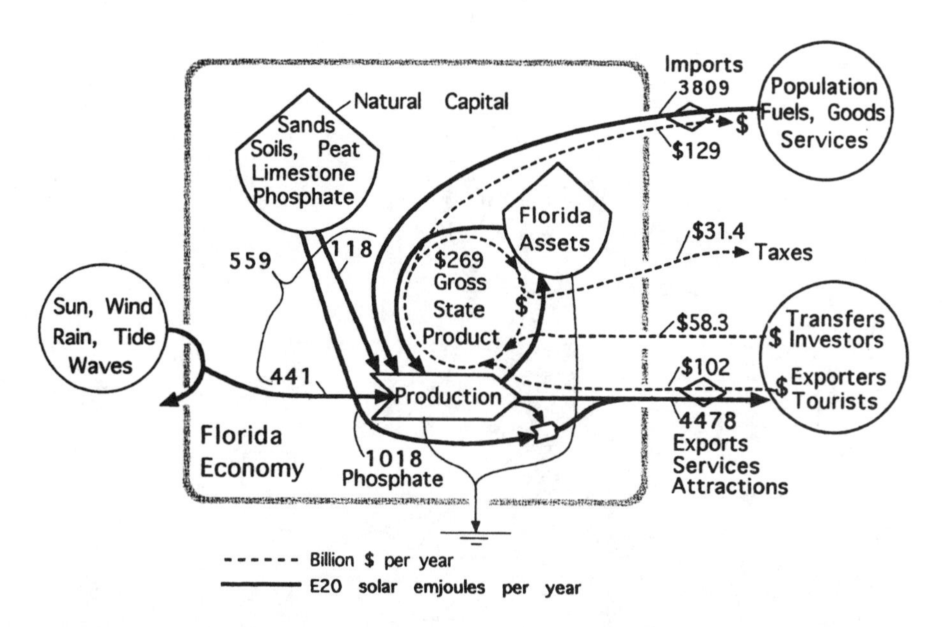

Figure 31.1 An overview of the economy of Florida and its environmental resource basis with annual emergy flows from Table 31.1. Dashed lines are the flows of money from Appendix B, Table B1.

the wealth of the economy. For example, if prices of foreign products drop, such as oil, then Florida can buy and use more for the same money. Then, there is more real wealth to use per person.

31.1.1 *More detailed view of Florida*

In Figure 31.2, Florida's system has been expanded to show the parts in more detail. On the left are the environmental and rural systems that receive the environmental inflow of sun, wind, rain, and geologic processes. These are areas of plant photosynthesis, either naturally growing vegetation or crops raised for harvest. On the right are the consumer sectors, mainly located in the towns and cities.

The coastal areas also receive waves and tides that generate many values in estuaries and beaches. With abundant rains there are wetlands, lakes, and waterways that are increasingly occupied with urban uses. Electric power use is high because of air-conditioning. Electricity is supplied by power plants that use nuclear, coal, oil, and gas fuels. Population inflows generate urban centers with commerce and industries. Some properties are unique, such as the phosphate industry, orange groves, vegetable crops, and tourism. Notice in the diagram how these are interrelated. Energy flows out in wastes and exports, and used energy dissipates as waste heat.

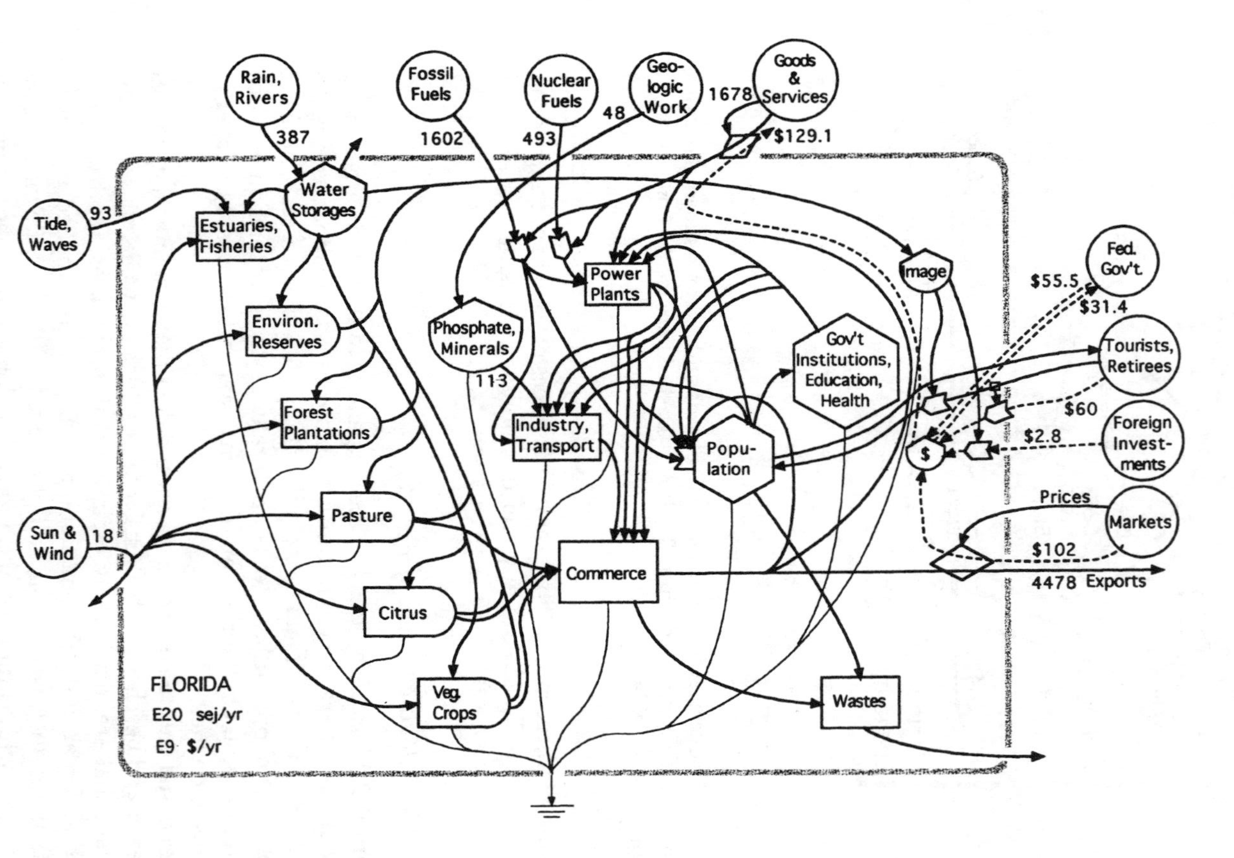

Figure 31.2 Systems diagram of Florida showing the main sectors of environment and society with annual emergy and money flows from Appendix B, Table B1.

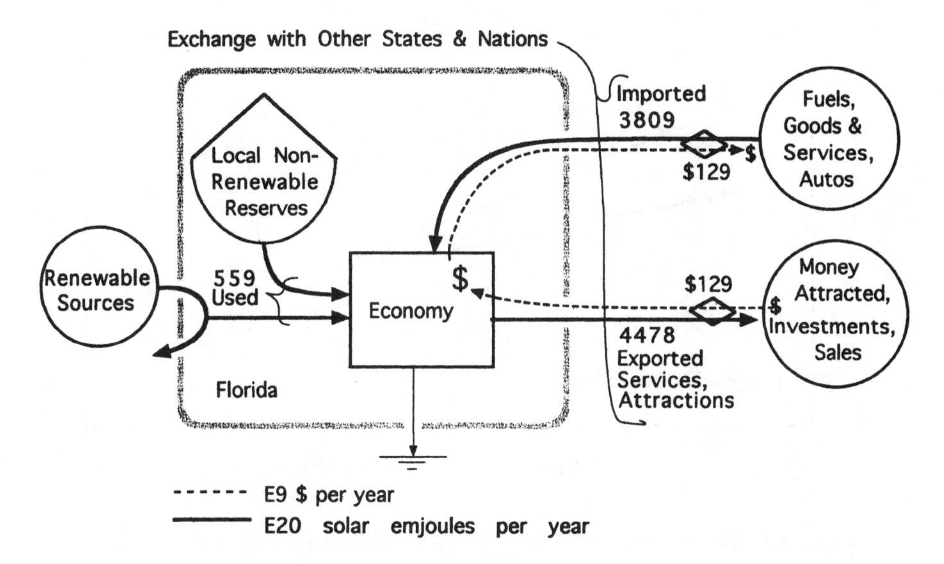

Figure 31.3 Diagram summarizing exchange with other states and nations showing the way the environment is essential for economic sales, services, and attractions.

31.1.2 Carrying capacity

Figure 31.3 summarizes the basis for Florida's environment and economy (using emergy evaluations from Table 31.1 and Figures 31.1 and 31.2). You can see that most of our wealth comes from imported fuels, goods, and services (3809 E20 sej/yr). If we had to run on the local environment alone (559 E20 sej/yr), we could support only a small percentage (about 15%) of our present population in the style typical of Floridians in the mid-1990s.

31.1.3 Attraction and image

People from outside Florida have a mental picture of what the state is like, such as oranges, warm temperatures, beaches, green wetlands, fishing, the space program, Disney World, and other theme parks. These mental pictures constitute the state's *image,* which influences people to visit or immigrate to Florida. The diagram in Figure 31.2 shows that the storage of image (in the minds of people thinking about Florida) is sustained by inputs of information from all the aspects of Florida, including the rural and environmental systems such as forests, sea coasts, farms, facilities for visitors, and information centers.

In Figure 31.2, the storage of image acts to increase the inflowing tourists and immigrants. The better the image of a state, the more attraction it has for people. Much of the southern part of the U.S., sometimes called the "sun belt", has a good image and is attracting more and more people. Coastal states not only have forests and rural areas, but also additional natural

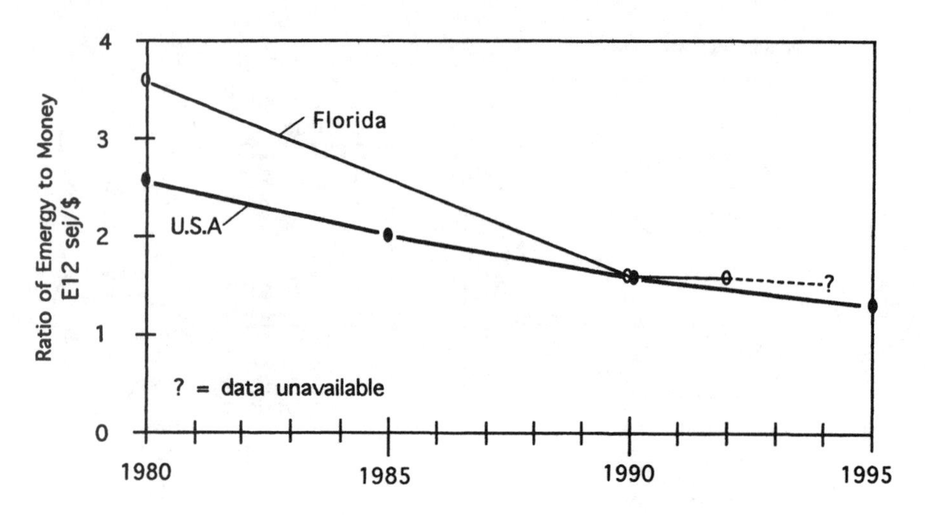

Figure 31.4 Recent trend of emergy-money ratios in Florida and the U.S.

resources such as estuaries and beaches that give them better images than some other inland states. Good image also attracts investment of money from outside sources. States with good images attract industries that are expanding or relocating.

As the attracted people, industry, and commerce add to the economy of a state, the image may decline. Crowding in cities, crime, the development of open space, and the loss of environmental values such as clean water and air cause the image to weaken. Other areas with higher images can draw tourists and investment away from those with a declining image.

31.1.4 *Emergy/money ratio*

As explained in Figure 29.2, the emergy/money ratio is the average quantity of emergy used for each dollar circulated. Each year the emergy/money ratio gets smaller (Figure 31.4), which means that a dollar buys less each year. This inflation is partly due to processing more money for the same emergy as the state becomes more urban. Earlier Florida had more real value for the dollar, but now it is about the same as the rest of the U.S.

31.2 *Comparisons between Florida and other areas*

Characteristics of the Florida system may be compared with other states and nations to help understand Florida's part in the world economy. A comparison of some emergy/money ratios in different states shows a wide range of values with rural states, Alaska and Maine, having the highest value (Table 31.2). The higher the ratio, the more emergy (wealth) is bought with a dollar. In rural areas, less money circulates, and people obtain many natural resources

Table 31.2 Emergy/Money Ratios

	E12 sej/$
Alaska	23.0
Maine	4.8
Texas	2.6
Florida	1.6
U.S.	1.4

(more emergy) directly from the environment (e.g., salmon, wood, wildlife, clean water). Wages are less. Consequently, a dollar buys more; however, costs are still high because more fuels and clothing have to be used during the long, cold winters.

Because Florida has become urban, with most people in cities, nearly everything has to be bought outside the city and transported in. It takes more money to live. More money circulates for the same wealth. A dollar buys less emergy in Florida than in less developed countries (Table 31.3). American tourists can buy goods very cheaply in Brazil, where the emergy/money ratio is much higher than that of the U.S. When Brazilians come to Florida, they find the prices of things very high.

31.3 Florida's trade

The exchange between Florida and other states and nations is summarized in Figure 31.3. The total circulation of money back and forth across the state line — including purchases and sales, transfers from the federal government, and

Table 31.3 Comparison of Emergy/ Money Ratios Among Nations

State or Nation	Emergy/Money (E12 sej/$)
Ecuador	8.7
China	8.7
Brazil	8.4
India	6.4
World	4.0
Cuba	3.5
Mexico	3.3
Italy	1.5
Florida	1.6
U.S.	1.4
Japan	1.4
Switzerland	0.5

financial transactions such as stock market earnings — is assumed to balance approximately ($129 billion per year, from Appendix B, Table B1). The sales figure for exports in Figure 31.3 came from general sales of oranges, tourist services, and vegetables ($23 billion); phosphate sales ($0.5 billion); and an estimate of sales of illegal drugs ($19 billion).

The real wealth exchange is uneven, however, mainly because of the export of phosphate — more real wealth goes out than comes in each year. The money received for phosphate is only for the work of mining and processing, and the emergy it can buy is much less than the emergy that nature contributed when it concentrated the phosphorus. Using the phosphate in state agriculture generates more real wealth than is obtained from selling it abroad. Not counting phosphate, Florida imports more emergy (especially in fuel) than is exported.

When a dollar circulates between Florida and another country as part of trade or investing, the benefit to Florida depends on the emergy/money ratios (Table 31.3). For example, if Brazil borrows a million dollars from a Miami Bank, this borrowed money has the buying power of the Florida emergy/money ratio of 1.6 E12 sej/$. Later, if Brazil pays back the loan with its currency converted to U.S. dollars, it gives the bank the buying power of its emergy money ratio of 8.4 E12 sej/$, 5.3 times as much. This is like paying 530% interest, in addition to the promised interest. As long as money is used as the basis for trade, less developed countries unfairly support developed areas such as Florida.

31.3.1 International influences

Not long ago, most of the income, population increases, tourists, and stimulus to the economy were obtained from other states to the north. In the last decades, however, there has been an increase in the roles of other countries. Cuban immigration that followed the Castro revolution as well as earlier settlements in Tampa gave south Florida Spanish-speaking populations. French-speaking immigrants came from Haiti. Miami is an international city serving as a trade center for a region much larger than south Florida, including Colombia, Venezuela, and Brazil. Many tourists are attracted from Europe. Although a reliable estimate cannot be found, many suspect that illicit drug trade through Florida has had a significant effect on the economy of south Florida.

31.3.2 Resources for the future

The economy of Florida (and that of most of the rest of the world) is presently dependent on fossil fuels. The future depends on the ability of foreign exchange to pay for continued imports of coal, gas, and oil. There is much uncertainty over when these fuels will begin to be less available and more costly. There is controversy over the usefulness of alternative energy sources and the future of Florida, as discussed in later chapters.

Questions and activities for chapter thirty-one

1. Define: (a) gross state product, (b) assets, (c) image, (d) emergy/money ratio.
2. Look for clippings in newspapers and current magazines concerning Florida's economy. Discuss the way the news items fit or do not fit the concepts given in the chapter as to the way the economy of Florida works. What are the current trends?
3. Discuss Florida's recent economic trends compared to the rest of the country. A good measure is the unemployment rate — is ours higher or lower? Speculate as to why.
4. From Figure 31.1 or Table 31.1, calculate what percent of total emergy use is from outside the state. From this measure, what can you say about Florida's self-sufficiency?
5. From Table 31.1, calculate what percent of the total emergy use is from the use of nonrenewable resources. How does this compare to your answer to Question 4? Why? In which states are there more nonrenewable resources within the state?
6. What percent of Florida's fuel budget would be cut off in the event of a war that stopped shipments from outside the country? (You'll have to research to find out how much we import from other countries.)
7. What percent of the total emergy budget is the flow of fuels from outside the state? What does this mean for our future?
8. What is happening to the image that people in other countries and other states have of Florida? What changes may affect that image? How will these changes affect growth? How will increasing population density affect the image?
9. When did your family move to Florida? What attracted you or your parents to this state? List the things that made Florida attractive to your family.
10. Compare Florida's emergy/money ratios for 1980 and 1992. What does the difference mean in the economy? What factors may have caused this change?
11. Compare the emergy/money ratios of the states in Table 31.2. What does this mean for investment in Florida compared to the other states?
12. In considering international trade, explain the difference between the balance of payments of *money* and the *emergy* flows they pay for. Give a Florida example.
13. In an even dollar trade between Florida and Ecuador, which country gets the most real wealth? Explain.
14. Locate a speaker with a different point of view about the future and limitations to growth. Discuss the points of difference.

chapter thirty-two

Florida's networks

Spreading over the landscape of Florida, the economy is organized by many networks. Before European colonization, people were mainly connected by waterways (Figure 1.2) and wildlife trails. Now the panorama is organized by dense networks of roads, railroads, pipe lines, and power lines joining towns and cities. The state is intricately controlled by telephone, radio, television, and the Internet, information networks that connect units of production, consumption, government, and business. In this chapter, we seek the understanding that comes from map views

32.1 Hierarchy of population centers

In Chapter 7 (on energy webs) and Chapter 27 (on cities) you learned that patterns of nature and those of the human economy were organized hierarchically in space. Rural areas converge products towards towns, and towns converge to cities, which are the hierarchical centers. When the landscape is healthy, these centers send back out to the towns useful services and high-quality products, and the towns do the same to the rural areas. Road maps show the way the towns connect to the cities and the inter-city highways. Aircraft and trucks converge on Miami, Orlando, Tampa, and Jacksonville, but they then turn around and diverge back outward to the smaller places. The closer you are to the high-emergy hierarchical centers, the more resources may be justified for transportation.

Cities first formed at river mouths or other places where more resources were spatially concentrated. Figure 32.1 shows Florida's big city centers of population that developed and the subordinate towns around them.

32.2 Transportation

Florida is covered with a dense network of roads, railroads, airports, boat channels, bikeways, and walkways — the means for transportation. Whereas the earlier Indian inhabitants used streams, wildlife trails, and beaches for

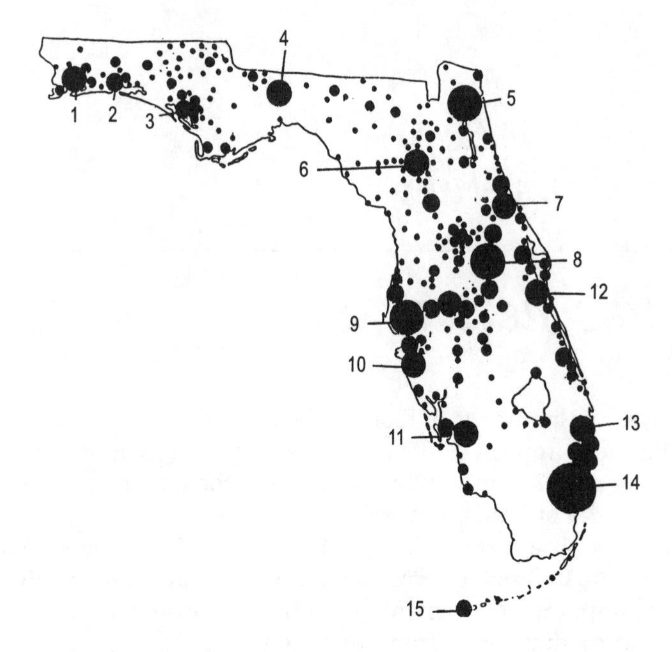

Figure 32.1 Map of Florida showing the hierarchy of cities and their location: 1, Pensacola; 2, Fort Walton Beach; 3, Panama City; 4, Tallahassee; 5, Jacksonville; 6, Gainesville; 7, Daytona Beach; 8, Orlando; 9, Tampa; 10, St. Petersburg; 11, Fort Myers; 12, Melbourne; 13, West Palm Beach; 14, Miami; 15, Key West.

their movements, modern society has used the abundant fossil fuels to build more elaborate transportation systems. Roads, bridges, rails, etc. are called the *infrastructure*. Infrastructure lasts about 50 years, which means that highways generally have to be replaced every half century. An energy systems diagram of the highway transportation system is given in Figure 32.2. Note that major contributions of emergy are for the liquid motor fuels used, the highway construction, the cars, and the people driving the cars.

The wealth stored in highways is much greater than would be inferred from the money spent in building and maintaining the highways. Money is only paid for the human service-labor contributions to road building, whereas very large quantities of emergy are stored in roads and railroads in the steel, concrete, and asphalt. Most of these materials were put in when the resources were abundant and cheap. It will be much more expensive in the future to replace and repair the infrastructure.

During our recent centuries of growth, building infrastructure was often the first step in development. It was a way to help the fossil fuels contribute to the economy. Because fuels had a high net emergy and were inexpensive, our whole pattern of life was influenced by this cheap transportation. Life in the U.S. became organized first around boats, then trains, and now around automobiles and highways. In the rush to make transportation efficient,

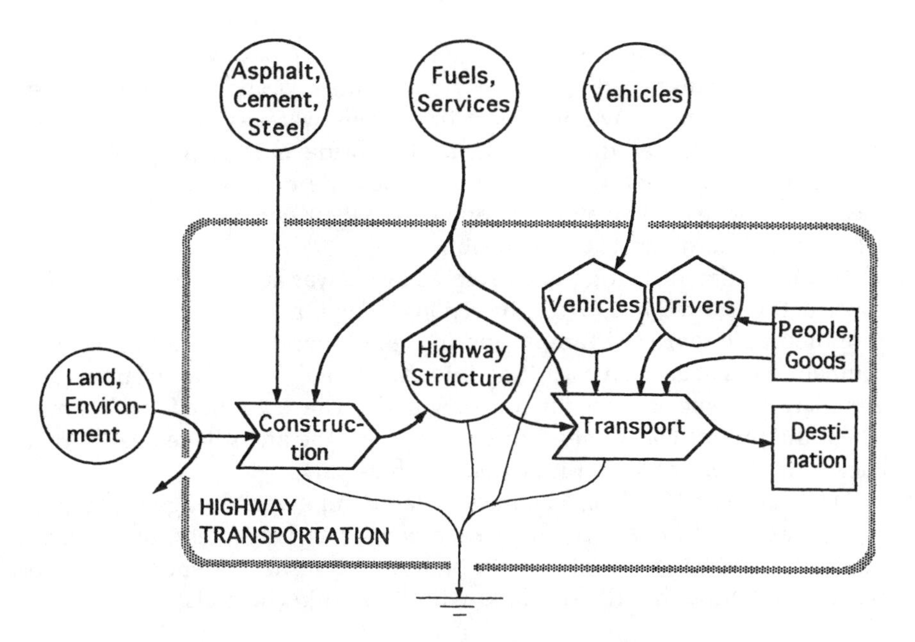

Figure 32.2 Systems diagram of highway transportation.

interstate highways were sometimes built right through the heart of cities, often destroying the patterns of city function that previously worked well.

Transportation has had major impacts on environmental systems because of the displaced land, rerouted water flows, blocked movements of wildlife, air pollution from exhaust gases, and piles of old cars. In recent years, public policies have improved conditions by eliminating lead from gasoline, and by requiring car exhausts to release less waste. The biggest waste is using larger cars than necessary, carrying extra weight back and forth, using up materials, fuels, and road surfaces.

32.2.1 Ships and barges

Boats were used first because they were able to pass through waters with minimal friction; they used wind-sail energy, tides, and currents; and they used the natural hierarchy of converging waters in the streams of the pioneer landscape. Steamboats were used on small rivers and lakes in Florida at the turn of the century.

An inland waterway was built by the federal government along the entire coastline of the U.S., mostly in the estuaries behind the barrier islands. The inland waterway was a 12-ft deep channel a few hundred feet wide, intended for smaller boats and small barges not suitable for the open sea with its frequent storm waves. The inland waterway cut through many of the natural patterns of tidal and river exchange, changing salinities, often reducing circulation when the dredged sediments (spoil) were left in small islands in the estuaries. But, this was many years ago, and natural processes have self-organized new patterns to fit the new water system. Exotic Australian pines colonized the spoil islands. In Florida, most of the boats on the inland waterway have ended up being pleasure craft. Concentrations of these boats have made waters eutrophic from their sewage. Speedboats also kill manatees and cause shore erosion.

Construction of a barge canal across Florida was authorized in 1942, and construction started in 1964, with sections being built on the west end of its proposed path, north of Tampa, and in central Florida. Because of its environmental impacts, such as cutting into the normal groundwater flows and eliminating scenic streams for canoeing and because the need for it diminished, the project was finally canceled in 1986. The areas have been turned back to the counties for multiple uses such as parks.

Harbors with 38-ft deep channels for large ships were dredged for the big cities, through Tampa Bay, at Jacksonville, Everglades City, and Miami. These channels have seawater on the bottom and fresher water on top. Saltier waters were brought farther inland with these ship channels.

32.2.2 Transportation alternatives

It is sometimes said that time is energy. By having abundant facilities and fuel for transportation, time of people is saved, and when the saved time is applied to productive purpose, the economy is enriched. However, when the price of fuels was high during the Arab oil embargo of 1973 (due to a temporary price-raising monopoly), efforts were made to save fuel by reducing speed limits. There were fuel savings, less accidents, and less wear on the highways, but more emergy was lost in work time because more time was spent on the highways.

Whether calculated as energy or emergy, the resources used per mile of transportation vary enormously. The least energy is used with walking and bicycles, and the most energy is expended by aircraft. As the intense noise of jet planes might indicate, airports are among the places with most concentrated energy use. As fuels for flying and for manufacturing aircraft become scarce and more costly, questions may be increasingly raised: does this trip contribute as much emergy to society as it takes away from the economy? Compared to traditional cars, electric cars are inefficient because three quarters of the fuel is necessarily lost converting fuel to electricity, increasing pollution.

As fuels have become more expensive, communication is replacing transportation. People are using phones, faxes, conference calls, and computer networks as means to work together without actually getting together. In planning for the future, we may need to make more emergy evaluations of transportation alternatives to see which have the best net benefit.

Other incentives and regulations may be required to make society more efficient and thus generate more real wealth for the same resource use.

32.2.3 Ecological engineering of transportation

Gradually, over the years, practices are developing to make Florida's highways and railroads contribute more as aesthetic greenbelts, sound and pollution screens, and wildlife corridors. Plantings and natural vegetation are being left more and more on the edges of highway rights-of-way for wildlife to travel along. Underpasses under Alligator Alley across the Everglades have been successful in keeping panthers and many other animals from being killed by cars. Fewer pesticides are being used in the fertile ditches so that fishes and wading birds can prosper there. Highways through wetlands are built up high to keep them from draining the wetland.

32.2.4 Florida and international networks

International trade is a major part of Florida, especially through Miami, which is an international city to which much of South and Central America converges. As long as fuels for transportation are inexpensive, international trade increases, with each area of the world specializing in the work most suitable to its climate, resources, and culture. Later, if fuel costs are much higher, it will not make as much sense to ship steel and cars back and forth across oceans. States such as Florida may become more self-sufficient, with more kinds of products aimed at more local markets.

32.2.5 Energy supply networks in Florida

Shown in Figure 32.3 is the network of main electric power transmission lines and power plants. The state is connected together in an "electrical grid" so that the demand for electricity can be met. When the system is interconnected in this manner, areas that have a surplus of electricity at any given time can send their surplus to areas that may be having a shortage. Brownouts and blackouts can be avoided in this way.

These interconnections, though, sometimes cause widespread power failures. When an overload in one plant causes it to cut off to prevent damage to equipment, the next plant becomes overloaded and cuts off and then the next. Loads are greatest in the afternoons of hot summer days when air-conditioning is at a maximum and people are cooking evening meals. Cold days also cause maximum loads during the winter when people use electric heaters.

Coal is distributed in Florida by railroads, especially to power plants. Fuels for cars, trucks, and airplanes are brought to Florida from refineries by ocean-going tankers that distribute them to tank farms, from which they are delivered by oil trucks to local distributors. As natural gas has become the cheaper and more efficient fuel, the gas pipeline from the panhandle has been extended to more and more Florida communities.

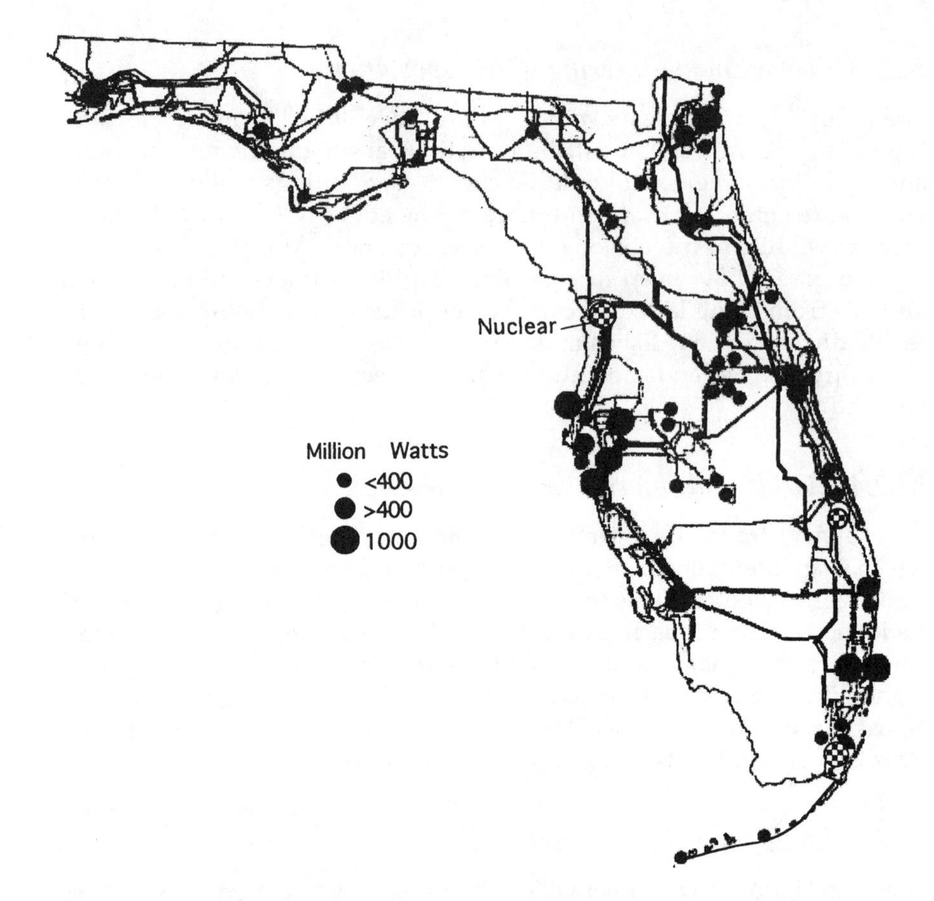

Figure 32.3 Network of electric power lines in Florida. (Modified from Fernald, E.A., Ed., *Atlas of Florida*, Florida State University Foundation, Tallahassee, 1981, p. 254.)

32.2.6 Information networks

Centers of information processing are found in the cities, in centers of government, and in universities and colleges. Long-term storage of information is required for an economy to be competitive.

Questions and activities for chapter thirty-two

1. Get a road map of your county and study the way the roads form a hierarchy connecting rural areas with the largest centers.
2. Obtain the latest atlas for Florida from the library or bookstore. Identify the maps that represent networks of flow of material, energy, people, and information.

3. Observe highways in your area and compare the appearance and wildlife of those that are mowed regularly with those that are allowed to develop a more varied vegetation.
4. Estimate what percentage of the emergy of transportation would be saved if people had only one car per family, could live within a bike ride of their work, and used public transportation for long trips.
5. Discuss what would make you decide to give up using your car. If you do not drive a car, discuss why not. Suggest ways to operate buses that would be efficient.
6. Discuss why the authors do not predict widespread use of electric cars. Why do you agree or disagree?
7. Research the abandoned barge canal across Florida. Discuss its history and current plans for use of the land.
8. Consider the future of Florida's transportation. Will there be more or fewer electric trains, highways, buses, Amtrak trains, airplane routes, river barges, personal cars?
9. Visit a high-voltage powerline and consider what is good use of the long corridors of land in regard to succession, wildlife, and human housing.
10. Discuss how Florida's networks may change as fossil fuels become less available and higher priced.
11. Count the number of each size dot in Figure 32.1. Plot the number of each on a graph where the horizontal axis goes from 1 to 6, representing the six sizes of cities shown. Make the vertical axis the number of cities in each size class. Does the shape of the graph suggest a hierarchy?

chapter thirty-three

Evaluation of energy sources

Much of Florida's information-rich, urban society depends on nonrenewable fossil and nuclear fuels (Chapters 29 and 31) to run its industries, cars, trains, trucks, airplanes, homes, and power plants; therefore, anticipating the future means evaluating present and future energy sources. A good energy source generates much more emergy than is required for its processing. For times ahead when the fossil fuels are less available and more expensive, are there good alternative sources to substitute? This chapter considers the net emergy yield of energy sources.

33.1 Power from heat engines

Much of our economy is based on the work of machinery using the heat from fossil and nuclear fuels. Learning the basic principles of energy conversion will give you a realistic view of what is possible.

33.1.1 Optimum efficiency for maximum power

One traditional way of evaluating the potential for an energy source to do useful work is to calculate the efficiency of the process (Chapter 7). Shown in Figure 33.1a, *efficiency* is the percent of the input heat energy that is transformed into an output of useful work. No real process is 100% energy efficient because some energy is always dispersed according to the second law (Chapter 7).

In science and engineering, the word *power* means useful work per time and is measured in energy units per time (joules or kilocalories per time). Speed of a process affects power and efficiency. Running very slowly is more efficient, but the work is done too slowly (power is small). If an engine is run too fast, the efficiency is too low as most of the energy is dispersed without being transformed into work. To convert heat at maximum power, only half

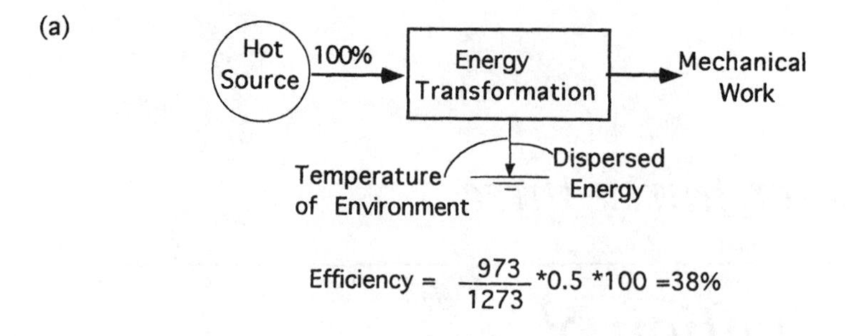

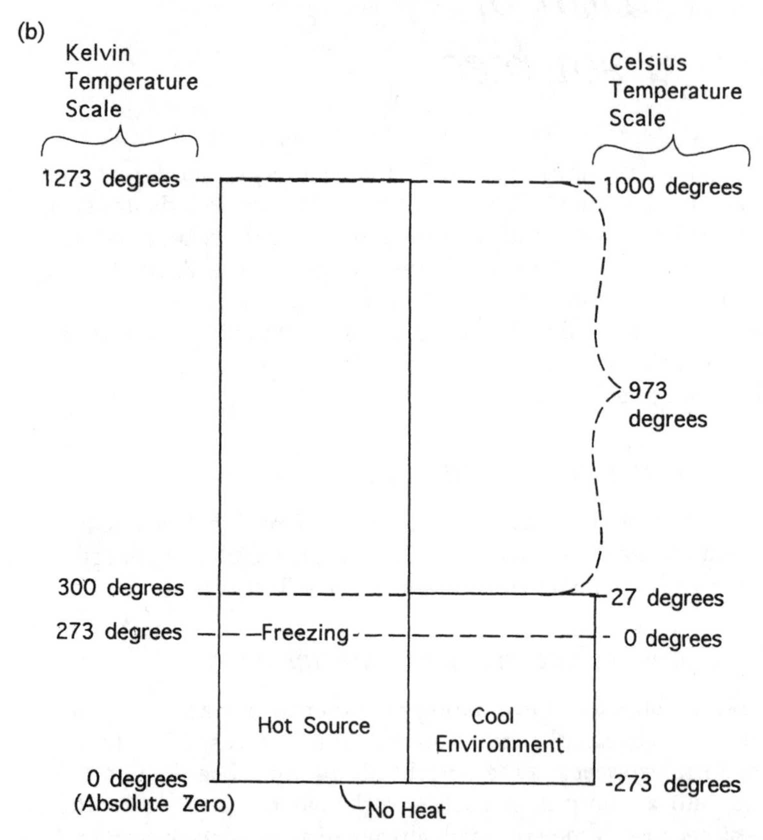

Figure 33.1 Diagrams for evaluating the work available from a source of heat. (a) Efficiency of converting heat to work; (b) bar graph for calculating the efficiency from Kelvin temperatures.

is converted and the rest is dispersed into the environment. There is an optimal speed that converts the available energy at a maximum rate. Cars and power plants are designed to generate maximum power and thus to run at about half the possible efficiency. There is a trade-off between power and efficiency.

33.1.2 Temperature measurements

To understand conversion of heat to work we have to put temperatures in degrees Kelvin. Temperature is measured in three scales: Fahrenheit, Celsius (centigrade), and Kelvin. Fahrenheit is commonly used in the U.S. — water freezes at 32°F and boils at 212°F. In most countries and in science, where metric measurements are used, the Celsius scale is applied — water freezes at 0°C and boils at 100°C.

The Kelvin scale given in Figure 33.1 has a value of 0 when there is no heat at all (called absolute zero) and 273 degrees at the freezing point of water (Figure 33.1). The Kelvin temperature is the Celsius temperature plus 273.

33.1.3 Efficiency of heat engines

Wherever there is a difference in temperature, there is a source of energy that can be converted into work. For example, train steam engines convert heat differences into the power of running wheels. When fuels such as coal or oil are used to do work, they are burned to make concentrated heat. Furnaces are at a high temperature, 1000°C or more. In power plants (Figure 33.1) and many other industries, concentrated heat is transformed into mechanical energy (turning electrical generators, etc.). Such processes are called *heat engines*. Storms in the atmosphere are heat engines.

The bar graphs in Figure 33.1b show the high temperature in the boiler of a heat engine and the lower temperature of the surrounding environment. The heat that can be transformed to mechanical work is only that fraction of the heat that is warmer than the environment. The percent of heat flow that can be converted to mechanical work is figured by determining what percentage of the temperature of the heat source is represented by this temperature difference, with temperatures given in degrees Kelvin.

For example, the heat source in Figure 33.1 is at 1273 degrees Kelvin when the environmental temperature is at 300 degrees Kelvin. The difference (1273 – 300) is 973 degrees. The fraction of the heat that can be converted into useful work is 973/1273 = 0.76. Convert this fraction into percent by multiplying by 100 to get 76%.

To operate at maximum power means adjusting the loading of the machinery to that speed which generates half of the possible work. In this example, the optimum efficiency for conversion to mechanical work at maximum power is 38% (0.5 × 76% = 38%).

Fuels vary widely in the concentration of heat they can generate per unit weight. Fast grown, low-density wood burns at low temperatures compared to natural gas (predominantly methane) and hydrogen, both of which generate higher temperatures. However, whether one fuel contributes more to the economy than another also depends on how much went into getting and preparing it.

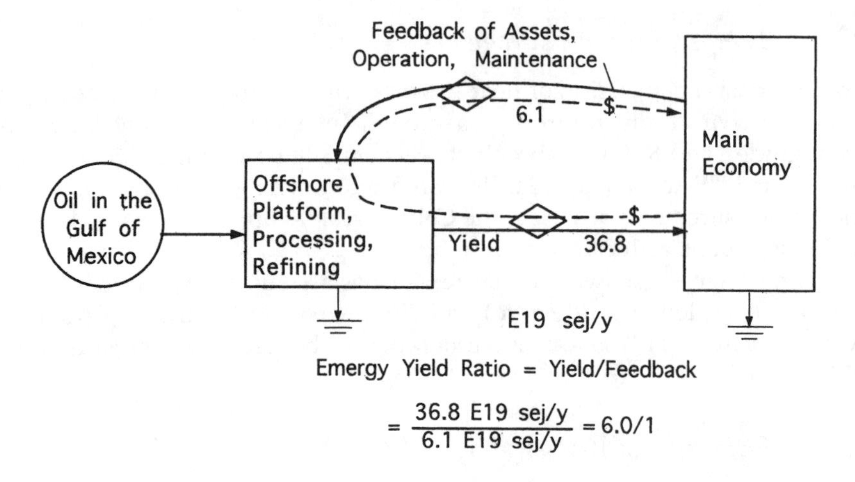

Figure 33.2 Diagram of the emergy yield ratio of a cluster of oil wells in the Gulf of Mexico in 100 feet of water. (From Odum, H.T. et al., Net energy analysis of alternatives for the United States, in *U.S. Energy Policy: Trends and Goals.* Part V. *Middle and Long-term Energy Policies and Alternatives*, 94th Congress, 2nd Session, Committee Print, prepared for the Subcommittee on Energy and Power of the Committee on Interstate and Foreign Commerce of the U.S. House of Representatives, 66-723, U.S. Government Printing Office, Washington, DC., 1976, pp. 254–304.)

33.2 Net emergy evaluation of fuels

Next we will compare the kinds of fuels available for use in Florida, evaluating them according to the net contribution of real wealth as measured in emergy units. A process has a net emergy yield if it contributes more emergy than it requires from the economy.

33.2.1 Emergy yield ratio

A useful way to compare different sources is to calculate how much yield is obtained from a source compared to the *feedback* efforts — in other words, by calculating a ratio of the yield to feedback, with both expressed in emergy units. This ratio is the *emergy yield ratio* (Figure 33.2). If the ratio is one, all the energy yield is going right back into the getting with no net benefit. A source good enough to run the rest of the economy yields many times more emergy to operate the economy than is fed back into the process from the economy.

In the example of oil drilling offshore in the Gulf of Mexico (Figure 33.2), yield is the flow of oil to the economy from the wells. Feedback includes the emergy of all the inputs to the processing: drilling rigs, ships, secretaries, office buildings, inspectors, drillers, divers — all the people and things that go into getting that oil from the ground, from prospecting to drilling to

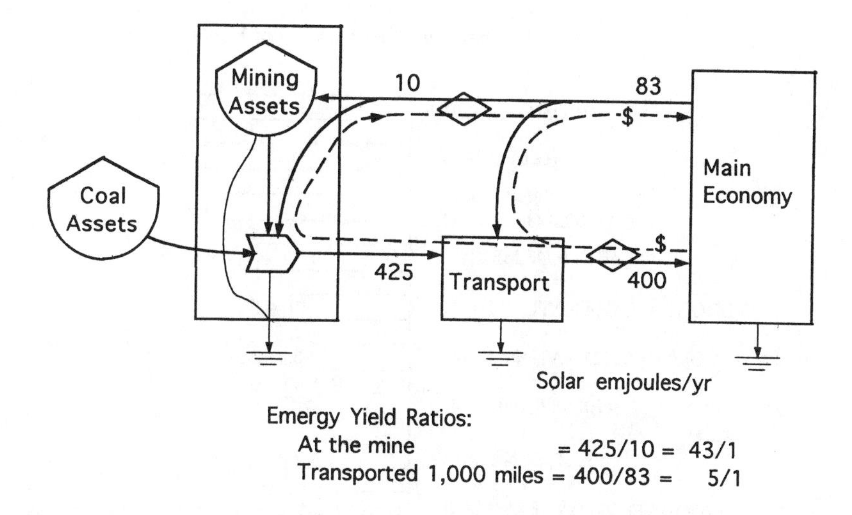

Figure 33.3 Diagram of emergy yield ratio of coal strip mining before and after transport to eastern cities. (From Ballentine, T., A Net Energy Analysis of Surface Mined Coal from the Northern Great Plains, M.S. thesis, Environmental Engineering Sciences, University of Florida, Gainesville, 1976.)

transportation to selling. Environmental impacts should be included in net emergy evaluations.

The 6/1 emergy yield ratio of Gulf of Mexico oil in Figure 33.2 is typical of fuel sources in the 1990s, although it is considerably less than earlier values. The U.S. economy was highly stimulated in the 1950s and 1960s when a typical source was 40/1 (40 emjoules obtained from each one emjoule expended in the efforts to find and process the fuel energy). The emergy yield ratios have declined as fuel energy has gotten more difficult to find and more and more energy is being spent in prospecting, transporting, and processing.

By changing equipment, fuel users (power plants, industries, homes) can switch from one heat source to another, depending on the prices. In general, energy sources with higher emergy yield ratios are cheaper because costs of processing are less. Thus, free markets tend to shift users to the source with the highest emergy yield ratio. This ratio can be used to predict if and when a proposed source will be economical.

33.2.2 *Effect of transportation*

Many fuels that have good emergy yield ratios when used near their source have much lower emergy yield ratios at distant points because of the energies used up in transportation. For example, the emergy yield ratio of coal mined in Wyoming at the mine site is greater than 40/1 (Figure 33.3). However, the transportation to bring it to eastern cities increases the feedback energy so

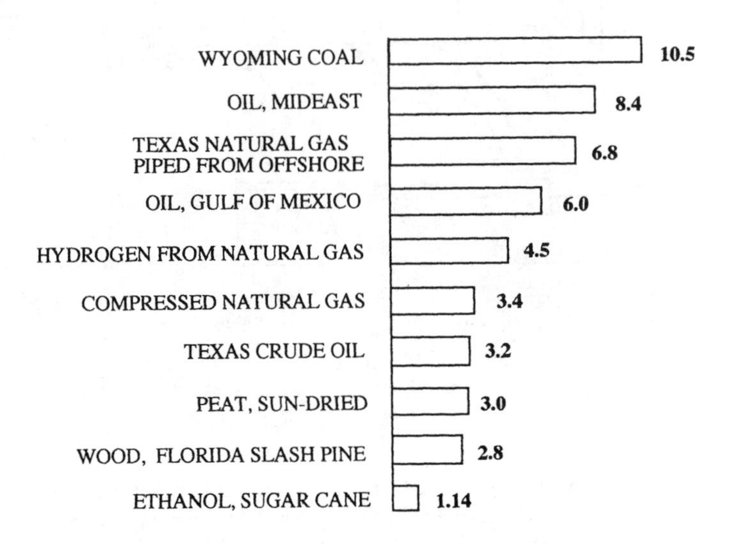

Figure 33.4 Comparison of emergy yield ratios of fuel sources important to Florida. For comparison with electric power sources, see Figure 33.5.

that the emergy yield ratio in Chicago is 5/1. The coal mines in West Virginia are much closer to Florida, and so compete better for Florida's markets than those from farther west. About 10% of the net emergy yield of oil from the Middle East is used up in transporting oil to refineries in the U.S.

33.2.3 Emergy yield ratios of foreign oil

When the U.S. buys oil on the world market and pays with dollars, the oil supplier can use that money to buy products from the U.S., thus receiving emergy according to the U.S. emergy/money ratio. The benefit to the U.S. is indicated by the emergy yield ratio of the purchase (emergy of fuel received divided by the U.S. emergy going to the seller). As inflation and worldwide prices change, so does the price of a barrel of oil. Before 1973, the emergy yield ratio in purchasing a barrel of oil was 40/1. In 1980, it was about 4.5/1; in 1990, 8/1; and in 1996, 7/1. The economy of the U.S. is stimulated more when the emergy yield ratio is higher.

33.2.4 Comparison of fuels

Figure 33.4 compares the emergy yield ratios of various fuel sources important to Florida. Each is based on a real example. The fuels at the top predominate in 1997 because their ratios are as high as that obtained from importing oil from abroad. Those with yields less than 5/1 do not now compete well as suppliers of heat.

Substantial reserves of *peat* are found in Florida. Peat is partially decomposed plant matter in marshes and swamps. It forms slowly a few millimeters a year. Its energy is intermediate in concentration between green plants and wood. To yield net emergy (Figure 33.4), it has to be naturally dried using dry winds and sunlight. Peat deposits are usually part of wetlands that serve the economy by increasing water quantity and quality, supporting wildlife, providing soil for agriculture, etc. (see Chapter 18). After public discussion, the peat in the swamp at the headwaters of the Santa Fe River was not mined for fuel; instead, the swamp was kept for its public value to the landscape. A date determined by the radiocarbon method indicated that the 10-ft (3-m) deep deposit took 1500 years to form, representing a natural contribution of 1.2 billion emdollars.

33.2.5 Transportation fuels

Abundant liquid motor fuels — diesel oil, jet aircraft fuel, kerosene, and gasoline — have been the basis for the design of cars, trucks, trains, and planes in this century. Petroleum out of oil wells is a mix of hydrocarbons (compounds of carbon and hydrogen in chains of different length). The mixture is separated into the various useful products in oil refineries. The petroleum is heated and converted into vapor. Depending on the size of the molecules, these vapors condense back into liquids, each at a different temperature. By controlling the temperatures, short-molecule hydrocarbons such as methane (one carbon per molecule) and propane (three carbons per molecule) are separated from gasoline (six or seven carbons per molecule), as well as heavier fuels with seven more carbons per molecule which are used in aircraft, trains, and diesel trucks.

There are alternative sources of motor fuel, but none give as much net emergy to the economy as gasoline made from inexpensive petroleum did in the past. Natural gas can be compressed into cylinders carried by a truck or bus; however, the pressure is so high that the cylinders have to be made of heavy steel and are costly to transport. Hence, the emergy yield ratio of compressed natural is less (Figure 33.4). Propane and butane are liquids at ordinary temperatures and do not require such heavy tanks to carry them, but like all inflammable vapors and gases, they require great care to prevent leaks.

Corn, sugar cane, and the other organic materials can be processed to manufacture alcohols (methanol and ethanol) to run automobiles. After adding the extra goods, services, equipment, fuels, and electricity requirements for this manufacture, the emergy yield ratios are close to one. This means that fuels can be made from agricultural and forestry production, but the processes do not compete nor stimulate the economy. Some biomass plants set up in Florida to supply alcohol as an auxiliary fuel failed because their emergy yield ratio was too low (Figure 33.4).

Liquid fuels can be made from coal, but about 45% of the coal is used up in the process. Early in the century, some cars were operated with small steam engines, like the trains of that day, but the engines required to control the pressures were very heavy.

Hydrogen is the lightest and most flammable of all fuel gases and is a very high-quality fuel. Because of its cleanliness and high energy content, hydrogen gas is important for special uses such as sending rockets into space and reacting with oxygen in fuel cells that make electricity in satellites. With the smallest of all gaseous molecules, hydrogen can leak rapidly if connections are not perfect and can react explosively with air. Hydrogen is made from natural gas with processes that use up some of the available energy. This makes it have a lower emergy yield ratio (Figure 33.4). Cars can be run with tanks of hydrogen, but the net emergy yield is less than running them with tanks of natural gas.

33.3 Emergy yield ratios of electric power production

Electric power is a high-quality energy that supports much of our human society because of its flexibility. As shown in Figure 7.6, additional energy must be used up in transforming fuel to electricity. About 4 emjoules of fuel emergy are required for each emjoule of electric power; therefore, electricity produced from fuels has a lower emergy yield ratio than the fuels before conversion. Electric power should not be used as an energy source where fuels are satisfactory. Various ways of generating electric power from other types of energy are evaluated in Figure 33.5. Electric power is often given in watts; a watt is a joule per second. The cumulative use of electricity is the product of power in kilowatts (1 kilowatt = 1000 watts) and the time in hours, which is kilowatt-hours.

33.3.1 Nuclear power

Nuclear power plants convert nuclear fission fuels (enriched uranium) into concentrated heat and then into electricity. The emergy yield ratio of nuclear plants (evaluated by Lapp, 1991) was about 4.5/1, better than the ratio for electricity from some fossil fuels. His calculation included an estimate of losses due to accidents and the plan for the wastes to be sealed up within the concrete shells after the plant was decommissioned (plant lifetime of about 35 years). This estimate was almost twice as high as one made 20 years earlier when nuclear technology was just getting started. Florida has two nuclear fission power plants at Turkey Point near Miami, two at St. Lucie, and one at Crystal River on the Gulf coast. The long-range future of nuclear power may depend on the availability of low-priced nuclear fuels.

Many planners hoped for cheap energy to come from nuclear fusion and nuclear breeder reactors; however, after 20 years of research, there is little practical evidence that fusion here on Earth can generate net emergy. The

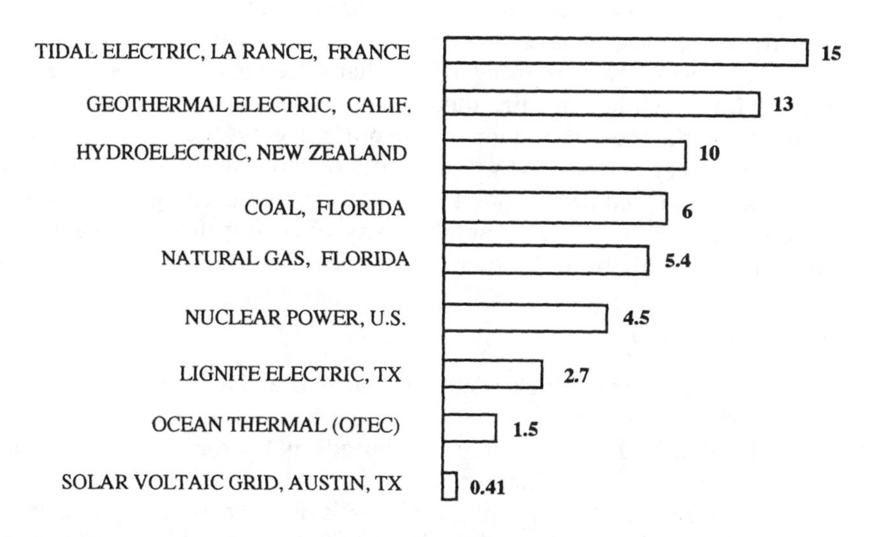

Figure 33.5 Emergy yield ratios for electric power sources.

process requires a temperature of 50 million degrees and equipment of enormous cost and complexity to generate a few seconds of reaction.

In the breeder reactor, the processing of uranium produces plutonium as a by-product. Since plutonium itself is a nuclear fuel, its production makes the original uranium go further; however, plutonium is extremely dangerous. It is toxic, causing cancers of the bone. The great costs of processing the intensively radioactive breeder wastes, so as to use plutonium safely, make the emergy yield of the breeder doubtful. Since plutonium can be made into atomic bombs by terrorists who obtain stolen nuclear wastes, the breeder projects were stopped in the U.S. Breeder reactors are operating in France, and we will have to wait for an emergy evaluation to learn if there is a net benefit.

33.4 Emergy evaluation of renewable energy sources

As the nonrenewable sources of energy that support our economy begin to dwindle, extensive research has been done to find renewable energy sources to replace the nonrenewable fuels. Evaluating the emergy yield ratios of alternative energy sources identifies which ones will support and stimulate the economy and not take more emergy from the economy than they give back.

Some alternative energy sources that are proposed for the future have emergy yield ratios that are less than one. Others have ratios that are much less than current sources of energy that support our economy. If an energy source has an emergy yield ratio that is less than one, it consumes more energy than it produces and is therefore not a source, but a consumer. Sources like these can exist only when there are rich supplies of other energies with which to subsidize them.

Sunlight is the most abundant energy but it is very dilute (low quality). Much of its energy is used up by the process of concentrating it for use. As Table 7.1 shows, it requires about 40,000 joules of solar energy to produce about one joule of coal. This is another way of saying that it takes about 40,000 joules of sunlight to do the work of 1 joule of coal.

33.4.1 Solar energy use through biomass

Biomass is a quantity of living or dead organic matter. Human societies have always used various kinds of biomass for food, fuels, clothing fiber, and housing. In Florida, human society was based on solar energy biomass in Indian and colonial times (Chapter 30).

Although the quantity of sunlight that falls on the country every day is quite large, natural vegetation of the biosphere uses up much of the solar energy received just to further concentrate solar energy into higher quality products such as wood and organic soils. Sunlight must be caught by leaves, transformed many times, converged, and accumulated. The efficiency observed (0.1 to 2%) may be the best that can be done to convert sunlight energy into organic matter without the help of other energy sources.

The emergy yield ratio of *biomass* depends on how intensively it is managed. When wood develops in a Florida forest for some years without much human effort other than harvesting, the emergy yield ratio is moderate (about 3/1). However, when short-term agricultural crops are harvested for energy purposes, the emergy yield ratio is low because much fuel and labor has to go into the crop each year. The emergy yield ratio declines as the intensity of management increases.

In our times, using solar energy to grow wood and agricultural products (foods, fodder corn, hay, etc.) is the main way that solar energy is brought into our economy, but these are not used as primary fuels and need not have a net emergy contribution. Everything is subsidized directly and indirectly by our heavy use of fossil fuels.

Whether from Texas, Louisiana, or foreign imports, coal, natural gas, and oil yield more than any renewable fuels. In the future, when these sources are unavailable, renewable fuels such as wood from organic production may be the only alternative. However, there may be strong competitive demands on the same land to produce food, clothing, building materials, and household goods, as well as fuels. Recall that the carrying capacity of Florida on solar energy alone is about 15% of the present economy (see Section 31.12).

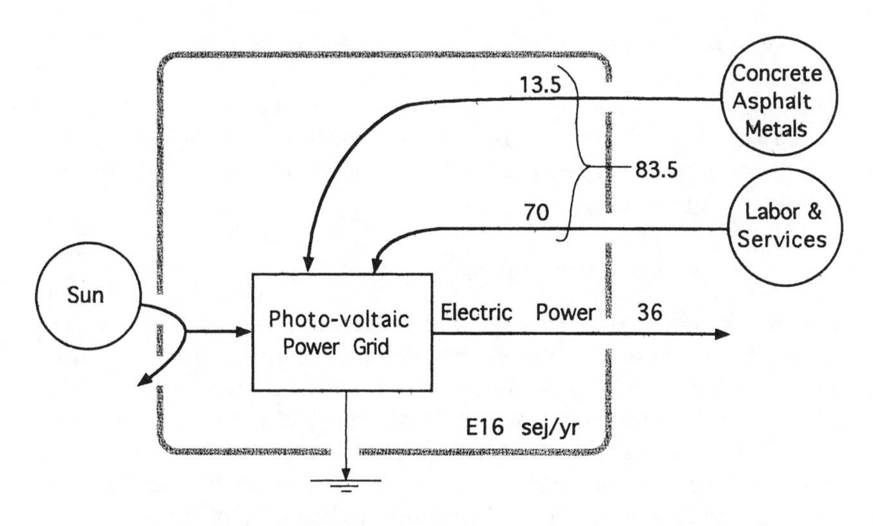

Figure 33.6 Emergy yield of a power grid of solar voltaic cells in Austin, Texas. (From King, R. J. and J. Schmandt, *Ecological Economics of Alternative Transportation Fuels*, report to the Texas State Dept. of Energy, Lyndon B. Johnson School of Public Affairs, University of Texas, Austin, 1991.)

33.5 Solar technology

With solar technology, devices are used to convert solar energy more directly to mechanical and electrical energy in one step, without going through the series of steps that generates biomass. However, using several steps is apparently more efficient (higher emergy yield ratios) than trying to skip from low-quality sunlight to higher quality mechanical and electrical energy in one step.

33.5.1 Solar voltaic cells for electric power

Solar voltaic cells generate electricity directly from sunlight. The green chloroplasts in plants are solar voltaic cells that start the photosynthesis process by first generating electricity in the biochemical systems of the plant cells. Much of the landscape of the world is already covered with green solar voltaic cells. Plant cells may have already achieved the maximum electric yield that is possible.

Solar voltaic technology uses metal cells of silicon with about the same efficiencies and power output as the green ones, but vastly greater inputs are required from the economy. Figure 33.6 shows the inputs and electric power yields observed for a grid of solar cells generating electric power in Austin,

Texas. When all the indirect solar emergy in goods and services is considered, there is no net emergy yield.

Very light electrical cars have been run on solar voltaic cells across Australia where there are no clouds, but solar cells do not yield net emergy, and the heavy batteries required to store electricity take up most of the energy for maintenance and transport.

33.5.2 Solar energy hot water heaters

Solar energy is widely used in sunny climates to heat panels of pipes in which water gets hot because of the blackened surfaces absorbing the sun. This hot water is then stored in a tank from which it is used directly as hot water or circulated to help heat the house. These *solar hot water heaters* are expensive because wide surfaces of costly glass, plastics, and metals are required.

Solar water heaters are not energy sources, but are consumer devices, all of which use more emergy from the economy than they produce. However, solar water heaters use less fuel than electric or gas heaters, so they are a way of saving energy. Two water heaters used in Miami were compared: one solar, the other gas. Both systems used fossil fuel indirectly to supply and maintain the equipment. The solar heater took more initial investment in equipment but used fuel only on cloudy days. The gas heater took less equipment but required continual fuel use. Solar water heaters are still in use in Florida, as they have been for decades, but they do not yield net emergy. They are a way of using excess capital to reduce fuel and dollar expenditures. Their use is a form of energy conservation.

Solar energy helps all economies and is essential to them through its support of plant production, heating, generation of winds, evaporating water, and driving the hydrological cycle. However, its ability to run economies directly through the use of *solar technology* is very limited. Emergy yield ratios are low.

33.6 Other renewable energy flows

Many of the energy flows of atmosphere, ocean, and geological systems can be diverted to human technology, but in all cases this means diverting it from its present work of maintaining a healthy planet.

33.6.1 Wind

Windmills have a net emergy yield ratio when they are located in areas of very steady moderate winds. However, in Florida, with average winds less than 7 miles per hour (15 km per hour), there is little net emergy yield. Small windmills may be useful for pumping water for watering stock or for irrigation in areas far from electricity. Sailboats yield net emergy if large sail areas and low emergy materials are used.

33.6.2 Geothermal power and ocean thermal electrical conversion

Small natural temperature differences are used in many of the Earth's processes. The atmosphere makes natural heat engines that produce wind from differential heating of the earth and atmosphere. Tapping the heat of the earth for man's industrial processes has been economically successful only in the vicinity of volcanoes (in Iceland, New Zealand, and California), where temperatures are high near the surface.

One proposed source of energy is the gradient between the hot surface water (27°C) of tropical seas and the cold bottom water a thousand feet below (2°C). Ocean thermal electrical conversion (OTEC) was much discussed in regard to Florida's Gulf Stream, but considering the relatively small temperature difference (25°C) from the surface to the deep water and high costs due to the threat of hurricanes, evaluations show only a small emergy yield ratio.

An engineering firm in Gainesville discovered that several thousand feet down in the sediments below Miami there are temperatures near freezing (2°C), whereas temperature usually increases with depth in the ground. Apparently, cold, deep seawaters exchange with these rock strata. Although considered for an OTEC geothermal installation, it was not authorized, as pumping water in or out of strata might cause earthquakes or land subsidence.

33.6.3 Hydroelectric power

Hydroelectric power is generated by water in mountains running downhill. Water in elevated lakes is allowed to fall through pipes under the force of gravity, turning turbines which generate electricity. Since there are no mountains in Florida, there is not much hydroelectric potential in the state. Florida produces only 0.1% of its electricity from the Jim Woodruff hydroelectric plant on the Apalachicola River. The disruption that dams cause to other river values, such as fish migrations and boat movements, may be greater than the small amount of power which can be generated.

33.6.4 Waves and tides

Wave energy coming ashore along coasts of the world is large in total quantity, and it does much daily work maintaining beaches, attracting tourists, etc. (Chapter 13). However, wave energy is difficult to use as a general energy source because it is spread out along a great length of coastline and is variable, with large energy one day and nearly none the next.

The rise and fall of water levels due to tides have been used to make electricity with emergy yield in several places in the world where tides are 20 ft (6 m) or greater. There are few areas with such great tides and none in Florida.

33.6.5 Using an appropriate energy quality

An economy is wasteful if it uses a higher quality energy source (higher transformity) than is necessary. For example, a water heater using electricity made from natural gas is wasteful compared to a natural gas water heater, because unnecessary transformations are required for the electricity.

33.6.6 Future sources for Florida

An examination of fossil fuels and alternative sources does not show any abundant new sources with high emergy yield ratios This means that as fossil fuels get used up economic growth is not to be expected considering what is now in sight.

Questions and activities for chapter thirty-three

1. Define: (a) energy sources, (b) emergy yield, (c) emergy yield ratio, (d) feedback, (e) ethanol, (f) peat, (g) solar technology, (h) solar voltaic cells, (i) OTEC, (j) nuclear fission, (k) uranium.
2. Explain why a source with an emergy yield ratio lower than another source will not compete until the richer source is used up and its ratio declines. Give an example.
3. Discuss what you think will happen to transportation by car as imported oil becomes less available. Consider both the possibility of gasoline being produced from natural gas, coal, or other materials temporarily and the possibility of none of these alternative processes resulting in much emergy yield.
4. Florida used 276 E20 sej of coal in 1993 (Appendix B, Table B1). If its emergy yield ratio was 5/1, what was the emergy of the feedback costs? What did that feedback cost in 1993 dollars? (You need to check Figure 29.2 for the emergy/dollar ratio.)
5. Make a collection of pictures and articles about alternative energy sources. (A group of students could work together on each source and give a report to the class.) Evaluate the articles. Do the data from different articles agree? Do the conclusions agree?
6. Take magnifying glasses outside on a sunny, dry day. Try to light a piece of paper or leaf by directing the sun at one spot. (Cigarette papers are easier than other paper.) Discuss the implications for solar technology.
7. Discuss possible futures if no new sources are found.
8. When the supply of fossil fuels is much smaller, which would you choose to use land for: food, sheep and cotton for clothes, wood for heat and to build houses, or fuel to run your car?

9. In Florida, agriculture runs more on fossil fuel than it does on the sun and other natural emergy. What will have to be done to maintain the same yields when there is less fertilizer, fewer machines, and fewer varieties of crops?
10. Why do solar technology devices have poor yield ratios? What are the proven ways of supporting human beings on solar energy?
11. Why have we largely replaced the direct use of coal and wood with electricity in our modern economy?
12. Name five problems that make nuclear power controversial. Discuss the one you think is most serious.
13. Find some articles on breeder and fusion nuclear power. Compare the conclusions in the articles with those in this chapter.

chapter thirty-four

Wastes and recycle

With the development of large urban populations and industry, accumulating wastes have become a major problem for society. Ideally, as explained in Chapter 5, a good system has no wastes, but instead passes its by-products to other uses or reuse. Our landscape of urban economy and environment is in transition, perhaps being organized towards the ideal. This chapter considers the wastewaters, air wastes, and solid wastes and ways of recycling for reuse or beneficial return to environment.

34.1 Background

Earlier in this century when oil and other resources were still very cheap, there was a rush to develop new kinds of chemical substances, such as chlorine-based plastics, with an emphasis on long-term durability. Yet, often these products are not used for long and are thrown out quickly. Many of the products and their wastes are not biodegradable. It is convenient for marketing and advertising to package products without regard for the needs of the whole system. No one takes responsibility for making it all work together. Since the environment is not paid for its services, the economic system has no mechanism to organize good reuse and recycle between the economy and the environment. Thus, wastes accumulate until concentrations become toxic, poisoning groundwaters, and they take valuable land out of good use to become dumps. As populations and economic development increased in Florida, conditions began to become hazardous and detrimental to the economy, decreasing the standards of living, interfering with tourism, and requiring increased taxes to deal with the mess.

Some of the initial efforts to deal with hazardous wastes failed because they did not consider the whole landscape. To solve air pollution, substances in smoke stacks were collected and dumped into streams. To solve stream pollution, substances in water were extracted and burned, putting wastes back into the air. Finally, as people have begun to understand that they have to work together to make a better system, programs for recycling garbage are being developed, economic incentives for better use of by-products are being tried, and regulations are being implemented where wastes are dangerous.

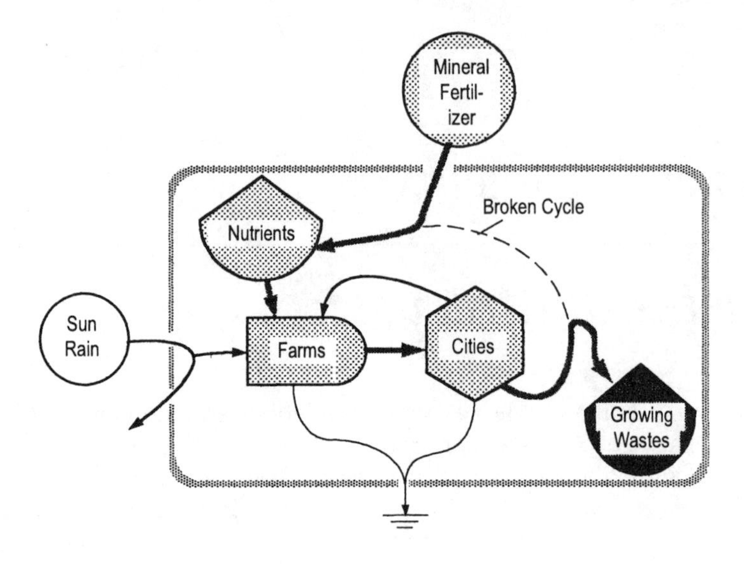

Figure 34.1 System of agriculture and towns and the processing of materials. Dashed line shows the broken cycle.

34.2 Principles of reuse and recycle

34.2.1 Closing broken cycles

Simpler agrarian society recycled the wastes of human society back to agriculture; some farms in China still carry human wastes (called night soil) from villages back to the fields. However, our intensive industrial agriculture uses fertilizer from geological deposits such as Florida's phosphate deposits and bird guano deposits found on some islands in dry climates. Guano is the accumulated waste of animal colonies rich in nitrogen and phosphorus. As Figure 34.1 shows, a closed cycle was replaced with a one-way flow from geological deposits into waste accumulations in our landscape.

34.2.2 Converging products and dispersing recycle

Generally, when materials are collected to make products, work is done to bring required materials to one place. Materials converge (Figure 34.2). As we explained in Chapter 7, materials recycled to environmental systems need to be dispersed back to broad areas again, so that the ecosystems can utilize them at low concentration. A study made of Paris in 1860 showed that hay was carried into the city for the horses that were used for transportation, then the horse manure was collected and dispersed out over the surrounding farms, making possible very fertile agriculture. In our society, our economy often pays to bring products into the city, but does not pay enough to

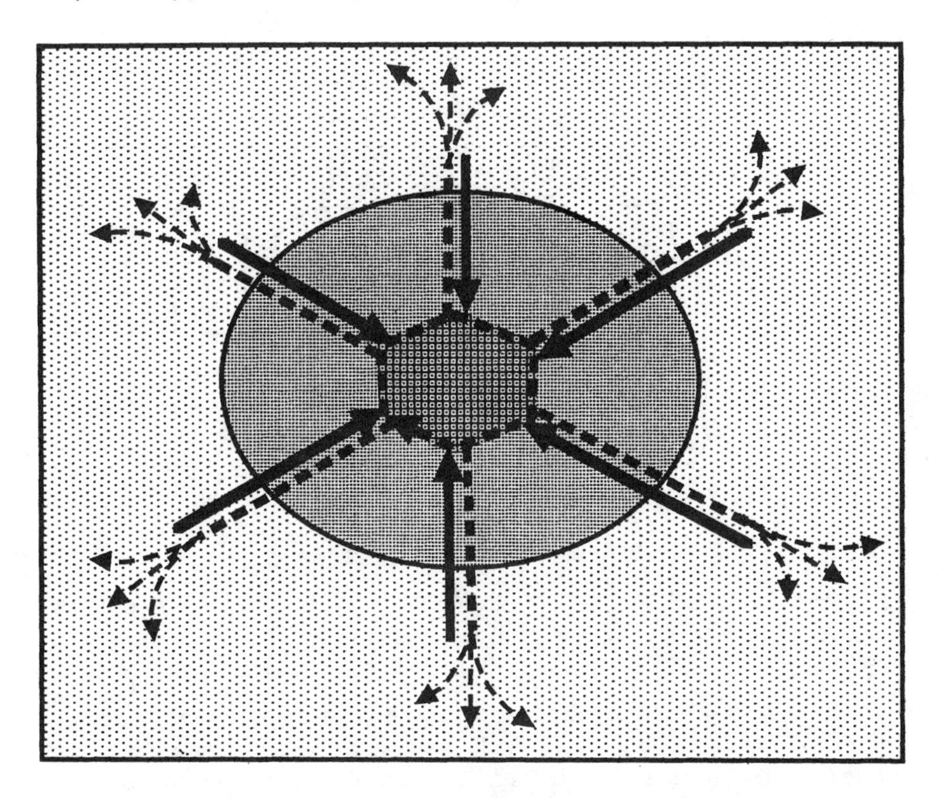

Figure 34.2 Pattern of materials converging in products going to town and the dispersal of the wastes to rural areas necessary for environmental recycle.

disperse materials over the landscape in a manner that the environmental systems can profitably use.

It has often been said, "The solution to pollution is dilution." One problem with following that slogan is that wastes dispersed into nature are sometimes immediately reconcentrated again. Wastes dispersed in streams are often reconcentrated when the streams converge to a lake. Radioactive fallout from the testing of nuclear weapons ended up in the Gulf of Mexico in 1962 and was immediately collected by *Sargassum* seaweed and dumped on Gulf beaches when these floating plants came ashore.

34.2.3 Economics and emergy evaluation of waste alternatives

For the economic system to process wastes, there has to be some profit; however, when estimated with emergy evaluations, waste processing is usually found to be a net benefit to society even if not economical. As with all environmental resources and problems, the economic free market does not necessarily do what is best for the public environment unaided. Some

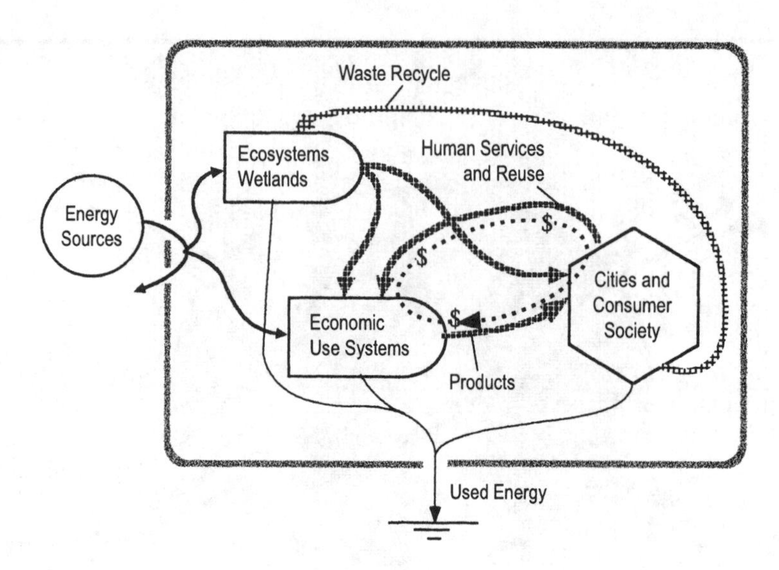

Figure 34.3 System of economy and environment with materials conserved over reuse and environmental recycling pathways.

kind of incentives have to be added to protect the public interest. For example, suppose emergy evaluation shows that recycling of paper is in the public benefit at a time when it is cheaper for the paper maker to cut fresh trees. Then, some incentive, such as tax write-off, can be given to the industry to benefit the public while keeping the private industry competitive. Some subsidy from public funds is justified.

34.2.4 Reuse and environmental recycle

Some wastes are so concentrated that they can be economically reused. These are usually too concentrated to fit the more dilute cycles of the ecosystems. For example, aluminum cans, glass bottles, and stacks of used paper are best processed for reuse (recycle within the economy). On the other hand, dilute wastes such as automobile exhausts, nutrients in storm-water runoffs, and grass cuttings are not economically collected and can be absorbed if dispersed into environmental cycles. Thus, there are two kinds of recycle: reuse and recycle to the environment. Figure 34.3 shows the dual pathways of waste recycling, part through reuse by the economy and part dispersed out to natural areas.

From the point of view of the whole system of economy and environment, nothing should be manufactured that cannot be reused or recycled, either to other parts of the economy or to the environmental life support systems. Things that go to the environment should be decomposable for use by the ecosystems. They are said to be *biodegradable*. Some plastics are not.

34.2.5 Environmental technology and industrial ecology

At first, people thought that wastes should be destroyed in treatment plants, which used energy, materials, and money. The idea was to make jobs with environmental technology; however, jobs that are not necessary and do not contribute to overall productivity are another kind of waste that reduces wealth and standard of living in the long run. Reuse and appropriate dispersal for natural uses allows the wastes to aid landscape productivity. Technological treatments should be used only for wastes that cannot be easily reused, such as toxic substances. Modern policies call for industries to include waste processing as part of their operations and costs. If a product and processing of its wastes cost more than the profit, then that product need not be made. Including more of the material cycles as part of an industry is *industrial ecology.*

34.2.6 Self-organization underway

With public approval, progress is being made in developing better recycling systems. As necessary materials are incorporated into new products, the wastes are released as the products are used. In the process of self-organization, wastes and by-products may accumulate temporarily, but usually the wealth that is represented by the storage attracts a specialist that can use it beneficially. For example, when an industry has a waste product that is accumulating, eventually it may pay some entrepreneur to develop a business to use it for something else, perhaps separating the most valuable materials for the economy while routing the remainder to nature in a way that helps the ecosystems. Similarly, for many kinds of chemical substances, there are special kinds of microorganisms somewhere capable of consuming them.

34.3 Wastewaters

34.3.1 City wastewaters

Shown in Figure 34.4, the sewage from cities and towns is collected by pipes and carried to sewage plants, which may have one or more treatment stages. In *primary treatment,* the solids are allowed to settle and are removed as *sludge.* This sludge contains most of the viruses and toxic metals. There is some risk if used for fertilizer. Sometimes this sludge is burned; sometimes it is buried.

Waters remaining after primary treatment have large quantities of organic matter dissolved or present as small particles. If released into public waters, these high concentrations of organic matter would cause so many bacteria to grow that the dissolved oxygen, normally 8 parts per million, might be reduced to less than 2 parts per million, causing fish to die.

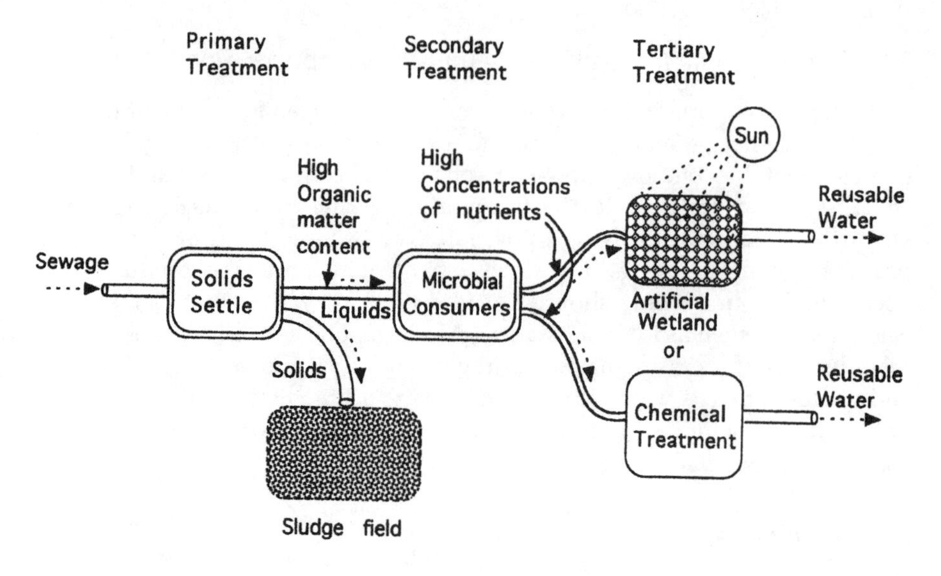

Figure 34.4 Processes of treatment of sewage from cities and towns.

Secondary treatment removes most of this organic matter. In the trickling filter method, waters are sprayed on a bed of gravel rocks, which have populations of tiny consumer microbes and small animals living in the spaces between the rocks. By the time the waters come out the bottom of the filter bed, the organic matter has been consumed. The consuming process releases the nutrients.

Another method of secondary sewage treatment takes less space than the trickling filter. It is the *activated sludge* process. The organic rich waters are seeded with bacteria and bubbled with air so that the microbes work to consume the organic matter very quickly. The bacteria use the oxygen as soon as it is pumped in. Only the rich nutrients remain.

Secondarily treated sewage waters have very high concentrations of phosphorus and other nutrients. If these are released to public waters, the algae are so stimulated that the waters become green. The overgrowing by a few species eliminates others, reducing biodiversity. Then, at night or on cloudy days, organic matter made by the algae decomposes, reducing the oxygen, which kills the fish. In other words, waters receiving treated sewage are too eutrophic (Chapter 17).

Tertiary treatment (the third stage of treatment, shown in Figure 34.4) takes out the excess nutrients and any other undesirable substances still in the water. In the center of big cities, tertiary treatment may be done with chemicals, including charcoal, to adsorb organic substances such as pesticides. However, this technological tertiary treatment is expensive and produces waste products to be disposed of. In some states, ponds are used for

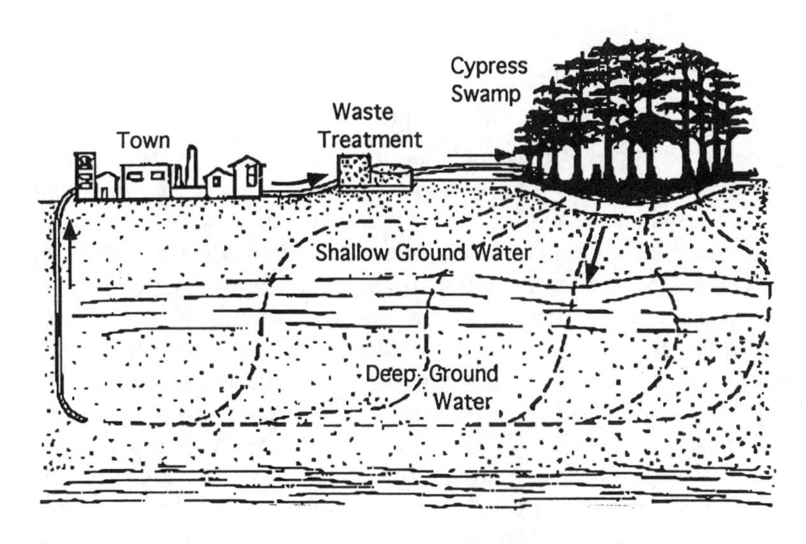

Figure 34.5 Recycle of treated sewage through a cypress swamp.

artificial ecosystems to consume the wastes; however, in Florida, where clay soils are scarce, pond bottoms are often porous sands, making it easy for waters to percolate into groundwaters before the wastes have been removed.

Florida has been a leader in ecological engineering measures to apply the nutrient waters from secondary treatment to the environment in a beneficial way. Artificial wetlands are constructed, seeded with many species of plants and animals, and allowed to organize a mix of species that can prosper by using the nutrients. Wetlands are a natural filter, some nutrients going into plants, some being taken up by microbes in the peat, and some being taken up by the soils underneath the wetland. Waters leaving these special wetlands are typical of public waters and can be released to help maintain water levels in lakes and groundwaters. Figure 34.5 shows recycling of wastewaters through cypress wetlands (see also Figure 23.4).

In many places, dumps and leaky underground gasoline tanks have polluted the upper groundwaters. Nitrates from excess agricultural fertilizers near the Suwanee River are making groundwater unsafe for drinking. Cleaning up contaminated groundwater is very difficult and expensive. Often the nearby people cannot use well water but must pay for special pipelines to bring water from elsewhere.

Some industrial wastes are purified in ways similar to sewage by passing the waters through a series of ecosystems. Wetlands help vaporize organic pollutants such as gasoline. Recent studies show wetlands are good at filtering out heavy metals. One swamp in Jackson County filtered out much of the lead released in a battery-washing industry.

34.3.2 Paper mill wastes

Paper mills are important in north Florida, where large areas are used to grow trees used in paper making (Chapter 22). Great quantities of water are used. After toxic chemicals have been removed, there remain large volumes of black wastewaters. The black color is mostly the natural lignin that is in wood, cementing the fibers together. When trees decompose in nature, brown lignin drains into streams, making them black. Since lignin decomposes slowly, natural black water is a normal part of southern waters, and aquatic life is adapted to it. Oxygen in such waters is about half saturated. Paper wastes are somewhat similar to the natural black waters, but if released into streams the organic matter decomposes more rapidly, using up dissolved oxygen. Preliminary studies show that wetlands can recondition these wastewaters so that waters coming out are like those of natural swamps. Perhaps paper mill wastewaters can be cycled back through the pine lands from which the wood comes, draining into the small dispersed swamps that occupy 10 to 20% of the pinelands.

34.4 Air wastes

Fifty years ago, the air of Florida and the eastern United States was usually clear with a visibility of 50 miles. Now, the air is filled with turbidity and waste chemicals whenever circulation is stagnant. Much of the pollution is from all the fuel combustion that runs our economy. This includes the cars, power plants, industries, home heaters, and gasoline-driven machines used at home and in industry. The air wastes from engine exhausts are a little like volcanoes, belching acid vapors. Just as smoking is bad for the lungs, so smoke from our economic processes is also unhealthy. When clouds form and release rain, many of substances carried in air produce acid rains (Figure 2.1), which go to the lakes and streams.

Technically, *acid* is the concentration of *hydrogen ions* (abbreviated H^+). Hydrogen ions react with *hydroxyl ions* (abbreviated OH^-) to form water (abbreviated H_2O). A base has more hydroxl ions than hydrogen ions. Acid is measured on a pH scale from the most acid, pH 1, to neutral at pH 7 when acid and base ions are equally concentrated. Waters are basic (alkaline) from pH 7 to the most alkaline at pH 14. The pH scale is logarithmic, which means that a one-unit change occurs when the concentration is ten times higher.

Rain is normally slightly acid (pH 5) due to normal carbon dioxide of air. Air that is full of acid vapors from volcanoes or from city combustion may be as low as pH 3 (100 times more acidic than pH 5). Central Florida around Tampa and Orlando sometimes has strongly acid rain.

If acid rain fills up a softwater lake that has no contact with limestone (calcium carbonate), the water remains acid and fish are killed. Some lakes in the sandhills of central Florida are softwater lakes that are easily damaged by acid rains. If the acids first run through limestones, the acid is neutralized (pH 7). Since much of Florida's surface has limestones, acid rain is less a problem in Florida than elsewhere.

Fluorine is naturally concentrated in phosphate deposits, and when phosphate is made into superphosphate (see Section 25.1), fluorine is released. Some years ago, fluorine from superphosphate plants was killing vegetation in central Florida, but now this element is captured for other use.

In order to reduce acid rains and other bad effects of air wastes, scrubbers can be used in smoke stacks to filter out these wastes, but this adds 10% or more to production costs. It would not hurt an industry or a municipal economy if all stacks were required to do this, but if the U.S. required it and competing economies in nearby countries did not, then the U.S. enterprises would not be able to compete, and jobs would be lost here.

As with most environmental impacts, it is the high levels of economic and population growth that overload the environment. If there were fewer combustion processes, the environment could absorb and buffer the air wastes, helping make industry competitive. When the environment can no longer maintain air quality adequately, jobs move to areas less congested.

34.4.1 Ozone

Oxygen is almost 21% of the atmosphere and is normally in molecules consisting of two joined oxygen atoms, abbreviated as O_2. High in the stratosphere, the ultraviolet light of the sun causes the oxygen to form *ozone,* which has three atoms of oxygen joined together and is referred to as O_3. Ozone is involved in air pollution issues in several ways.

The abundant ozone in the stratosphere forms a layer that catches much of the ultraviolet light that would otherwise reach the surface of the earth. Ultraviolet light contains the wavelengths of light too short to be seen by the eye, with twice as much energy per photon as ordinary light. It is capable of damaging the genes in the skin cells of an exposed human body. Cells with damaged genes can cause skin cancers.

One type of substance released to the air in our society in recent years is chlorofluorocarbon, abbreviated CFC. These substances have been accumulating in the atmosphere and are now destroying the ozone in the stratosphere, removing the protective filter and allowing ultraviolet radiation to increase at the surface. This effect showed up first over the south and north poles in winter because the stratospheric ozone layer is lowest there. Treaties have been developed to ban the use of CFCs worldwide. Increasing ultraviolet light is of great concern to Florida with its large population of bathers on beaches.

Whereas ozone in the stratosphere is good for health, ozone in air near the ground is not healthy for the lungs or plants. When air is filled with byproducts of combustion, there are chemical reactions aided by sunlight that create ozone in our atmosphere at ground level. Ozone is sometimes used as an index of air pollution.

34.4.2 Greenhouse gases and climate change

There is now worldwide concern over the increases of carbon dioxide (CO_2), methane (CH_4), and other gases which may be changing world climate. Normally much of the sun's energy that comes to Earth goes back out as infrared heat radiation (light with wavelengths too long for the human eye to sense). Carbon dioxide, methane, and other gases absorb some of this outgoing radiant energy so that the atmosphere becomes warmer. This is called a *greenhouse effect* because the glass of greenhouses lets sunlight pass in but catches the infrared radiation going out so that the greenhouse gets warmer.

As you learned in Chapter 4, carbon dioxide is a normal by-product of the respiration of life, volcano fumes, and the usual fires of nature. Normally, increases in carbon dioxide increase photosynthesis which uses up carbon dioxide excesses. The carbonates in the ocean also take up some of the excess carbon dioxide. Because the worldwide consumption of fuels is so great and forests have been reduced, the carbon dioxide excess is so large that it may be causing temperatures to rise, at least in the tropics.

Methane and nitrous oxide gases are normally released from wetlands and animal guts, places where oxygen is limited and consumption of organic matter is incomplete. Normally, these gases are used up in reactions with oxygen. But now, with additional releases from cars and industry, they are being produced faster than the normal processes of decomposition can use them.

At the international environmental congress in Rio de Janeiro in 1992, the concern about global change was so great that there were movements to restrict world fuel consumption to reduce the production of carbon dioxide and other greenhouse gases.

34.5 Solid wastes

In earlier times, there was not much solid waste because most solids from households were biodegradable. Garbage could go to pigs and chickens on farms where there was plenty of land. Wastes could be reduced by burning some of it and dispersing the rest into soils. But then people formerly on the farms moved to the cities to jobs in industry and business. Now there is little space to process wastes, and it is necessary to collect garbage and do something with it. Since foods and other goods have to be transported into the cities, packaging has become a necessity, and the solid wastes are full of containers.

In many countries, the solid wastes are spread out on the land without covering, so that the atmosphere and the animals can help with decomposition. Sometimes this is done with incinerator ashes. Such areas are favorite places to see birds and mammals. There are swarms of sea gulls, vultures, egrets, and many kinds of hawks that come to eat the rats and mice that are an abundant part of this kind of decomposition system. In the early days of national parks, hotel garbage was spread out for the bears, and the tourists would go every evening to watch them. This practice was stopped as wastes became unsuitable and because it led to an unnatural concentration of animals.

Now, solid waste is a major problem since it is full of plastics, glass, cans, paper, and toxic substances. Many cities are using their trucks to pick up tree limbs and leaves from residences. All this is stored in landfills, holes in the ground that are covered over with a layer of dirt by bulldozers. Often these landfills drip toxic solutions into groundwaters, which become unsafe to use for farming or housing. Near Naples, one such landfill is being unearthed again to remove items of some value to reprocess them.

The first plastics were readily decomposed by the ultraviolet rays of the sun, but when the chlorine-based plastics were developed everything changed. You cannot safely burn chlorine-based plastic in ordinary incinerators because the chlorine compounds in the smoke spread over the landscape, where they can be toxic for people as well as the ecosystems. Such plastics can be safely burned in special high temperature furnaces, but these are expensive and have to be carefully regulated. Great controversy has accompanied plans to put in regional incinerators for toxic plastics. Chemical industries are now advertising that they can reuse these plastics in making more plastic products, but the costs are not yet known, and the amount of plastic recycling is still small.

As the public became environmentally concerned, the concepts of simplicity and cleanliness were linked to the concepts of protecting nature. People interested in improving the environment organize clean-up days to pick up manmade litter that is scattered through forests, fields, and beaches. The result is a big stack of solid waste that has to be processed.

People are encouraged to keep as much of their organic substances on their lots as possible, developing small compost piles for their leaves, limbs, and some food garbage, thus developing rich soil for their lawns trees, and gardens.

Better ways of processing solid wastes are now being used in Florida. Citizens help by separating out some major components to aid recycling: aluminum cans go back to the aluminum plants to be reused; old cars go back to the steel industry for reuse; and much paper is separated out to be recycled for newspaper.

Some furnaces have been designed to burn solid waste for heat energy, although this still leaves a lot of ash, some of which may be toxic. These systems do not yield net emergy, but they may reduce the emergy required for waste processing.

In an experimental forest in Gainesville, Dr. Wayne Smith of the University of Florida planted a crop of pine trees in a foot of shredded solid waste. The pine tree growth after 16 years was as good as any in the state. The solid waste forest looked entirely normal. You can dig with a shovel in the soil and occasionally find a piece of a can or plastic, but these items are so dispersed that they are not noticeable and may eventually decompose.

High concentrations of mercury are found in fish and peat in the Everglades. Some of the mercury apparently comes from the atmosphere, where global levels of mercury from urban sources have increased.

It is a very creative time, with many potential solutions being tried. What works in one community may be copied by others. This is the normal process of self-organization (Chapter 8). Developing good designs of the economy with environmental systems so that all work cooperatively is sometimes called ecological engineering (Chapter 18). This uses the self-designing tendencies of ecosystems along with some technology and construction by humans to make the whole system beneficial.

Questions and activities for chapter thirty-four

1. Define: (a) waste, (b) solid waste, (c) hazardous waste, (d) converge, (e) acid rain, (f) CFC, (g) by-product, (h) plastic, (i) biodegradable.
2. Describe a waste problem where incentives might be used to encourage private industry to solve it. Be specific: name the problem, the possible solution, and incentives.
3. Describe the use of wetlands to give tertiary treatment to sewage. Find a local example, if possible. What are the advantages? Disadvantages?
4. Collect rainwater on several days. Test it for acidity (pH) with a pH meter or litmus paper. Compare the pH with that of tap water. Discuss whether acid rain may be a problem in your area.
5. Explain the roles of ozone in the stratosphere and near the earth. What else can you think of that is "good" in one situation and "bad" in another?
6. What projects could you do in and around your house or apartment to minimize the use of materials that you do not recycle?
7. Discuss the idea that spreading solid wastes around the countryside is all right. For example, old cans provide homes for small animals; garbage stimulates plant and animal growth. Justify your reaction.
8. Make up your own example for Figure 34.2. Be specific as to what comes into and what spreads out of the town.

chapter thirty-five

Population and carrying capacity

Standard of living depends on resources and how many people use them. We have already discussed Florida's natural and economic resources; in this chapter, we will relate Florida's economy and its population.

35.1 Population and standard of living

With the rapid increases in population that Florida has experienced in the last three decades, the state is becoming one of the most densely populated areas of the nation. Records of population growth shown in Figure 35.1 indicate that the population has grown almost five times in the 45 years since 1950.

This growth in population is a result of people migrating to Florida from other states and other countries to the south. They are attracted to Florida for its reasonable combination of environmental quality and economic development. Many people come with an image of a rural Florida that was true when they last saw it, but which is not true in many sections now.

Figure 35.2 traces gross state product (roughly dollar income) and population. Economic production and population have increased together. Apparently the gross state product increased faster than the population between 1961 and 1990. As a state becomes more urban, more money circulates. Things have to be done and paid for in cities that are not required in more rural living.

If the population of a region increases and the amount of resources driving the economy does not, then the resources per person decreases. Sometimes the amount of resources per person is called the *standard of living*. One measure of this is emergy per person, which may be a better measure of individual wealth than income as it includes those resources used directly from the environment or from other people without paying money.

From 1980 to 1993, Florida's population grew from about 10 million people to about 13.6 million, a 36% increase. During that time Florida's total emergy use increased from 3283 E20 sej to 4369 E20 sej, a 33% increase. This

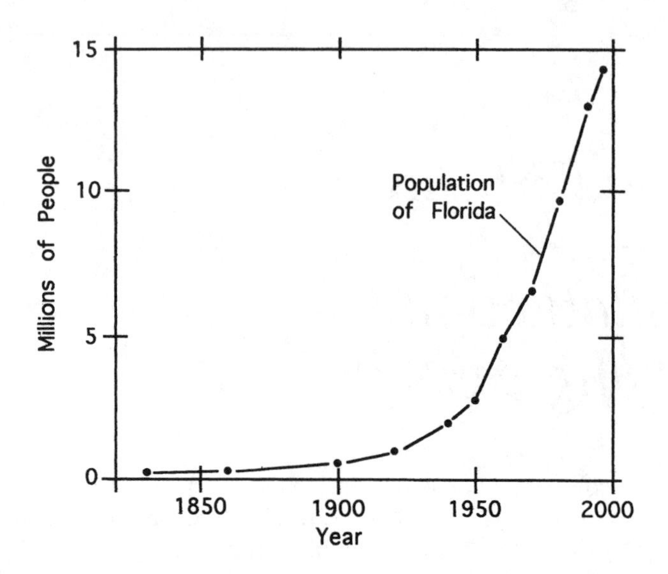

Figure 35.1 Record of population growth in Florida since 1830.

means the average standard of living, measured in emergy per person, was little changed. Will our emergy per person increase? Have we reached a steady state, or will the emergy per person decrease as migration into the state continues and the resource base levels or declines?

Table 35.2 shows some comparisons among countries and Florida. Notice the differences in total emergy use per year, total population, and the ratio of the emergy per person (a measure of the standard of living). The U.S. has a very high emergy use per person; whereas, countries such as India with enormous populations and moderate resources have a much lower standard of living.

35.2 Carrying capacity

Florida's carrying capacity for people is dependent on all resources and how they are used, but ultimately on the supply of water. Water is the limiting factor (Chapter 6). Humans can live on a much lower quantity of goods and services, but they cannot survive without water. Concern about availability of water shows up in legislative proposals to pipe water from the Suwanee River to the Tampa area, plans for desalination plants, and sporadic controls on sprinkling lawns and washing cars.

Many people believe that the growth of Florida's population will continue for many years to come because of the numbers of retired people who wish to move to the state. Others feel that the climate and lower cost of living will attract increasing numbers of people of all ages, especially if the nation should enter a period of lower resource availability. However, recent trends

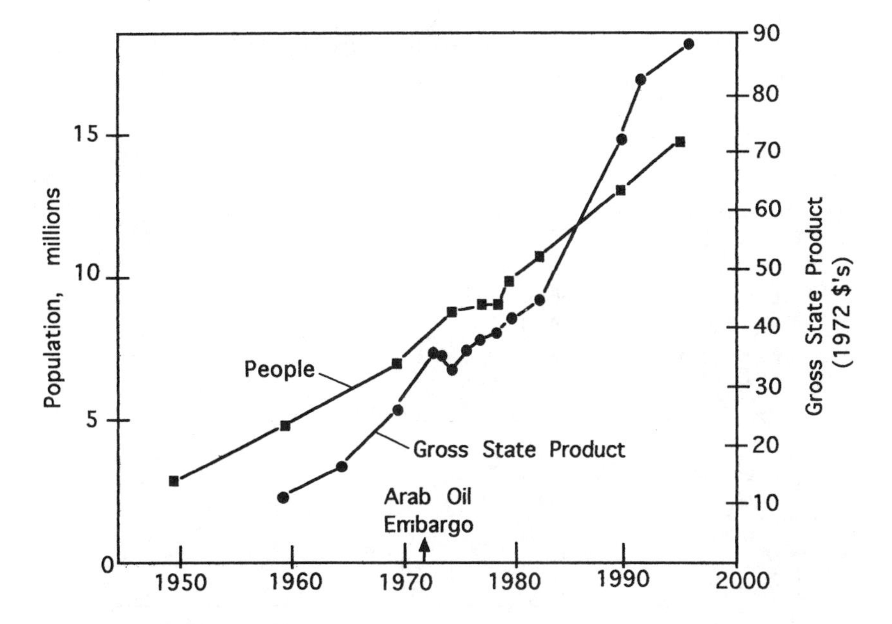

Figure 35.2 Graphs of population increase and change in gross state product since 1950.

suggest that in the long run, Florida's carrying capacity may not be able to sustain increased populations without an increasing resource base.

Because the economy of Florida is so strongly tied to that of the nation and the world, any turn-up or turn-down of these larger economies is felt even more within the state. For example, 20% of the economy of the state in 1993 was due to tourism dollars. People can only take vacations if they have been able to save excess money throughout the year, or they can only retire to Florida if they can sell their homes elsewhere.

Exactly how Florida's population growth responds to the future expansion and contractions of the world economy remains to be seen. However, one thing is certain. The long-term wellbeing of the population that lives in Florida depends on a healthy economy that includes both environment and developed areas.

Table 35.1 Emergy per Person in Florida

Year	E15 solar emjoules (sej) per person
1980	33
1990	27
1993	32

Table 35.2 Standards of Living for Various Countries and Florida, 1990

	Emergy (per year) (E22 sej)[a]	Population (E6 people)	Emergy (per person per yr) (E15 sej)
Australia	88.5	15.0	59
Sweden	41.1	8.5	48
U.S.	660.0	240.0	28
Florida	35.0	13.0	27
Netherlands	37.0	14.0	26
Italy	126.5	57.5	22
Brazil	178.2	121.0	15
Japan	153.0	121.0	13
Ecuador	10.3	9.6	10
Poland	33.0	34.5	10
Mexico	61.4	81.1	8
China	719.0	1100.0	7
World	2024.0	5250.0	4

[a] This includes environmental energies and fuels, all expressed in solar emjoules (sej).

35.3 The effect of declining resources on population

There is much controversy among population scientists about what the response of birth rates, death rates, and migration trends will be as economies begin to contract. The effect of declining carrying capacity on populations is open to some question; one answer is given by the simulation of a model in Figure 35.3.

In this model, economic assets control public health (births, deaths, epidemics), and public health controls population. As world assets increase, population growth increases, as well as the demand for nonrenewable fuel resources. Since the nonrenewable fuels are being used but not replaced, the assets begin to decrease. Population rapidly follows the downfall of the assets. Whether this model is an accurate picture of the future remains to be seen. If it is true for Florida, it would mean that population would continue increasing even after assets had declined. Then, later, population would decline, also.

Questions and activities for chapter thirty-five

1. Explain the difference between the standard of living measured as emergy use per person and measured as income per person.
2. Define the concept of carrying capacity as it pertains to a human system.

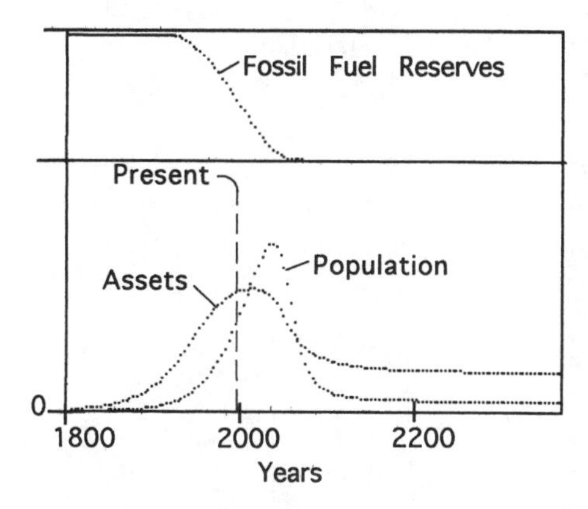

Figure 35.3 Results of simulating changes in world population based on changes in resources and economic assets. The model diagram and computer program are given in Appendix A, Figure A12.

3. Considering water as the limiting factor for population in Florida, discuss policies for more efficient use of water supplies.
4. Explain how declining resources affect a population.
5. Estimate how much land an average Floridian has to support his/her personal carrying capacity. The area of Florida is 151,570 km^2 (58,560 $miles^2$), and the population was 14 million in 1995.
6. Considering the data on emergy per person in Table 35.2, in which directions do you think migration is most likely? Why?
7. If we were to strive toward the goal of equality of emergy for all people in the world, how would the present emergy patterns have to change? How might this affect the U.S.? Florida?
8. Are you planning to have more, less, or the same number of children your parents had? Why?

chapter thirty-six

Simulating the future

The economy of Florida may grow, crest, or decline depending on the renewable and nonrenewable resources that become available elsewhere in the world. This chapter uses two simple computer simulation models to show how the future of Florida and the world may be affected by the environmental resources in Florida and fuel sources of the world. Simulation considers the alternative of continued world growth and a world scenario of rise and decline with available fuel reserves. The computer programs for the models in this chapter are given in Appendix A.

36.1 Simulation of a world mini-model

The global mini-model (Figure 36.1a) is based on resources. Soils and wood represent environmental resources. The model has world economic assets drawing inputs from the nonrenewable reserves of fuels and minerals and from the soils and wood that are "renewed" by the steady inflow of global solar energy. The economic assets are the buildings, roads, machines, and goods produced by the world economy. In the model, these stored assets feed back and interact to increase the energy use. The stored assets are also available for trade.

Results of simulating the world model are given in Figure 36.1b. It starts over 300 years ago with high storages of fuels, soils, and wood, but with low economic assets. The program first runs with the pathway from fossil fuel to the world economy turned off. Growth is based on use of the soils, wood, and related environmental resources. After some growth based on renewable sources, the assets reach a modest steady state.

Then the fuel pathway is turned on, representing the start of the industrial revolution. The abundant fuel resources cause the model to develop a great pulse of world economic growth, building assets, using up fuel reserves, and drawing down environmental resources. At a later stage, when fuel reserves are smaller, assets depreciate faster than they are being generated. In the simulation (Figure 36.1b) world assets crest and decrease.

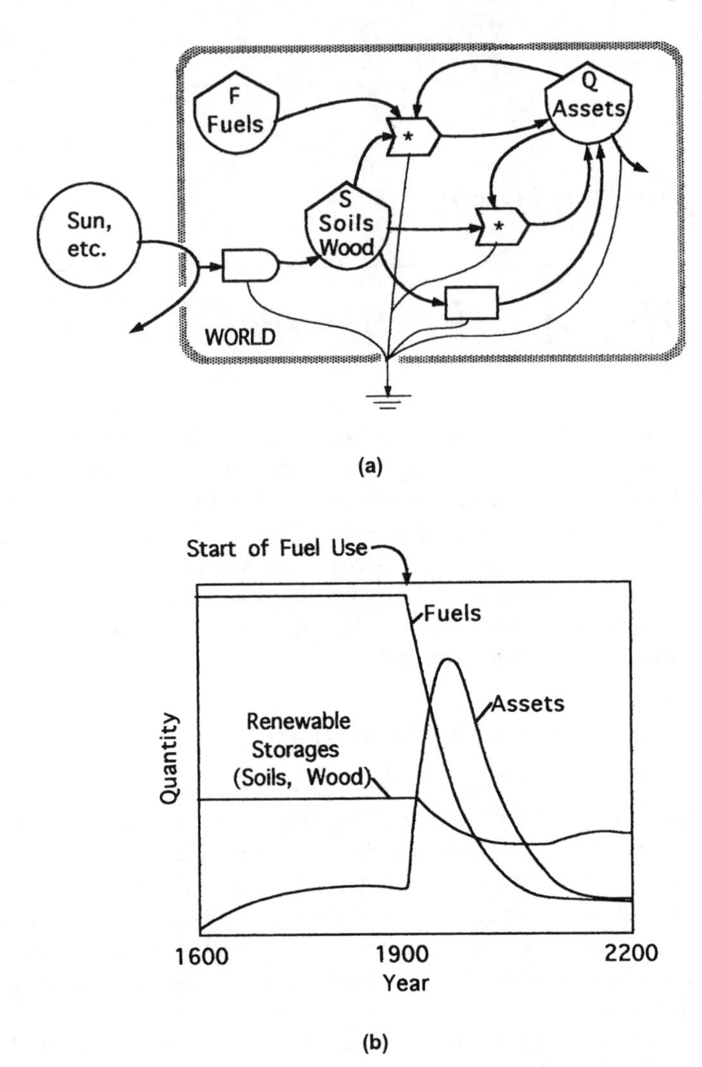

Figure 36.1 Simulation of a global mini-model. (a) Systems diagram of the world assets based on renewable resources (soils, wood) and fuel reserves; (b) simulation of growth with massive use of fuels beginning in 1900.

Leveling may be the general trend for many countries as soon as the net emergy of available fuels and minerals declines worldwide. It is difficult to visualize the real world economy doing anything very different in the long run; however, the exact shape and timing are beyond the scope of mini-models such as this. Overview models do not have enough detail to produce the small-scale up and down changes that may dominate a given decade. Just when the crest of the world's assets will occur remains to be seen. Meadows

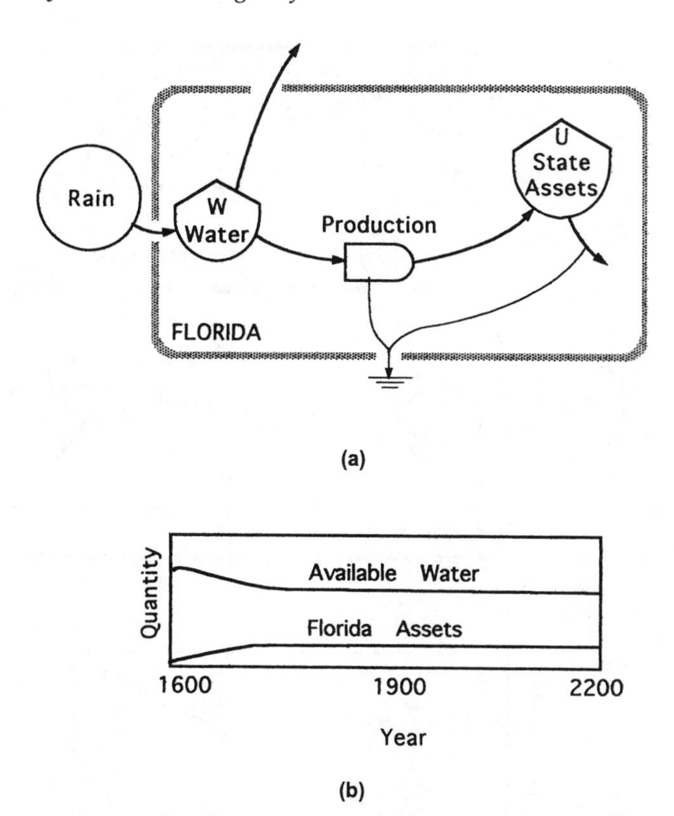

Figure 36.2 Simulation of a mini-model of Florida assets growing only on renewable resources, which are represented by the water available for development. (a) Systems diagram; (b) results of computer simulation of growth.

(1992), in a simulation of a much more complex model, suggests a cresting in the early part of the next century.

36.2 Simulating a Florida mini-model without exchange

In Figure 36.2, Florida is represented by assets based only on environmental resources within its borders. For the most part these are renewable resources, and water is used as an indicator of the various essentials necessary for human populations.

The first simulation of Florida using the model in Figure 36.2a was made without trade, without inputs from the world. The result was a simple growth and leveling off. Perhaps this represents the pattern during Indian times before Columbus.

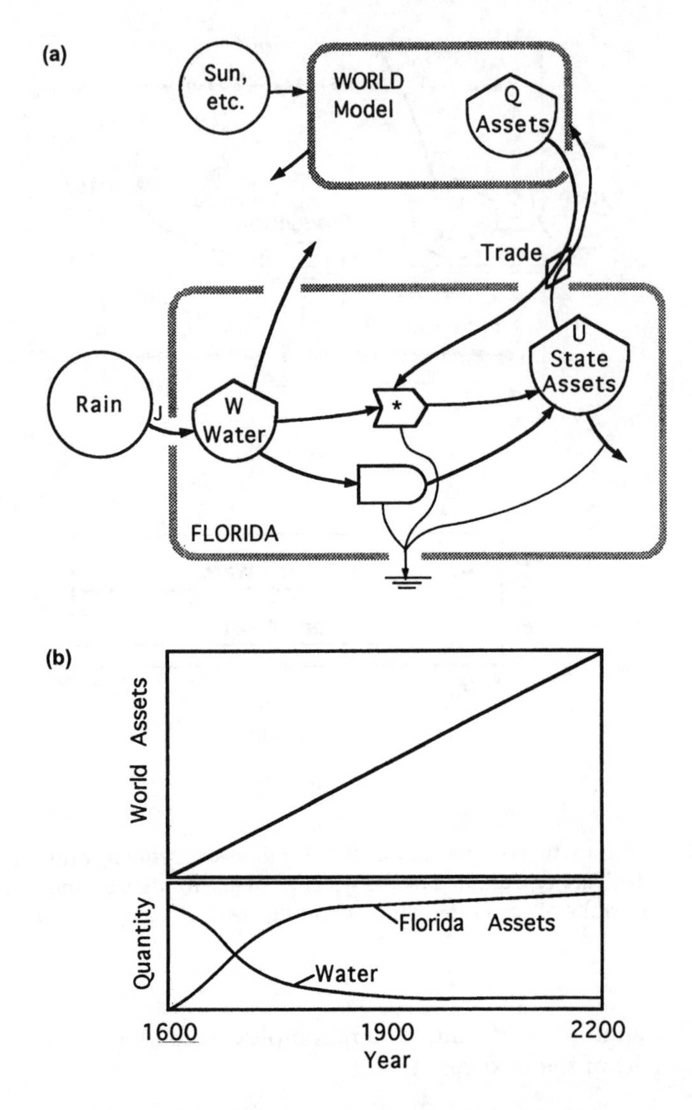

Figure 36.3 Simulation of the Florida assets mini-model including exchange with the rest of the world. (a) Systems diagram of the Florida assets mini-model showing trade exchange and nonrenewable sources. (b) Results of computer simulation of Florida assets if there were a constant growth in world assets. (c) Results of computer simulation of Florida assets if the world growth follows the pattern of rise and decline in Figure 36.1.

36.3 Simulation of Florida assets with world trade

Next we connect the Florida mini-model to the world mini-model by trade (Figure 36.3a). The lower pathway from the left, with a production symbol, represents the work of nature. The upper pathway is based on purchases

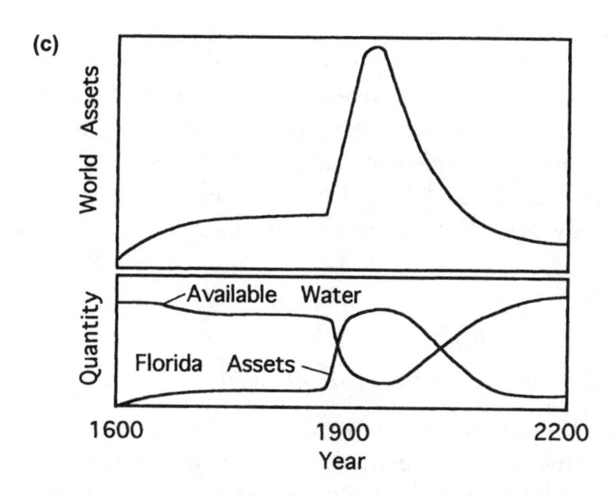

Figure 36.3 (continued)

brought into the state from elsewhere in the U.S. and the world. Both the internal and external pathways depend on environmental resources. In Florida, water represents environmental resources.

Oranges, vegetables, manufactured goods, and tourist services are exchanged for fuels, cars, goods, and services. The imports to Florida are in proportion to the world assets. In this way, the world model is made to drive the state model. Fuels, materials, goods, services, equipment, information, etc. are brought into Florida in exchange for sales and services that are sent out. In the model, these sales and services are represented by the pathway from assets that goes to the right through the exchange symbol marked trade.

The model relates Florida's growth to the assets of the world. The assets of the global economy depend on the availability of renewable and nonrenewable resources worldwide. When the world's resources are abundant, the things that people in Florida want to buy are inexpensive.

Like the diagram in Figure 31.3, the Florida model depends on the local resources to attract matching outside resources. The inflow of outside assets and fuels also depends on their availability. This availability is changing, as driven by change in world assets.

36.3.1 Simulation of state growth with steady increase in world assets

Many people believe that the growth of world assets will continue for a long time into the future. For this simulation, the world assets are arranged to perpetuate steady growth (upper part of Figure 36.3b), providing inputs to Florida in exchange for exports. The lower part of Figure 36.3b is the result of simulation of the Florida mini-model including trade with the rest of the world under these conditions.

In this scenario of continued world growth, the simulated Florida's assets do not continue indefinitely even though more and more investments of fuels and goods and services are received from outside. This is because the environmental resources (in this simulation represented by water) are drawn down to such low levels that they limit further growth of the economy. Growth has already been limited in some areas of Florida that are short of water.

36.3.2 *Simulation of Florida growth driven by the world pulse of growth*

In the simulation (Figure 36.3c), the world assets grow and decline as in the simulation in Figure 36.1. The growth of world assets makes available increased supplies of resources to the state's economy after the fuels are switched on. Thereafter, the state's assets (the bottom graph in Figure 36.3c) grow rather sharply in response to the increased availability of resources from the world market.

As world growth crests and declines, the availability of outside resources decreases, and the state's growth declines to a level sustainable by the renewable sources. Water resources are restored to a level that was characteristic of earlier steady-state times before the pulse of growth.

Summarizing, we used a simple model of the world economic system to consider what may cause growth of the world to crest and decline. Trade was arranged between the world and Florida models to dramatize the relationships between a state and the world. The simulations show the important role of fossil fuels in the world's future economic growth. The models show when a sustainable economy is possible.

Questions and activities for chapter thirty-six

1. Define: (a) simulation, (b) steady-state economy, (c) environmental resources, (d) trade.
2. List some renewable and some nonrenewable resources that influence the future growth of Florida.
3. Discuss what is meant by the assets of a state.
4. Discuss what happens to the environmental resources of Florida as the economic assets increase.
5. Describe what would happen to the growth of Florida if the amount of fossil fuels from the world were unlimited.
6. Run the world trend and combined Florida-world computer programs described in Appendix A. Try some "what if" changes to fit other possible futures.
7. Explain why you think soils and wood were used to represent the renewable resources of the world and water was used to represent the renewable resources of Florida.

8. If a model of the U.S. were substituted for the world model, what differences do you think this would make to the simulations of the Florida model? Explain.
9. Find one or two other predictions of world future trends from magazines, books, or people. Discuss how they compare to these predictions.

chapter thirty-seven

The future

As a part of the global economy, Florida's future partly depends on what happens elsewhere; however, principles from previous chapters should help us consider possible scenarios and the policies that would keep the partnership of society and environment prosperous. In this chapter, we consider four regimes that could develop (Figure 37.1): (1) growth, (2) steady state at present levels, (3) descent to a lower level, and (4) steady state at a lower level. If running out of nonrenewable resources is the predominant factor affecting the future, all of these scenarios might occur in 1-2-3-4 order. As we enter the twenty-first century, how should each of us prepare? What public policies should we support?

37.1 A scenario of continued growth

In the first scenario, Florida continues its economic growth, which is favored by some of the following conditions:

1. The world price of fuels does not rise enough to reduce the emergy yield ratio.
2. The U.S. economy remains prosperous, thus supporting domestic tourism.
3. The further development of economies in the Caribbean and Central and South America encourages their use of Florida for its centers of information and trade.
4. Florida industries help lead the wave of development of information and computer related industries — which may depend on increasing education of young people of Florida.
5. Environmental contributions such as those of groundwaters, beaches, and natural areas are not diminished by further growth.
6. Policies develop to prevent the harmful effects of further concentrations of population, cars, wastes, and unemployed.

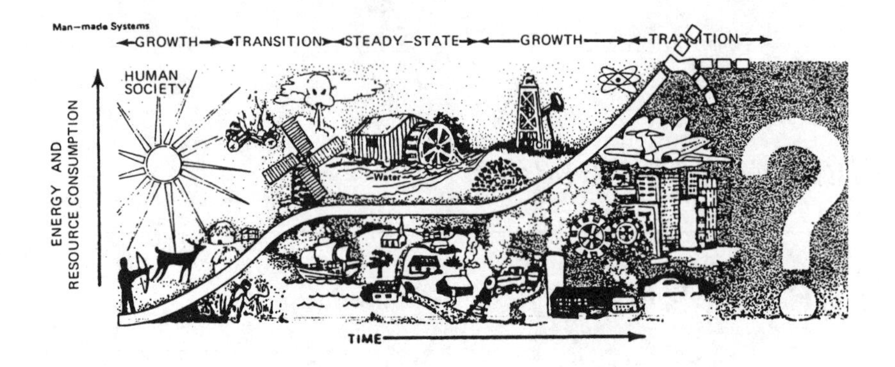

Figure 37.1 Past, present, possible futures. (From Browder, J. et al., *South Florida: Seeking a Balance of Man and Nature,* Bureau of Comprehensive Planning, Division of State Planning, Tallahassee, FL, 1977.)

7. Florida remains economically competitive by eliminating its waste and increasing its efficiency as quickly as its competitors do. Inefficiencies include unnecessary number and size of automobiles, large population supported on unearned incomes, unused housing spaces in second homes, inefficient use of energy in appliances and home heating and air-conditioning.
8. Present accumulations of air, water, and solid wastes are replaced with reuse and recycling.
9. Ways are found to prevent the further increase of eutrophic waters and unproductive land use.

Nearly everyone agrees that there are eventual limits to growth because the earth's resources are limited, but few are confident in predicting when and where growth will stop. Real growth can be recognized by a rising annual use of real wealth from foreign and domestic sources. An increase in the standard of living requires that the increase in emergy use be faster than the increase in population.

37.2 *A scenario of sustaining the present economy*

Sustainability is now a popular theme among planners and futurists. This ideal means continuing the present level of economic activity and standard of living. It is argued that even if further growth is possible, it would not be sustainable. To maintain a steady state economy for Florida at its present level will require the following measures to keep renewing each aspect of the Florida system to keep up with depreciation and disasters:

1. Shift jobs, initiatives, and taxes from building new developments to maintaining and replacing roads, houses, power plants, schools.

2. End population increase by decreasing immigration and/or reproduction. Encourage a stable age distribution (one in which the numbers in each age group are steady).
3. Arrange enough feedback from consumers to agriculture to keep it competitive with that in other countries.
4. Evaluate in advance new developments to ensure that they have a net benefit compared to what they are proposed to replace.
5. Change majority public opinion and political emphasis from growth to efficiency, from competition to cooperation.
6. Manage water demands and wastewater recycle to keep groundwater and lake levels in steady state (allowing the short-term fluctuations necessary for healthy vegetation).
7. Match the gradual decrease in net emergy yield of fuels with a corresponding increase in efficiencies of energy use (see item 7 in Section 37.1).
8. Develop policies on insurance and building limits in hurricane-risk areas that are sufficient to keep up with each year's hurricane repairs and replacements.
9. Meet the need for fertilizer/nutrients in agriculture, forestry, and aquaculture by shifting from mining and imports to reuse of wastewaters.
10. Sustain fisheries by not overfishing; maintain forestry production by operating lands in renewable cycles of growth and cutting; maintain soils by rotating land uses; maintain biodiversity by putting abandoned land in ecological succession.
11. Convert the relatively flimsy, short-lived housing of Florida to more permanent, long-lasting buildings such as those in Europe.

37.3 A scenario of descent

If and when a decreasing economy is required as fuels are used up and become more expensive, the big question is, "Can the economy decrease gradually, or will there be some kind of crash requiring a new start?". The following are measures that may allow a gradual decrease without loss of quality and standard of living:

1. Reduce population at the same rate as the decrease in net emergy yield of fuel resources.
2. When downsizing is required, cut salaries of employees instead of firing them. In this way, productivity is maintained without increasing costs of unemployment, welfare, and increased crime.
3. Develop public pride and purpose in projects that reorganize the landscape for increased quality of life that comes from less crowding (for example: starting a neighborhood gardening plot and recreation center).

4. Develop public works programs that refit released lands and buildings while sustaining employment.
5. Find ways of redirecting capital investments to projects of reduction, simplification, and conversion of some urban lands to agriculture.
6. Develop incentives for substituting bicycles for cars and living near jobs.
7. Arrange public transportation with routes in straight lines from outside to inside like spokes of a wheel, and between centers. Run small frequent buses. Let people get to the bus lines by walking and biking.
8. Develop courses and curricula that prepare students for reorganization, adapting to change, and making descent efficient.
9. Encourage local cooperation and self reliance to replace the decreasing strength of central governments.
10. Organize abandoned lands into parks and areas with ecological restoration.
11. Adapt housing (see item 13 in Section 37.4, below), clothing, and style of life to use less fuels.

37.4 A scenario of steady but smaller economy

When fuel resources on world markets are scarce (it may be a century or more), the economy of Florida has to operate at a much lower level. Within the U.S., coal reserves will still be available but will be more expensive. The net emergy yield to Florida of purchased resources will be lower than now, and the carrying capacity of the state for imported fuels less. In Chapter 31, it was estimated that 15% of the present economy could be supported on local renewable resources alone (sun, wind, rain, marine energies, and geologic processes). If the population could be reduced as much as the net emergy yield of the resources, the standard of living (emergy per person) could be sustained. The following are some characteristics of a reasonably prosperous Florida at a lower energy level.

1. The total activity of society is much reduced and the Florida landscape much different from its present organization. Social and political activities move to smaller centers. Differences in local conditions are expressed as different customs and cultures.
2. The present spatial organization around automobile use is replaced with people moving within bicycle range of their jobs or public transport. The present trends to replace transport with communication and computer work at home continues.
3. The great cities decrease not only because there are fewer people but also because more labor is needed on farms that have to operate with less energy and energy-based inputs.
4. With fuels expensive for transport, trade is more localized with homegrown and processed food products and goods. The economy diversifies.

5. Some agroecosystems are those of an earlier era which required less special fossil fuel and technological support and more human labor and yielded less. More land is needed for agricultural ecosystems, and cultivating practices do more at restoring, building, and protecting soils without expensive chemicals. With less transport and demand for specialized products, Florida agriculture production shifts away from winter vegetables and oranges.
6. Wasteful excesses and luxuries tend to disappear as priorities have to go to necessities.
7. With an understanding of the reorganization that is necessary, public attitudes and policies discourage growth and competition.
8. Use of remaining phosphates is directed to home agriculture rather than international sale, although large-scale mining is less because of high costs of fuels and electricity.
9. Nuclear power continues, and plants are replaced if nuclear fuels are still available, since electric power is essential to a high-level, information-based society.
10. With less demand, natural patterns of waters and wetlands are restored, increasing water quantities and qualities available at less cost.
11. There are great changes in the way the population treats its wastes. Dumps are obsolete as all materials are recycled and reused. Sewage is recycled on the landscape, increasing the productivity of farm and forest alike. With less fertilizer available, any materials that have nutrient value are composted and the nutrients returned to the landscape instead of being buried or shunted to the sea.
12. With smaller cities and fewer cars, less land is covered by buildings, roads, and parking lots. Vegetation is restored by punching holes in hard surfaces of parking lots and covering with a foot or two of soil. In the cities open space and lawns may give way to productive gardens and fishponds. Forestry practices would be less mechanized and more labor intensive.
13. Fuels and electricity are too costly for air-conditioning, and houses have to be better adapted to the climate. By building roofs sloped enough to protect from sun in the summer but designed to be opened in the winter, nature is used more for heating and cooling. Perhaps the "cracker" houses of Florida at the start of the century are a good, low-energy design (Figure 37.2). Deciduous trees are planted to shade in summer and open the houses to the sun in the winter.
14. With fuels and cars costly, tourism could decrease, although those now traveling abroad might come to Florida instead.
15. National budgets are less for military and space expenditures. Some of Florida's military bases and NASA facilities are closed.
16. Governmental activities decentralize with fewer laws, less control, and more participation by people in the decision-making process.

Figure 37.2 Using nature and "cracker" building style for heating and air-conditioning your house. (From Brown, M.T. and Palmer, C., *How Things Work; Why Things Work; Systems and Symbols; Both Sides Now,* four multimedia film cassettes, National Park Service, U.S. Department of the Interior, U.S. Government Printing Office, Washington, D.C., 1979.)

17. The best products of our technological society are retained where the energy cost of maintenance is not large, including vaccines, drugs, smaller computers, radios, television, bicycles, and books.
18. Agricultural practice require more diversity of crops. More crop rotation is necessary, with fallow stages to reduce weeds and insects and to allow successional vegetation to extract fertilizer elements from rocks and rains. By leaving vegetation strips around fields for habitat, birds and other insect predators are used in place of pesticides to offset the high costs of chemicals.
19. The struggle to maintain natural environments continues because demand will increase for land for agricultural and forestry production. However, with more urban lands becoming available, natural areas are retained while also increasing agriculture area.
20. Ways of living that involve more dispersed populations, more rural patterns of life, and fewer differences between the richest and poorest are regarded as progressive. Conditions for individuals are good if the emergy per person is kept large by the decline of population as emergy use decreases.

For generations we have been accustomed to thinking of progress as more, larger, more complex, and more technological. The anticipation of a lower energy world which is smaller, simpler, and less technological may seem like anti-progress. However, humans have been in lower energy societies for much of their history. We may be better adapted to the lower energy regime than to the intense urban world of the present.

Whatever the scenario, human democratic society through public discourse is very good at recognizing changed conditions, developing a unified zeal to adapt to new opportunities with invention and pleasure. Individuals can prepare themselves best by getting a broad education so as to be prepared for whatever is ahead.

Questions and activities for chapter thirty-seven

Consider the following questions for each of the four scenarios:

1. Describe how your community disposes of water, sewage, garbage, trash, metals. How would this change?
2. Describe the origins of the components of your most recent meal. How would meals change?
3. Discuss features of your community that would disappear.
4. Describe how you get to school. How would this change?
5. Discuss features that your community would add.
6. List three jobs that would not exist.
7. List three new jobs that would be available.
8. Try a week of a diet with small amounts of protein (i.e., 1/2 pound of high-quality meat) and no packaged foods. Think of other ways to experience future scenarios and try them.
9. What will you miss most from your present life? What least?
10. What sort of retraining should be given to Floridians to prepare them for these scenarios? How should formal schooling be changed?
11. Collect pictures of low-energy and high-energy activities, and make a collage of each. Be prepared to use them to explain your ideas about the changes to the class.
12. Convene a discussion group to think of additional ways to adapt. Hold class debates on the suggestions in this chapter.

Appendices

Appendix A

Programs for computer simulations

This Appendix contains the simulation models discussed in Chapters 6, 9, 10, 11, 35, and 36. Figures A1 to A14 include the diagrams, rate equations, and listings of the programs in BASIC, as well as cross-references to the models and discussions in the main text.

Changing programs for Macintosh computers

These programs are written for PC computers. To change a program for a Macintosh, delete the CLS statement, SCREEN statement, COLOR statement, last number in the LINE statement, and last comma and number in each PSET statement.

Running programs

- Load BASIC or QUICKBASIC.
- Open the program you want.
- Run it; the graph should look like the one in the text.
- Print the graph.

Each kind of computer has different ways to print the graphs. Here are two of the common ones. You may have to check with your computer manual.

For Windows95, when the graph appears, press Print Screen on the keyboard. This will put the graph on the clipboard. You can then paste the graph into a word processing or drawing program and print it from there.

For a Power Macintosh, when the graph appears, simultaneously hold down the command key, the Shift key, and the 3 key. This will put a picture of the screen on your desktop. Double-click on this picture icon and select Print from the File menu.

You can now do "what if" experiments like those suggested in questions after the chapters. Think of an example of this system that is different from the text. Predict what would happen if you changed one of the variables. Change the variable and run the program. If it does not do what you expected, figure out why. If it does do what you expected, try another change. (It is best to change only one variable at a time so you can tell which variable has produced a particular result.)

Environmental decision-making modules

In several chapters, we suggest running our symbol-based (object-oriented) computer exercises that have been published on CD-ROM. *Environmental Decision-Making (EDM)*, by E.C. Odum, H.T. Odum, and N.S. Peterson, contains computer programs which run with EXTEND. The programs teach ecological systems concepts suitable for general biology, ecology, and environmental science. Icons are connected by mouse, values are entered in dialog boxes, and simulations result. A teaching manual is available. Models include food chains and fishing in a pond (Chapter 17, Question 17), growth and fire in grassland (Chapter 19, Question 9), and logging, management, and sales in a forest (Chapter 22, Question 11). General systems icons are also included for readers to assemble their own simulations.

Environmental Decision-Making is part of a CD-ROM and booklet in the BioQUEST Library, which includes many simulation modules on topics across the bioscience curriculum. The BioQUEST CD-ROM and additional information are available from:

BioQUEST Curriculum Consortium
700 College Street
Beloit College
Beloit, WI 53511
Voice: 608-363-2743
Fax: 608-363-2052
e-mail: bioquest@beloit.edu
Web site: http://www.beloit.edu/~bquest

The simulation program EXTEND is available from:

Imagine That
6830 Via Del Oro, Suite 230
San Jose, CA 95119-1353
Voice: 408-365-0305
Fax: 408-629-1251
e-mail: extend@imaginethatinc.com
Web site: http://www.imaginethatinc.com

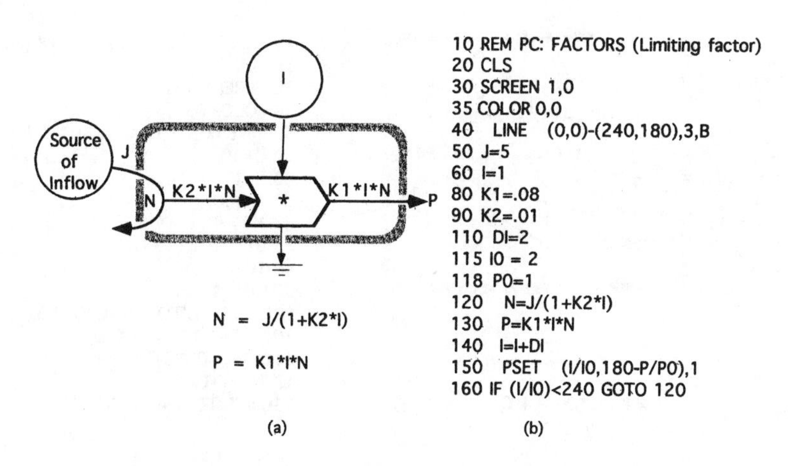

Figure A1 Model of limiting factors, FACTORS (see Figures 6.3 and 6.4). (a) Diagram; (b) computer program.

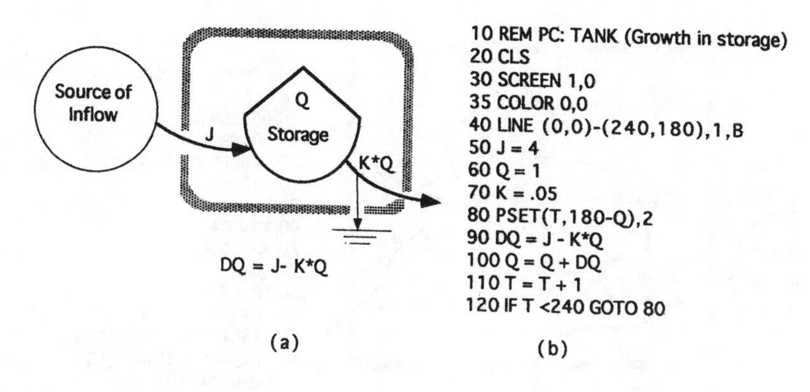

Figure A2 Model of growth in a storage, TANK (see Figure 9.1). (a) Diagram; (b) computer program.

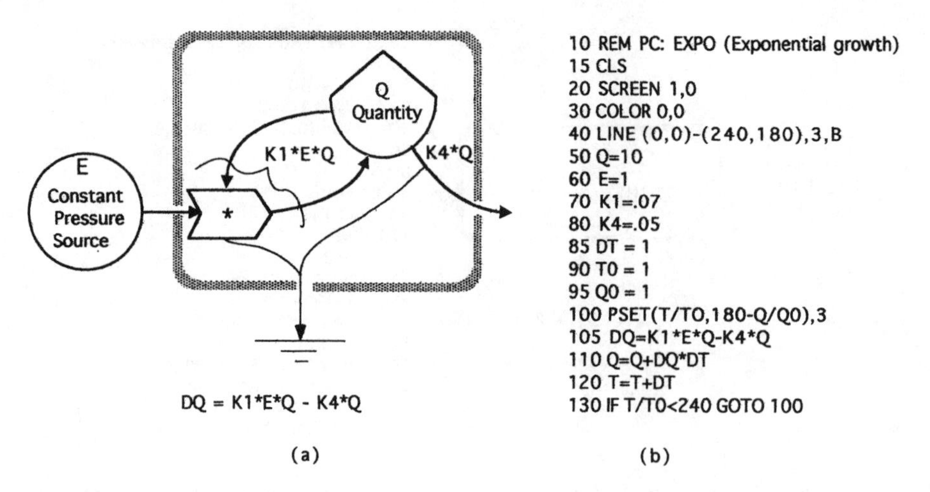


```
10 REM PC: EXPO (Exponential growth)
15 CLS
20 SCREEN 1,0
30 COLOR 0,0
40 LINE (0,0)-(240,180),3,B
50 Q=10
60 E=1
70 K1=.07
80 K4=.05
85 DT = 1
90 TO = 1
95 Q0 = 1
100 PSET(T/TO,180-Q/Q0),3
105 DQ=K1*E*Q-K4*Q
110 Q=Q+DQ*DT
120 T=T+DT
130 IF T/TO<240 GOTO 100
```

Figure A3 Model of exponential growth, EXPO (see Figure 10.1). (a) Diagram; (b) computer program.

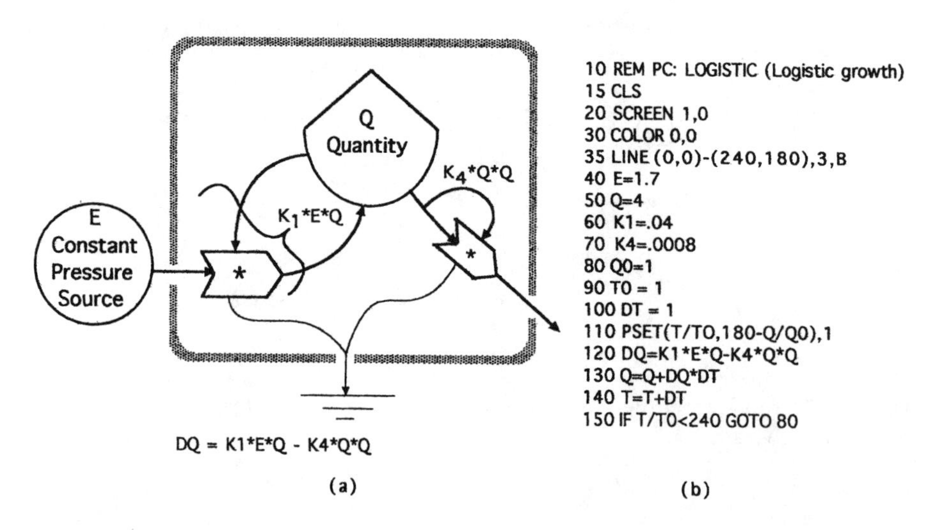


```
10 REM PC: LOGISTIC (Logistic growth)
15 CLS
20 SCREEN 1,0
30 COLOR 0,0
35 LINE (0,0)-(240,180),3,B
40 E=1.7
50 Q=4
60 K1=.04
70 K4=.0008
80 Q0=1
90 TO = 1
100 DT = 1
110 PSET(T/TO,180-Q/Q0),1
120 DQ=K1*E*Q-K4*Q*Q
130 Q=Q+DQ*DT
140 T=T+DT
150 IF T/TO<240 GOTO 80
```

Figure A4 Model of logistic growth, LOGISTIC (see Figure 10.2). (a) Diagram; (b) computer program.

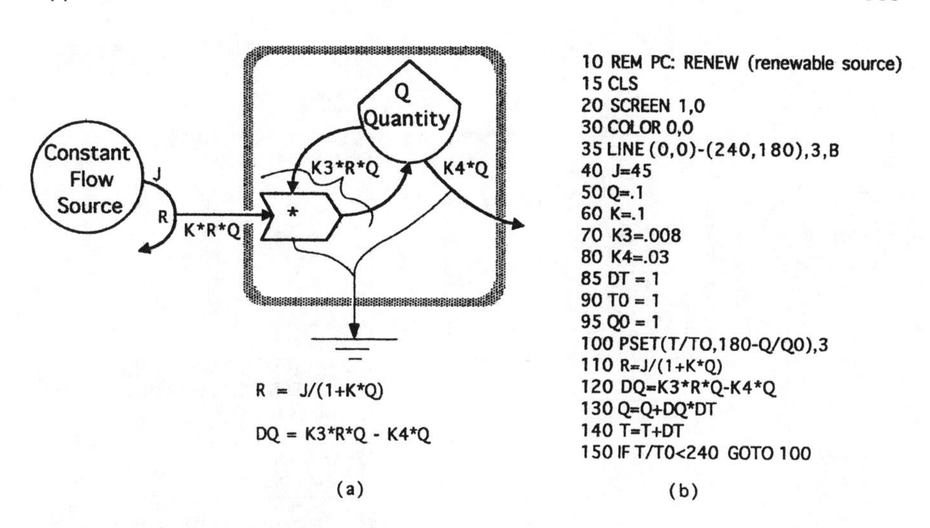


```
10 REM PC: RENEW (renewable source)
15 CLS
20 SCREEN 1,0
30 COLOR 0,0
35 LINE (0,0)-(240,180),3,B
40 J=45
50 Q=.1
60 K=.1
70 K3=.008
80 K4=.03
85 DT = 1
90 T0 = 1
95 Q0 = 1
100 PSET(T/T0,180-Q/Q0),3
110 R=J/(1+K*Q)
120 DQ=K3*R*Q-K4*Q
130 Q=Q+DQ*DT
140 T=T+DT
150 IF T/T0<240 GOTO 100
```

Figure A5 Model of growth on a renewable source, RENEW (see Figure 10.4). (a) Diagram; (b) computer program.

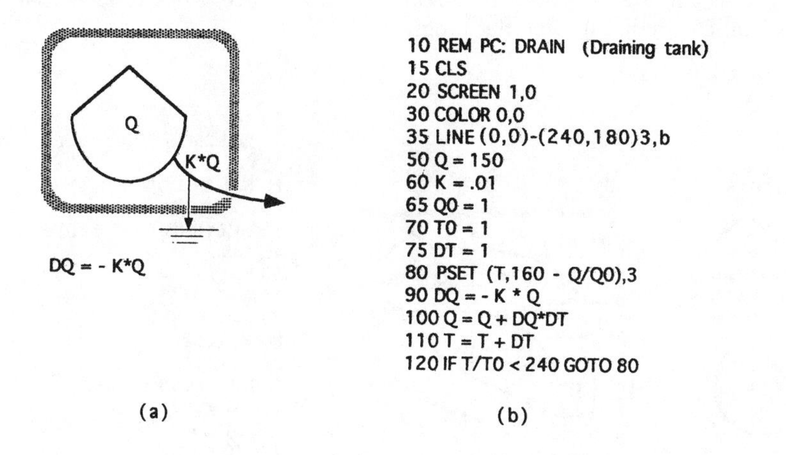


```
10 REM PC: DRAIN  (Draining tank)
15 CLS
20 SCREEN 1,0
30 COLOR 0,0
35 LINE (0,0)-(240,180)3,b
50 Q = 150
60 K = .01
65 Q0 = 1
70 T0 = 1
75 DT = 1
80 PSET (T,160 - Q/Q0),3
90 DQ = - K * Q
100 Q = Q + DQ*DT
110 T = T + DT
120 IF T/T0 < 240 GOTO 80
```

Figure A6 Model of draining tank, DRAIN (see Figure 10.5). (a) Diagram; (b) computer program.

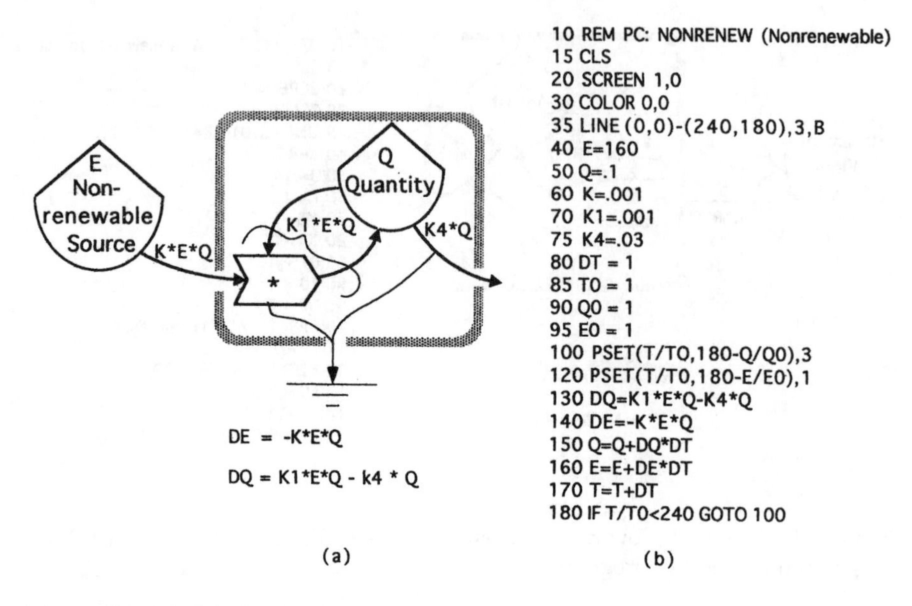


```
10 REM PC: NONRENEW (Nonrenewable)
15 CLS
20 SCREEN 1,0
30 COLOR 0,0
35 LINE (0,0)-(240,180),3,B
40 E=160
50 Q=.1
60 K=.001
70 K1=.001
75 K4=.03
80 DT = 1
85 TO = 1
90 QO = 1
95 EO = 1
100 PSET(T/TO,180-Q/QO),3
120 PSET(T/TO,180-E/EO),1
130 DQ=K1*E*Q-K4*Q
140 DE=-K*E*Q
150 Q=Q+DQ*DT
160 E=E+DE*DT
170 T=T+DT
180 IF T/TO<240 GOTO 100
```

Figure A7 Model of growth on a nonrenewable source, NONRENEW (see Figure 10.6). (a) Diagram; (b) computer program.

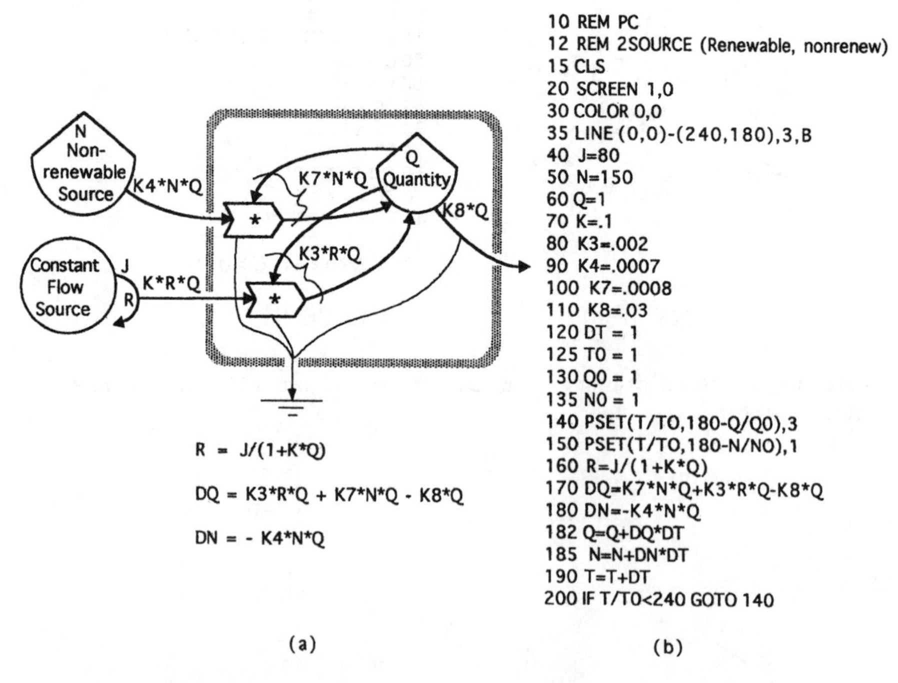


```
10 REM PC
12 REM 2SOURCE (Renewable, nonrenew)
15 CLS
20 SCREEN 1,0
30 COLOR 0,0
35 LINE (0,0)-(240,180),3,B
40 J=80
50 N=150
60 Q=1
70 K=.1
80 K3=.002
90 K4=.0007
100 K7=.0008
110 K8=.03
120 DT = 1
125 TO = 1
130 QO = 1
135 NO = 1
140 PSET(T/TO,180-Q/QO),3
150 PSET(T/TO,180-N/NO),1
160 R=J/(1+K*Q)
170 DQ=K7*N*Q+K3*R*Q-K8*Q
180 DN=-K4*N*Q
182 Q=Q+DQ*DT
185 N=N+DN*DT
190 T=T+DT
200 IF T/TO<240 GOTO 140
```

Figure A8 Model of growth on two sources, 2SOURCE (see Figure 10.8). (a) Diagram; (b) computer program.

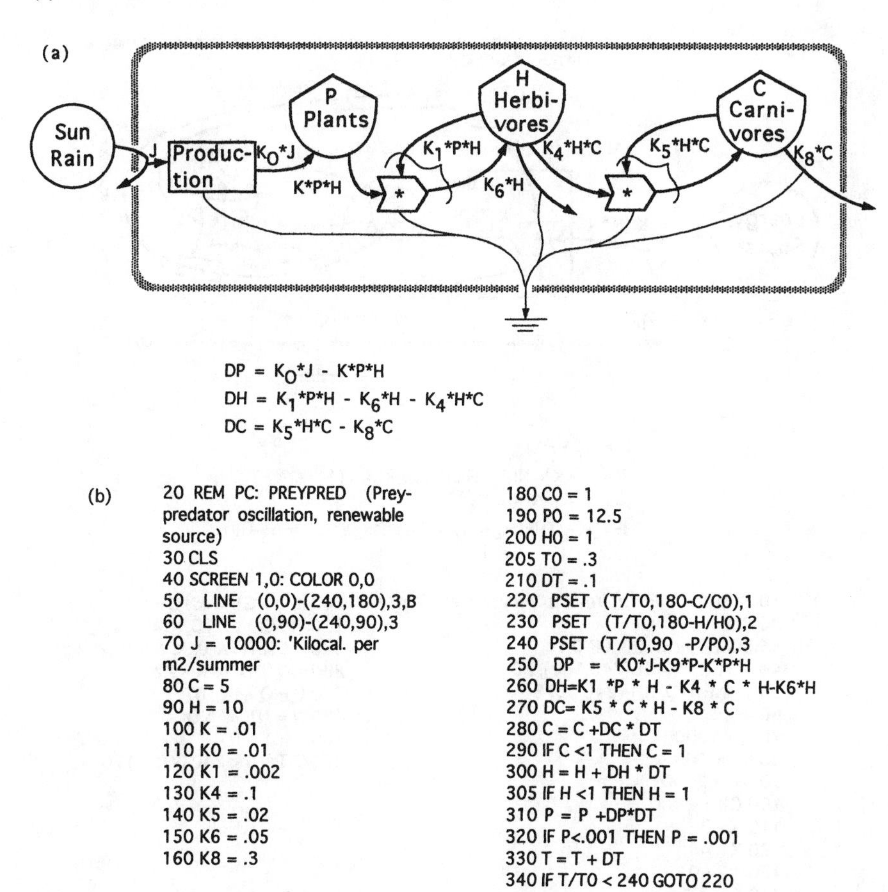

$$DP = K_0*J - K*P*H$$
$$DH = K_1*P*H - K_6*H - K_4*H*C$$
$$DC = K_5*H*C - K_8*C$$

(b)

```
20 REM PC: PREYPRED (Prey-
predator oscillation, renewable
source)
30 CLS
40 SCREEN 1,0: COLOR 0,0
50  LINE  (0,0)-(240,180),3,B
60  LINE  (0,90)-(240,90),3
70 J = 10000: 'Kilocal. per
m2/summer
80 C = 5
90 H = 10
100 K = .01
110 KO = .01
120 K1 = .002
130 K4 = .1
140 K5 = .02
150 K6 = .05
160 K8 = .3
180 CO = 1
190 PO = 12.5
200 HO = 1
205 TO = .3
210 DT = .1
220  PSET  (T/TO,180-C/CO),1
230  PSET  (T/TO,180-H/HO),2
240  PSET  (T/TO,90 -P/PO),3
250  DP  =  KO*J-K9*P-K*P*H
260 DH=K1 *P * H - K4 * C * H-K6*H
270 DC= K5 * C * H - K8 * C
280 C = C +DC * DT
290 IF C <1 THEN C = 1
300 H = H + DH * DT
305 IF H <1 THEN H = 1
310 P = P +DP*DT
320 IF P<.001 THEN P = .001
330 T = T + DT
340 IF T/TO < 240 GOTO 220
```

Figure A9 Model of prey-predator oscillation, PREYPRED (see Figure 11.2). (a) Diagram; (b) computer program.

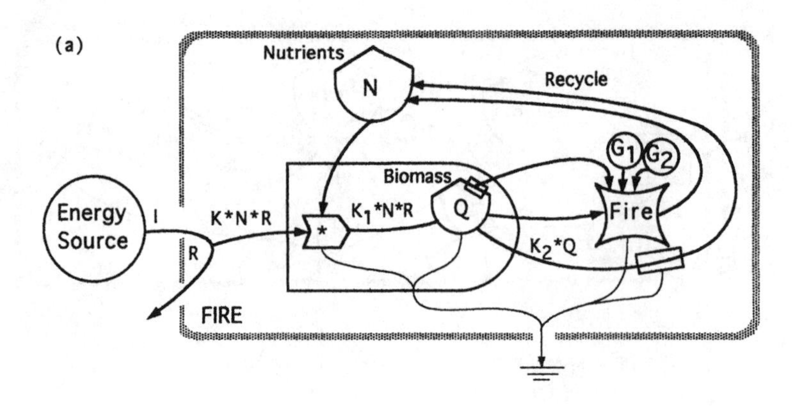

$N = N_t - F*Q$

$R = I - K*N*R$ Therefore $R = I/(1 + K*N)$

$DQ = K_1*N*R - K_2*Q$

IF $Q > G_1$ THEN $Q = G_2$ (Fire change : $DQ = G_1 - G_2$)

(b)

```
10 REM PC: FIRE  (Fire with recycled nutrients)
30 CLS
35 SCREEN 1,0: COLOR 0,1
40  LINE  (0,0)-(240,180),3,B
50  LINE  (0,90)-(240,90),3
60 I=10
70 Q = 1000
80 NT = 100
90 G1 = 5000
100 G2 = 2000
110 K=.9
120 K1 =8
130 K2=.01
140 F = .01
150 Q0 = .01
160 N0=.9
163 T0 = 1
165 DT = 1
170  R=I/(1+K*N)
180 IF Q > G1 THEN LINE (T/T0,180-G1*Q0)-(T/T0,180-G2*Q0)
190 IF Q > G1 THEN LINE (T/T0,90-(NT-F*G1)*N0)-(T/T0,90-(NT-
F*G2)*N0)
200 IF Q > G1 THEN Q = G2
210  PSET  (T/T0,180-Q*Q0),1
220  PSET(T/T0,90-N*N0),2
230 DQ =K1*N*R - K2* Q
240 Q = Q + DQ*DT
250 N = NT - F * Q
260 T = T + DT
270 IF T/T0 < 240 GOTO 170
```

Figure A10 Model of oscillation with fire, FIRE (see Figure 11.4). (a) Diagram; (b) computer program. Q = quantity of grass per m^2; G_1 = threshold amount of grass to turn fire on; G_2 = threshold amount of grass to turn fire off; NT = total nutrients in system; N = nutrients in soil; F = proportion of grass biomass that is nutrients; I = energy source (sun, wind, rain).

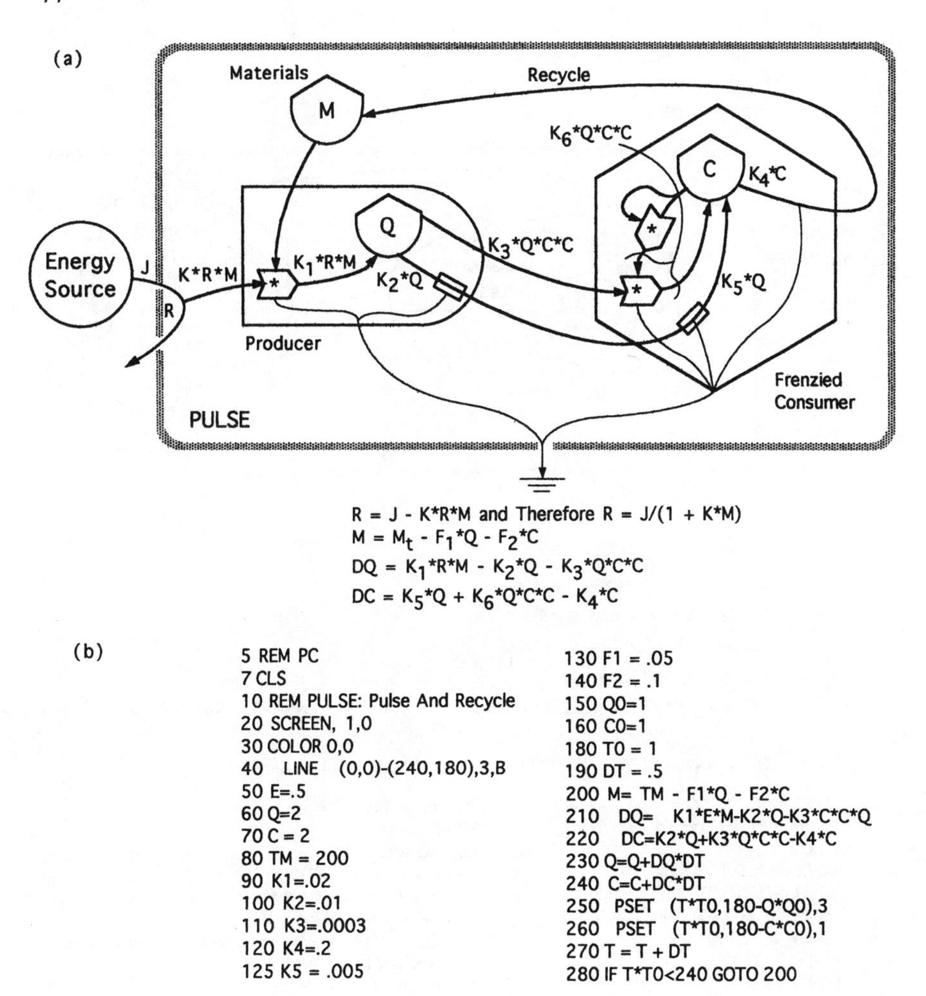

Figure A11 Model of pulsing system, PULSE (see Figure 11.5). (a) Diagram; (b) computer program.

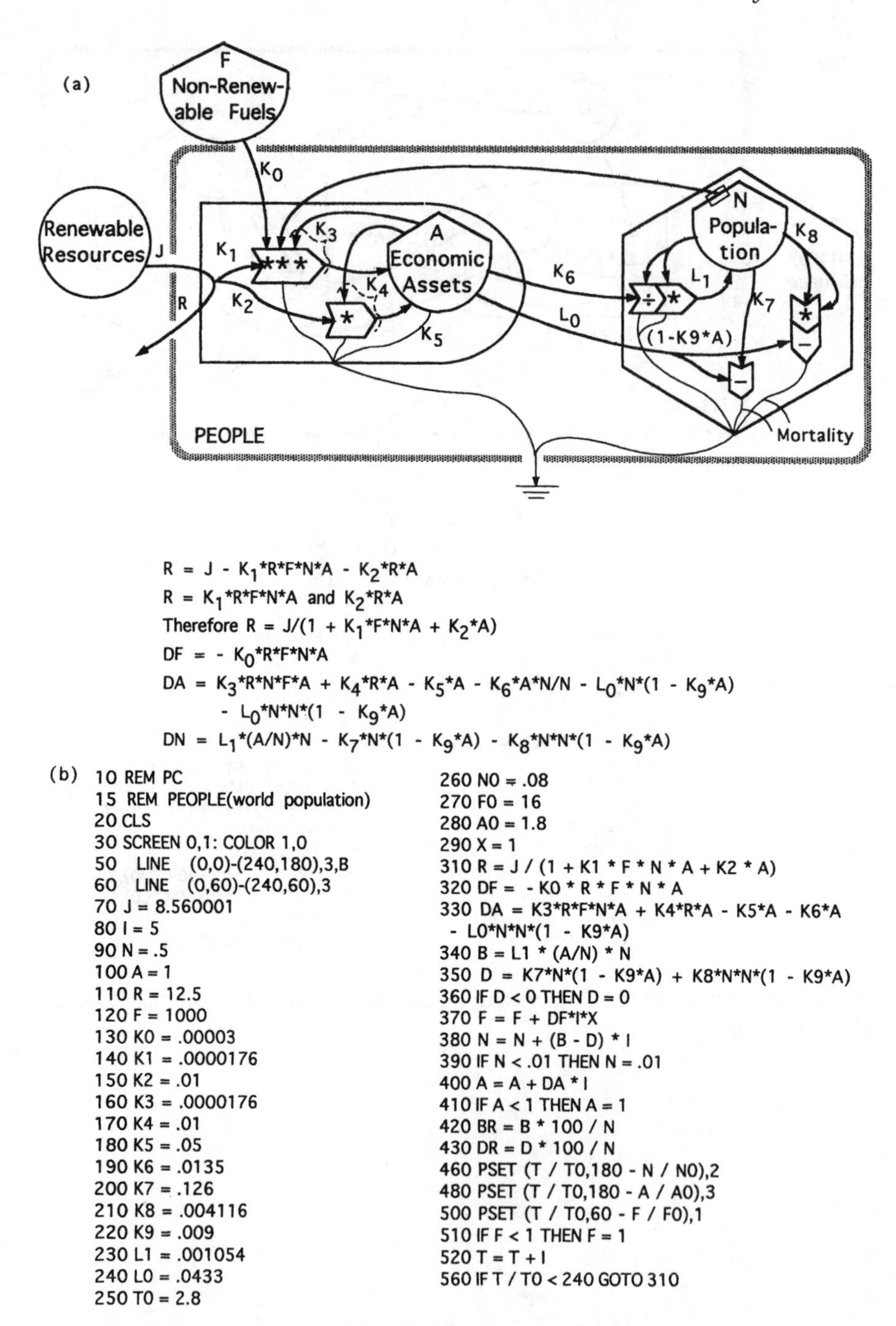

$R = J - K_1{*}R{*}F{*}N{*}A - K_2{*}R{*}A$

$R = K_1{*}R{*}F{*}N{*}A$ and $K_2{*}R{*}A$

Therefore $R = J/(1 + K_1{*}F{*}N{*}A + K_2{*}A)$

$DF = - K_0{*}R{*}F{*}N{*}A$

$DA = K_3{*}R{*}N{*}F{*}A + K_4{*}R{*}A - K_5{*}A - K_6{*}A{*}N/N - L_0{*}N{*}(1 - K_9{*}A) - L_0{*}N{*}N{*}(1 - K_9{*}A)$

$DN = L_1{*}(A/N){*}N - K_7{*}N{*}(1 - K_9{*}A) - K_8{*}N{*}N{*}(1 - K_9{*}A)$

(b)

```
10 REM PC
15 REM PEOPLE(world population)
20 CLS
30 SCREEN 0,1: COLOR 1,0
50  LINE  (0,0)-(240,180),3,B
60  LINE  (0,60)-(240,60),3
70 J = 8.560001
80 I = 5
90 N = .5
100 A = 1
110 R = 12.5
120 F = 1000
130 KO = .00003
140 K1 = .0000176
150 K2 = .01
160 K3 = .0000176
170 K4 = .01
180 K5 = .05
190 K6 = .0135
200 K7 = .126
210 K8 = .004116
220 K9 = .009
230 L1 = .001054
240 LO = .0433
250 TO = 2.8
260 NO = .08
270 FO = 16
280 AO = 1.8
290 X = 1
310 R = J / (1 + K1 * F * N * A + K2 * A)
320 DF =  - KO * R * F * N * A
330 DA = K3*R*F*N*A + K4*R*A - K5*A - K6*A
  - LO*N*N*(1 - K9*A)
340 B = L1 * (A/N) * N
350 D = K7*N*(1 - K9*A) + K8*N*N*(1 - K9*A)
360 IF D < 0 THEN D = 0
370 F = F + DF*I*X
380 N = N + (B - D) * I
390 IF N < .01 THEN N = .01
400 A = A + DA * I
410 IF A < 1 THEN A = 1
420 BR = B * 100 / N
430 DR = D * 100 / N
460 PSET (T / TO,180 - N / NO),2
480 PSET (T / TO,180 - A / AO),3
500 PSET (T / TO,60 - F / FO),1
510 IF F < 1 THEN F = 1
520 T = T + I
560 IF T / TO < 240 GOTO 310
```

Figure A12 Model for population change, PEOPLE (see Figure 35.3). (a) Diagram; (b) computer program. Each flow in the diagram is indicated by the coefficient written on the line.

(a)

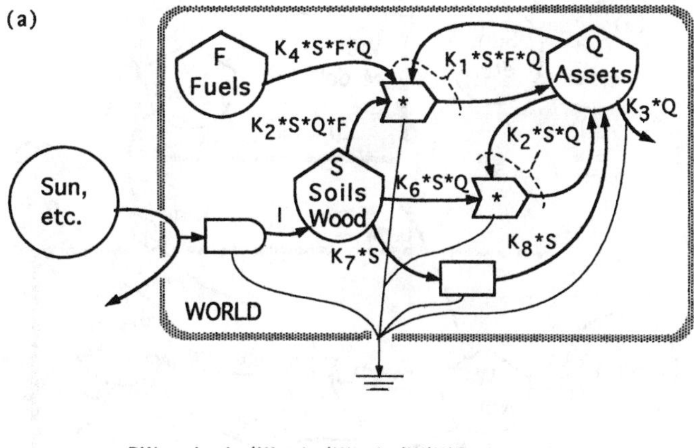

$$DW = J - L_3*W - L_5*W - L_1*W*U*Q$$
$$DU = L_3*U*Q*W + L_4*W - L_6*U - L*U$$
$$DS = I - K_7*S - K_6*S*Q - K_5*S*F*Q$$
$$DF = - K_4*S*F*Q$$
$$DQ = + K_8*S + K_2*S*Q + K_1*S*F*Q - K_3*Q$$

(b)

```
2 REM PC
4 REM  WORLD
8 SCREEN 1,0
10 COLOR 0,1
12   LINE   (0,0)-(320,180),3,B
15  S=60000!
20  SO=1000!
25 DT = 1
30 Q=1
40 Q0=2
50 F=0
60 FO=6
70  I=120
80 TO=2.2
90  K1=1E-09
100  K2=.0000005
110  K3=.04
120  K4=2E-09
130  K5=1.8E-08
140  K6=.000018
150  K7=.0018
160  K8=.00001
200 IF T=300 THEN F=1000
205 REM GRAPHING ASSETS
210   PSET(T/T0,180-Q/Q0),3
220  PSET  (T/T0,180-S/SO),1
222 If T <300 Then PSET (T/T0,12)
225 REM GRAPHING FUEL
230  PSET  (T/T0,180-F/FO),2
240   D1=K1*S*F*Q+K2*S*Q-K3*Q+K8*S
250   D2=I-K7*S-K5*S*F*Q-K6*S*Q
260   D3=-K4*S*F*Q
270 Q=Q+D1*DT
280  S=S+D2*DT
290  F=F+D3*DT
500 T=T+DT
510 IF T<320*TO GOTO 200
```

Figure A13 Mini-model of world growth, WORLD (see Figure 36.1). (a) Diagram; (b) computer program.

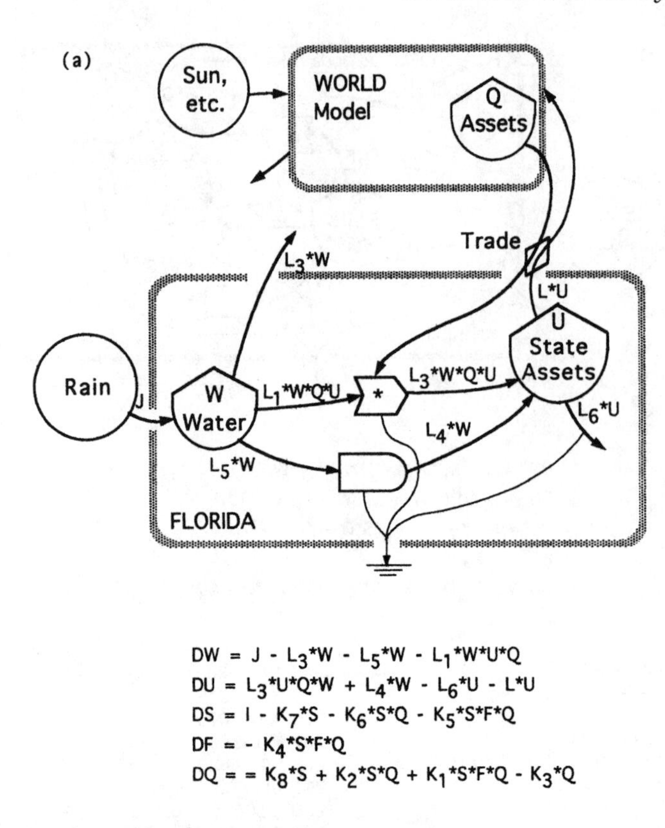

Figure A14 Model for growth of Florida, FLORIDA, driven by the mini-model of the world's growth as discussed in Chapter 36. (a) Diagram; (b) computer program.

(b)
```
20 REM PC: FLWORLD: (Florida model driven by world model)
30 CLS
40 SCREEN 1,0: COLOR 0,1
50 LINE (0,0)-(320,180),3,B
60 LINE (0,80)-(320,80),2
70 I = 120
80 J = 2
90 S = 60000!
100 Q = 1
110 W = 10
120 F = 0
130 U = .1
140 K1 = 1E-09
150 K2 = .0000005
160 K3 = .04
170 K4 = 2E-09
180 K5 = 1.8E-08
190 K6 = .000018
200 K7 = .0018
210 K8 = .00001
220 L = .1
230 L1 = .00016
240 L2 = .000266
250 L3 = .17
260 L4 = .01
270 L5 = .006
280 L6 = .04666
290 S0 = .001
300 Q0 = .3
310 F0 = .08
320 DT = 1
330 T0 = .45
340 U0 = 3
350 W0 = 3
360 REM WORLD IS TOP OF SCREEN
370 IF T = 300 THEN F = 1000
380 PSET  (T * T0,80 - Q * Q0),2
390 PSET (T * T0,80 - S * S0),1
400 PSET (T * T0,80 - F * F0),3
410 PSET (T * T0,180 - W * W0),1
420 PSET (T*T0,180 - U*U0),3
430 D1 = K1 * S * F * Q + K2 * S * Q - K3 * Q + K8 * S
440 D2 = I - K5 * S * F * Q - K6 * S * Q - K7 * S
450 D3 = - K4 * S * F * Q
460 D4 = J - L1 * Q * U * W - L3 * W - L5 * W
470 D5 = - L * U + L2 * U * Q * W + L4 * W - L6 * U
480 Q = Q + D1*DT
490 S = S + D2*DT
500 F = F + D3*DT
510 W = W + D4*DT
520 IF W < .1 THEN W = .1
530 U = U + D5*DT
540 T = T + DT
550 IF T * T0 < 320 GOTO 360
```

Figure A14 (continued)

Appendix B

Emergy evaluation of Florida

Table B1 Florida's Emergy Use (Discussed in Chapter 31)

Item	Data (units/yr)	Solar Emergy (sej/unit)	Solar Emergy (E20 sej/yr)
Direct sun[a]	1.7 E21 J	1.0	17.1
Wind[b]	5.3 E16 J	6.63 E2	0.64
Tide[c]	1.12 E17 J	4.9 E4	54.9
Waves[d]	1.23 E17 J	3.06 E4	37.6
Rain (including shelf)[e]	2.20 E18 J	1.54 E4	338.8
Elevated potential	3.6 E15 J	1.04 E4	0.37
River inflow (chemical)[f]	9.5 E16 J	5.0 E4	47.5
Hydroelectric power	2.3 E15 J	1.7 E5	3.9
Hurricanes[g]	1.5 E17 J	4.1 E4	61.8
Geologic cycle[h]	1.4 E17 J	3.4 E4	47.6
Fuels[i]			
Gas from within Florida	7.9 E15 J	4.8 E4	3.8
Gas from outside Florida	3.9 E17 J	4.8 E4	187.2
Imported coal	6.9 E17 J	4.0 E4	276.0
Imported petroleum	1.7 E18 J	6.6 E4	1139.1
Electricity from imported nuclear fuel	2.9 E17 J	1.7 E5	493.0
Total fuels used			2099.1
Total outside fuels			2095.3
Peat mined[j]	4.3 E15 J	3.5 E4	1.6
Phosphate mined[k]	29.0 E12 g	3.9 E9/g	1131.0
Used in Florida	2.9 E12 g	3.9 E9/g	113.1
Imported goods and services[l]	129.1 E9 $	1.3 E12/$	1678.2
Net population migration[m]	108.5 E3	3.3 E16/person	35.8
Exports			
Goods and services[n]	141.26 E9 $	1.3 E12/$	1836.4
Phosphate[o]	26.1 E12 g	3.9 E9/g	1017.9
Illegal drugs[p]	18.7 E9 $	1.3 E12/$	1598.0
Petroleum[q]	3.8 E16 J	6.6 E4/J	25.5
Total			4477.8

Table B1 (cont.) Florida's Emergy Use

Item	Data (units/yr)	Solar Emergy (sej/unit)	Solar Emergy (E20 sej/yr)
Total imports[r]			
Goods and services			1678.2
Fuels			2095.3
Population			35.8
Total			3809.3
Sources from within Florida[s]			
Nonrenewable			118.4
Renewable			441.2
Total			559.6
Total emergy use[t]			4368.9
Emergy/$ ratio = total use/gross state product = 1.62 E12 sej/$[u]			

Abbreviations: sec, second; yr, year; m, meter; g, gram; kcal, kilocalorie; J, joule; kwh, kilowatt-hours. *Area of Florida:* land = 1.4 E11 m^2; marine shelf — Atlantic = 0.34 E11 m^2, Gulf = 1.4 E11 m^2; total = 3.14 E11 m^2; gravity = 9.8 m/sec^2.

[a] Direct sun: (1.29 E6 SE kcal/m^2/yr)(3.14 E11 m^2)(4186 J/kcal) = 1.7 E21 J/yr.

[b] Wind kinetic energy: momentum diffusion from winds above, using data for Tampa (Swaney, 1978). Eddy diffusion coefficients: 2.82 (January) and 1.7 (July); mean, 2.26 m^3/m/sec; vertical velocity gradient, 2.6 E3 (January), 1.51 E3 (July); mean, 1.89 E5 m/sec/m. (3.14 E11 m^2)(1000 m)(1.23 kg/m^3)(2.23 m^3/m/sec)(3.154 E7 sec/yr)(1.89 E3 m/sec/m)^2 = 5.3 E16 J/yr.

[c] Tidal energy, assuming 90% absorbed on continental shelves (Campbell and Odum, 1997). *Atlantic,* 0.9 m: (0.5)(0.34 E11 m^2)(0.9)(706 tides/yr)(0.9)(0.9)(1.025 E3 kg/m^2)(9.8 m/sec^2) = 0.5 E17 J/yr. *Gulf,* 0.5 m: (0.5)(1.4 E11 m^2)(0.9)(706 tides/yr)(0.5)(0.5)(1.025 E3 kg/m^2)(9.8 m/sec^2) = 0.62 E17 J/yr. *Atlantic + Gulf = Florida tide:* (0.5 E17 J/yr) + (0.62 E17 J/yr) = 1.12 E17 J/yr.

[d] Wave energy absorbed, calculated as energy of waves multiplied by length of coastline and shoreward velocity: $(gz)^{0.5}$, where g = square root of gravity, z = water depth. *Atlantic coast,* 0.65 m, measured in 3-m water depth (z); 9.33 E5 m coastline facing wave fronts: (9.33 E5 m)(1/8)(1.025 E kg/m^3)(9.8 m/sec^2)(0.65)(0.65)(3.154 E7 sec/yr)((9.8 m/sec^2)(3.0 m))$^{0.5}$ = 8.46 E16 J/yr. *Gulf coast,* 0.45 m, measured in 2-m water depth (z); 1.23 E6 m coastline facing wave fronts: (1.23 E6 m)(1/8)(1.025 E3 kg/m^3)(9.8 m/sec^2)(0.45)(0.45)(3.154 E7 sec/yr)((9.8 m/sec^2)(2.0 m))$^{0.5}$ = 4.36 E16 J/yr. *Atlantic + Gulf = Florida waves:* 8.46 E16 J/yr + 4.35 E16 J/yr = 1.23 E17 J/yr.

[e] Rain, 1.4 m/yr, including shelf. *Chemical potential of rainwater:* (3.14 E11 m^2)(1.4 m)(1 E6 g/m^3)(5 J/g) = 2.20 E18 J/yr. *Geopotential energy of rain runoff,* mean elevation of 23 m: (1.4 m rain)(0.1 m runoff)(1 E3 kg/m^3)(1.14 E11 m^2)(23 m)(9.8 m/sec^2) = 3.60 E15 J/yr.

[f] River inflows, Apalachicola, 603 m^3/sec: (603 m^3/sec)(1 E6 g/m^3)(5 J/g)(3.154 E7 sec/yr) = 9.50 E16 J/yr. Hydroelectric power, 1993 (Floyd, 1996): (2.2 E12 btu/yr)(0.252 kcal/btu)(4186 J/kcal) = 2.3 E15 J/yr.

[g] Hurricanes (8% probability per year); hurricane energy: 4.8 E5 kcal/m^2/day; 1 day for passage of hurricane; 3% of energy; 10% of energy contributed to surface zone: (4.8 E5 kcal/m^2/day)(1 day)(0.08)(3.14 E11 m^2 area)(0.03)(0.10)(4186 J/kcal) = 1.5 E17 J.

[h] Geologic cycle, Earth energy share of uplift; Earth heat flux typical of old, stable areas, 1.0 E6 J/m^2/yr: (1.4 E11 m^2)(1 E6 J/m^2/yr) = 1.4 E17 J/yr.

[i] Fuels (Floyd, 1996). *(i)* Gas from within Florida, 1995, 7.17 E9 ft^3/yr: (7.17 E9 ft^3/yr)(1.1 E6 J/ft^3) = 7.9 E15 J/yr. *(ii)* Florida consumption of gas from outside Florida, 1994, 367.6 E9 ft^3/yr: (367.6 E9 ft^3/yr)(1.1 E6 J/ft^3) = 4.0 E17 J/yr; Florida consumption – Florida production = total outside gas: (4.0 E17 J/yr) – (7.9 E15 J/yr) = 3.9 E17 J/yr. *(iii)* Imported coal, 1993, 654 E12 btu/yr: (652 E12 btu/yr)(1055 J/btu) = 6.9 E17 J/yr. *(iv)* Imported petroleum, 1993, 1636.0 E12 btu/yr: (1636 E12 btu/yr)(1055 J/btu) = 1.7 E18 J/yr. *(v)* Nuclear electricity, 1993, 276.5 E12 btu/yr: (276.5 E12 btu/yr)(1055 J/btu) = 2.9 E17 J/yr. *(vi)* Total fuels: *i* + *ii* + *iii* + *iv* + *v*. *(vii)* Total outside fuels: *ii* + *iii* + *iv* + *v* (in solar emjoules, sej).

[j] Peat mined, 1994 (Floyd, 1996), 2.06 E5 tons: (2.06 E5 ton)(1 E6 g/ton)(5 kcal/g)(4186 J/kcal) = 4.3 E15 J/yr.

[k] Phosphate production, 1994 (Scoggins and Pierce, 1995): 29 E6 metric tons = 29 E12 g. 10% used in Florida (Florida Conservation Foundation, 1993).

[l] Imported goods and services; external expenditures assumed equal to estimated total $ per year from out-of-state minus taxes. *(i)* Total dollars into Florida equals sum of footnotes n (exported goods and services), o (exported phosphate), and p (drug sales): $141.16 E9 + $18.7 E9 + $0.53 E9 = $160.49 E9. *(ii)* Federal taxes paid, 1994 (Floyd, 1996): $31.4 E9; imported goods and services = total dollars into Florida – taxes: $160.49 E9 – $31.4 E9 = $129.09 E9.

[m] Population in-migration (Floyd, 1996): 1995, 14.2 E6 population; 1993, 108,462 net in-migration.

[n] Exported goods and services. *(i)* General export sales: $23 E9, calculated from graph of gross domestic product and exports (Brown, 1986) using the 1992 gross state product (Floyd, 1996): $268.6 E9. *(ii)* Tourists, 1995 (Floyd, 1996): $60.0 E9. *(iii)* Total transfer payments from federal government, 1994 (Floyd, 1996): $55.5 E9. *(iv)* Foreign investment, 1992 (Scoggins and Pierce, 1995): $2.76 E9. *(v)* Exported goods and services = *i* + *ii* + *iii* + *iv* = $141.26.

[o] Phosphate export, 1994: 90% of production (see footnote k) = 26.1 E6 metric ton; ($20.42/metric ton)(26.1 E6 metric ton) = $0.53 E9.

[p] Illegal drug export (Drug Enforcement Agency): 1992, $17 E9; 1994, figured as a proportion of GNP, $18.7 E9.

[q] Oil produced in Florida but sent out of state for refining and use, 1993 (Floyd, 1996), 5.6 E6 barrels: (5.6 E6 barrels)(6.9 E9 J/barrel) = 3.8 E16 J. Energy/$ ratio, U.S., 1994 (Odum, 1996): 1.3 E12 sej/$.

[r] Total imports; total outside fuels, goods, and services; and in-migration from out-of-state equals footnotes i(*vii*) + l + m.

[s] Sources from within Florida; several concurrent climatic inputs not included in totals to avoid double counting by-products. Renewable sources are equal to the sum of tide, chemical potential of rain, and river inflow: (338.8 E20 sej + 18.8 E20 sej + 47.5 E20 sej) = 405.1 E20 sej/yr. Nonrenewable sources are equal to the sum of phosphate, peat, and home fuels used: (113.1 E20 sej + 1.5 E20 sej + 3.8 E20 sej) = 118.4 E20 sej/yr.

[t] Total emergy use = renewable + nonrenewable + imports: (441.2 E20 sej + 118.4 E20 sej + 3809.3 E20 sej) = 4368.9 E20 sej/yr.

[u] Emergy/dollar ratio = total use/gross state product (1992 gross state product = $268.6 E9): (4368.9 E20 sej)/($268.6 E9) = 1.62 E12 sej/$.

Table B2 Florida's Stored Resources (Discussed in Chapter 29)

Item	Data (units)	Solar Emergy (sej/unit)	Solar Emergy (E20 sej)
Sands	?		
Soils	?		
Peat	?		
Crude oil[a]	68.5 E16 J	6.6 E4/J	451.8
Natural gas[b]	110.9 E16 J	4.8 E4/J	52.3
Phosphate[c]	1.0 E15 g	3.9 E9/g	39,000.0
Population[d]	14.2 E6 people	3.3 E16/person	4686.0

[a] (Tootle, 1991): (109 E6 barrels)(6.28 E9 J/barrels) = 68.5 E16 J.

[b] (Tootle, 1991): (99 E9 ft^3)(1.1 E6 J/ft^3) = 10.9 E16 J.

[c] (Bureau of Mines, 1992): (1 E9 tonnes)(E6 g/tonnes) = 1 E15 g.

[d] (Floyd, 1996): 14.2 E6 people in 1995.

Appendix C

Useful conversions

Metric conversions

1 kilocalorie (kcal) = 4186 joules (J)
1 millimeter (mm) = 0.001 meter (m) = 0.1 centimeter (cm)
1 centimeter (cm) = 0.394 inches (in)
1 meter (m) = 39.37 inches (in) = 3.28 feet (ft); 1 ft. = .3 m
1 kilometer (km) = 1000 meters (m) = 0.621 miles (mi); 1 mi = 1.6 km
1 liter (l) = 1.06 quarts (qt)
1 hectare (h) = 2.5 acres
1 gram (g) = .035 ounce (oz)
1 pound (lb) = 454 g
Degrees Celsius (centigrade) = 5/9 × (degrees Fahrenheit – 32)
Degrees Fahrenheit = 9/5 × (degrees Celsius) + 32
Degrees Kelvin = degrees Celsius + 273
1 barrel (bbl) of oil = 6.9 E9 J
1 watt = 1 joule per second
1 kilowatt = 1000 watts
1 kilowatt-hour = 860 kilocalories

Emergy conversions

Emergy/money ratio for the U.S. in 1995 = 1.4 E12 sej/$.
Emergy/money ratio for Florida in 1993 = 1.6 E12 sej/$.

A plastic template for drawing energy symbols is available from the Wetlands Center, Phelps Lab, University of Florida, Gainesville, FL, 32611.

Appendix D

Types of wetlands

Table D1 Types of Wetland Ecosystems in Florida

Freshwater Wetlands

Softwater, low-nutrient wetlands, with rainwater, little land runoff
- Bogs, with sphagnum moss and insectivorous plants
- Bays, with bay trees and other evergreen trees
- Dwarf cypress — dwarf trees growing on limestone and rainwater; predominates Big Cypress area east of Naples
- Cypress and gum ponds — dome-shaped tree line on high, flat plains receiving runoff waters from a small watershed; deciduous
- Elevated headwater swamps with deep peat beds — Santa Fe Swamp, Okeefenokee Swamp, headwaters of Suwanee and St. Mary's Rivers

Forested wetlands with more nutrient runoff from larger watersheds
- Cypress strand — flowing water swamp without distinct channels; strands from cypress ponds have pond cypress and black gum; strands with larger drainages tend to have bald cypress
- Hydric hammock — diverse forested wetland receiving groundwater seepage with nutrients; short hydroperiod; hardwoods, palms, cypress, gums
- Floodplain swamp — high diversity of hardwoods including ash, maple, wetland oaks, cypress, and gums; short hydroperiod; meandering channel at low water; seepage that carries nutrients
- Lake-border swamp — bald cypress and other wetland trees, depending on the fertility of the lake waters
- Tree islands in Everglades marshes — low-nutrient forests on peat accumulations, bays
- Custard apple swamp in south Florida, with flowing nutrient waters

Freshwater marshes, herbaceous and grass-like, rooted vegetation, successional or arrested by frequent fire
- Lake-border reeds and rushes where nutrients are low
- Lake-border cattails and eutrophic plants with high nutrients
- Stream-margin aquatic plants; springs and hardwaters; *Pontideria*, bulrushes, *Nymphea*
- Vegetation with slowly flowing, low-nutrient waters; sawgrass as in Everglades

Table D1 (cont.) Types of Wetland Ecosystems in Florida

Saltwater Wetlands

Salt marshes in northern Florida are predominately:
- *Spartina* — flushed regularly with tide; narrow salinity range
- *Juncus* — flushed less frequently; wide salinity range
- Mangroves in south Florida estuaries, where freezing is rare
  - Red mangroves (*Rhizophora*) — salinity range of soil is small
  - Black mangroves (*Avicennia*) — salinity range of soil is large
  - White mangroves — disturbed and successional conditions

Glossary

Glossary

Acid rain — precipitation with acidity more than normal, a low pH
Aerobic — containing oxygen or requiring oxygen for respiration
Agricultural — having to do with farming; cultivation of ground to raise food
Agroecosystem — agricultural system of planted fields
Agrarian city — a city based on farms and solar energy
Algae — a category of simple aquatic organisms that photosynthesize; mostly microscopic; phytoplankton, seaweed
Amplifier — a means of multiplying an effect
Amplitude — the distance from the crest to the trough of a wave
Anaerobic — without oxygen
Annual rings — rings in a tree cross-section indicating yearly tree growth
Aquaculture — cultivation of edible organisms in water
Arboreal — living in trees
Archeology — the study of past cultures, based on relics and remains
Arid — dry
Assets — wealth of nature and human development
Balance of payments — money going out of a country compared to the money coming in
Barrier island — narrow strip of sand and dune vegetation containing the beach shore
Barter — trade without money exchange
BASIC — Beginner's All-Purpose Symbolic Instructional Code; one of several languages that computers use
Beach — the often sandy or pebbly shore of a body of water
Benthic — the sediment water interface at the bottom of aquatic systems
Benthos — community of organisms living in or on the bottom of aquatic systems
Biodegradable — can be decomposed by living organisms
Biodiversity — number of different kinds of living organisms
Biomass — the amount of organic matter, living or dead, in a given area
Biome — type of ecosystem which lives in a climatic zone

Biosphere — the part of the Earth's surface and atmosphere where living things exist
Biota — living organisms
Blue-water ocean — low-nutrient plankton ecosystem of tropical seas and the Gulf Stream
Bog — wetland community characterized by peat accumulation
Boil in a spring — turbulent outflow of water from underground channels
Bottomland — floodplain of a stream
Buffer — barrier; in chemistry, a chemical that can neutralize acids or bases
By-product — material produced along with the primary product of the process
Calcareous — contains calcium carbonate
Calorie — heat energy required to raise the temperature of a milliliter of water 1° Celsius (this is a calorie, spelled with a small "c"; a Calorie, with a capital "C", is a kilocalorie)
Canopy — top layer of tree leaves
Carbon dioxide — a gas, the molecules of which are formed from the combination of one carbon and two oxygen atoms (CO_2); necessary for photosynthesis
Carnivore — animal that eats other animals
Carrying capacity — amount of animal life, human life, or industry that can be supported on available resources
Celsius — another name for the centigrade temperature scale in the metric system; freezing is 0°C and boiling is 100°C
Chaos — scientific term for an oscillation with shifting characteristics
Chemical potential energy — energy available from chemical substances
Chlorofluorocarbons (CFC) — substances used in air-conditioning which are reducing the ozone shield
Chlorophyll — green pigment in plants which absorbs sunlight for photosynthesis
Clay — class of minerals found as microscopic crystals in soils and sediments
Climate — average weather conditions of sun, temperature, rain, storms, and seasonal changes
Climax — a mature stage of succession in which the dominant species reproduce themselves and are able to maintain the system
Coefficient — a numerical factor that indicates the proportion that one system flow is of another
Colonial — organisms of the same type (species) living together in groups or colonies
Cold front — zone of weather at boundary of cold air mass pushing under a warmer air mass
Commerce — buying and selling of goods
Commodity — a useful product, such as corn or steel
Community — an interacting group of organisms of many species
Conifers — evergreen trees that bear cones as reproductive organs

Conservation — preservation of natural areas and ecosystems
Constant-flow source — a renewable source, supplies a steady input
Constant-pressure source — an energy source that is so large it is considered to supply constant pressure to a system
Consumers — organisms or people that use the outputs of production
Continental shelf — gently sloping land extending from the shoreline outward into the oceans from each of the continents
Control burn — planned burn of a forest to eliminate underbrush
Converge — come together, bring together
Coral reef — ecosystem of colonial jellyfish and symbiotic algae on limestone skeletons
Coriolis force — a force resulting from the Earth's rotation that causes motions in the Northern Hemisphere to be bent to the right
Cross-section — a side view that would result if something was sliced in two
Crown fire — fire that burns the whole forest not just the ground vegetation
Culture — shared information and values of a society
Currency — money
Cycle — an action that makes a circle and returns to the initial state
Cyclonic storm — low-pressure system that generates temperate latitude weather, with air circulating counterclockwise in the northern hemisphere (as viewed from above); also, tornado, hurricane
Deciduous — trees that shed leaves seasonally
Decomposer — organism that consumes dead matter and returns nutrients to the system
Decomposition — the process of breaking down organic matter to simpler nutrients
Debris — trash, as refuse washed up on the beach
Defoliation — the dropping of leaves
Demand, economic — willingness to purchase a product
Demise — death, ceasing to exist
Denitrification — process of microbes returning nitrogen contained in chemical nutrients back to the atmosphere as nitrogen gas
Density — weight of a unit volume of liquid, solid, or gas
Depreciation — loss of structure with the dispersal of material
Detoxify — to remove the effects of a harmful substance
Detritus — mixture of organic matter that is in the process of being decomposed and the microbes that are decomposing it
Difference equation — an equation for change over a time interval, as used in computer simulation programs
Dinoflagellates — a type of green plant cells that swim with flagella, important in plankton
Disperse — to move apart, becoming less concentrated
Diversity — different kinds; the condition of being different
Domesticate — change to live with humans
Dredge — dig up sediments under water

Drift line — high tide margin where floating debris collects
Drill — snail that makes a hole in the shell of an oyster and eats the insides
Dynamic equilibrium — a steady-state balanced system
Ecological engineering — environmental management, with part of the design resulting from human actions and part from attributes of nature
Ecology — study of populations, communities, and ecosystems
Economic web — the relationship of producers and consumers in an economic system which involves the exchange of money
Ecosystem — a community of organisms in interaction with the environment
Eddy — a current of air, water, etc., with a circular motion
Efficiency — ratio of output energies to input energies
Emdollar — dollar equivalent of emergy: emergy divided by emergy/dollar ratio
Emergy — the energy that was required and used to make a product or service; its embodied energy
Emergy analysis — calculation and comparison of emergy inputs and outputs of a system
Emergy investment ratio — ratio between the emergy of purchased inputs and that of local free inputs
Emergy yield ratio — ratio between emergy of yield and that of the feedback from the economy
Emjoules — unit of emergy
Endangered species — plants and animals close to extinction
Energy — the ability to produce heat
Energy quality — the ability of a particular type of energy to be used in work, measured by its transformity
Environment — the landscape on which people depend
Emergy/money ratio — the ratio of emergy use to money (dollars) flow
Epidemic — extensive spread of disease through a community
Epiphytes — plants that grow on other plants but are not parasites and produce their own food by photosynthesis
Equilibrium — a stable state without available energy for change; dynamic equilibrium, a steady state
Erosion — wearing away of land surface by water or wind
Estuary — partly enclosed coastal waters characterized by a mixing of fresh and saltwaters
Ethanol — ethyl alcohol; can be made from fermentation of biomass such as corn and sugarcane
Eutrophic — high in nutrients, productive
Evaporation — the change from liquid to vapor
Evapotranspiration — the loss of water to a system through the process of evaporation and transpiration
Everglades — "river of grass"; a large marshy wetland in the center of the southern part of Florida

Evergreen — trees and shrubs that do not shed their leaves seasonally; always green
Exotic — introduced; not native
Experiment — studies performed under controlled conditions to test an hypothesis
Exponential — rapid growth or decline, based on an autocatalytic function
Fahrenheit — traditional temperature scale used in the U.S.; freezing is 32°F; boiling is 212°F
Fallow — land removed from cultivation and allowed to start succession and soil restoration
Fecal pellets — units of animal excrement
Feedback — return of a portion of the output back to the source of input
Feedback loop — flows connecting two units so that the output of each stimulates the other
Fertility — ability to reproduce
Fertilizer — chemical added to stimulate plant production
Filter feeder — organism that gets its food by filtering the water that surrounds it, e.g., clams, oysters
Final demand — the ultimate or final consumers in an economic web
Fire climax — a forest of fire-resistant plants that is maintained by fire
Fire tolerant — resistant to fire
First energy law — energy can change form but cannot be created nor destroyed
Fisheries — enterprises managing, harvesting, and selling fish
Flood plain — shore area along a body of water that is sometimes covered with water; wetland along a river
Flow chart — diagram of the steps in a program (as for a computer)
Food chain — a description of a series of energy transfers in a living system, in which the organisms obtain their energy by eating others
Food web — feeding relationships between organisms in a community; includes a number of branching, interconnecting food chains
Foot-candle — the light falling on a surface 1 ft away from a candlelight source
Fossil fuel — coal, oil, and gas; fuels made from plant decomposition millions of years ago
Gene pool — information in the genes of a group of living organisms
Geological uplift — elevation of land as part of earth cycles
Geology — earth science
Geothermal power — electricity generated from the heat in the earth, near volcanoes
Global — of the whole world
Gram — metric unit of mass equal to 0.035 ounce
Grass flats — shallow underwater ecosystem of rooted plants
Grasslands — a landscape covered by grasses and herbaceous growth
Greenhouse effect — buildup of atmospheric heat caused by atmospheric carbon dioxide and water vapor which absorb outgoing heat radiation

Greenways — strips of vegetation connecting natural areas; intended for wildlife and recreation
Gross National Product (GNP) — flow of money through the household and government sectors of a national economy
Gross production — total output from a production process before use
Gross state product — flow of money through the household and government sectors of a state economy
Groundwater — water which flows under the ground, often pumped up for drinking water
Gulf Stream — strong, deep-water ocean current which flows from the Florida Keys up the east coast and across to Great Britain
Gyre — ocean current flowing in a circle
Habitat — place where an organism lives
Hardwater — water that is high in mineral (calcium and magnesium) content
Hardwood — broad-leaved tree, such as oak or hickory
Heat — energy of random motion of molecules, measured as temperature
Heat sink — the outflow of dispersed heat from the system
Herbaceous — a non-woody plant
Herbivore — an animal that gets its energy from eating plant products
Hibernate — to spend the winter in a resting state
Hierarchy — organization of objects or elements in a graduated series
High pressure — areas of greater atmospheric pressure; in weather terminology, a "high"
High-quality energy — energy that is concentrated or is of a special type; can do greater amounts of work; of high transformity
Holocene — geologic epoch from 10,000 years ago until the present time
Humidity — concentration of water vapor in the air
Humus — dark material found beneath the litter of a forest floor, formed by the partial decomposition of wood
Hunting and gathering — primitive human existence based on eating small animals, nuts, and fruits
Hurricane — tropical cyclone with winds of 74 mph or greater
Hydroelectric power — electricity generated using the pressure of water flowing from a height, such as a mountain stream
Hydrologic cycle — circulation of water
Hydrology — study of water
Hydroperiod — length of time that a wetland area is flooded by water
Hydroponic — growing in water without soil
Ice age — period in Earth's history when polar ice sheets covered continents down to 40 degrees north latitude
Icon — small picture on computer screen representing a program or file
Image — mental picture
Immigrate — move into an area
Industry — economic productive activity
Inflation — decrease in buying power, or worth, of money

Information — knowledge, facts, data; (special use) parts and connections of a system
Infrastructure — structures (building, roads) organizing society, including transportation, utilities, government
Inorganic — nonliving; not part of a living thing
Input-output table — chart of money exchanges between different sectors of the economy
Intensive agriculture — farming that uses large inputs of fuels, machinery, labor
Interaction — process that combines different types of energy or material flows
Intertidal zone — area between low and high tides
Intertropical convergence — weather zone near the equator where trade winds converge, causing clouds and rain
Inversion — a layer of warm air over a colder layer at the ground; occurs often at night
Invertebrates — animals without backbones; insects, lobsters, clams, etc.
Ion — chemical unit with an electrical charge (+ or –)
Isobar — lines of equal pressure
Iteration — the repetition of a step or event
Joule — the Système International d'Unités unit of energy; a kilogram of force exerted for a meter
Key (geographical) — small island made of old reefs; the Keys at the southern tip of Florida
Kilocalorie — Calorie, equal to 1000 calories
Kinetic energy — energy of motion
Landfill — dump for solid wastes
Landscape — a part of the Earth's surface, including land and water
Land rotation — alternating use of land for agriculture and for ecological succession to allow it to rest and replenish nutrients
Larvae — immature stages of animals such as insects, shrimp, lobsters, plankton, fish
Latitude — distance north and south from the equator measured in degrees, reflecting the 360 degrees of a circle (Earth)
Law of diminishing returns — a commodity has a decreasing effect as it becomes more abundant; law of limiting factors
Leach — dissolve, as through soil
Legume — a plant of the pea and bean family whose roots contain nodules of nitrogen-fixing bacteria
Lichens — a complex organism that develops from the symbiotic relationship between algae and a fungus
Light — visible part of the electromagnetic spectrum of radiation
Lightning — passage of large electrical charge between clouds or between clouds and ground
Limestone — rock made of cemented shells, mostly $CaCo_3$

Limiting factor — a factor, such as temperature, light, or nutrients, that limits the growth, abundance, or distribution of an organism
Litter — wastes such as sticks, leaves, and paper on the forest floor
Logistic growth — autocatalytic population growth that is limited by negative effects of crowding
Longshore current — current parallel to the shore from waves breaking at an angle
Low-energy agriculture — a simple farming system that uses the labor of humans and farm animals rather than fuels and machinery
Low pressure — an area of the Earth where the atmospheric pressure is lower than at other areas; in weather terminology, a "low"
Low-quality energy — dilute energy; type of energy not so useful in work
Macrophytes — plants large enough to be seen without a microscope; usually refers to aquatic or wetland plants
Maize — a type of primitive corn grown by Indians
Mangroves — saltwater wetland trees found where frosts are not frequent
Maritime — from the sea
Market value — dollars that can be obtained for a good or service in the economy of buying and selling; what people are willing to pay
Marsh — a wetland area covered mainly by grasses
Maximum power principle — principle that explains that the system designs that prevail are those that organize to use more energy
Megafauna — large animals
Megajoule — one million joules
Metabolism — processes of living organisms that utilize energy to maintain their structures and activities
Microclimate — selected area where climate conditions are different from the surrounding area, such as within a dense forest
Microcosm — small ecosystem in a container
Microeconomics — small-scale economics; production, sale, and use of basic products
Microorganisms (microbes) — living units so small that they can only be seen with a microscope
Migrant labor — workers who move from one job to another during the year, especially in agriculture
Migrate — to move from one place to another
Model — a diagram that shows important relationships in a simple way
Molecule — smallest unit of a substance that keeps its properties
Money transaction — energy and materials flow one way and money flows in the opposite direction to pay for them
Monoculture — growing only one type of crop over a large area
Monsoon — periodic wind of the Indian Ocean and southern Asia, associated with heavy seasonal rains
Mosaic — a patchwork of various kinds of areas
Native — having its origin in a particular region and found living there

Neap tide — smallest tidal range, when sun and moon are at right angles
Net emergy — amount of emergy remaining from production after using up some emergy in the production process
Net production — gross production minus consumption over a certain time period
Niche — way of life of an organism, where it fits in the ecosystem
Nitrogen cycle — flow of nitrogen through an ecosystem (as from the air, soil, plants, animals, decomposers) back to the soil
Nitrogen-fixation — conversion of atmospheric nitrogen to nitrates (and other forms of usable nitrogen) by bacteria
Nomadic — population that periodically moves
Nonrenewable — resource found in a definite amount that cannot be replaced in a given time period
Nuclear breeder reactor — uses plutonium (a very poisonous carcinogen and a by-product of nuclear fission) to produce electricity
Nuclear fission power plant — electricity produced by the energy produced from breaking apart the nucleus of the uranium atom
Nuclear fusion power — energy produced from bringing together hydrogen atoms (the same reactions found on the sun); used to produce the hydrogen bomb, but not currently controllable enough to produce electricity
Nursery — place where animals spend early part of life, grow rapidly, and mature, as in an estuary
Nutrient — substance required by living things for basic life processes; used here as inorganic, mineral chemicals
Ocean — body of saltwater that covers 70% of the earth
Ocean thermal electrical conversion (OTEC) — electricity generated from the difference in heat between the surface and the depths of the ocean
Oceanography — science of the ocean
Oligotrophic — low in nutrients, unproductive
Organic matter — material that is living or once was living
Organism — living plant, animal, or microbe
Oscillation — repeated pulses, quantities going up and down
Overfishing — catching fish in quantities that are not sustainable
Oxidize — a process in which another substance chemically combines with oxygen
Oxygen — one of the 92 natural elements; released by green plants in photosynthesis
Ozone — O_3; a very reactive form of oxygen formed by sunlight in the stratosphere; occurs as a pollutant near the ground
Peat — solid matter that results from the partial decomposition of plants in water
Pelagic — open ocean
Peninsula — land extending out into the sea like a finger
Percolate — to move or trickle down through a permeable substance, as water through soil

Periphyton — algae and small animals that grow attached to surfaces, such as rocks and sticks in lakes and streams
Permafrost — soil that remains always frozen
Phosphorus cycle — flow of phosphorus through an ecosystem
Photon — unit of light
Photosynthesis — process in green plants that uses light energy, water, nutrients, and carbon dioxide to produce organic matter and oxygen
Phytoplankton — plant life suspended in water, algae, seaweed, mostly microscopic
Pioneer — the first species to move in and colonize an area
Plankton — organisms suspended in water, moving with the water
Plankton-benthos ecosystem — waters of medium depth with part of the animals on the bottom and part suspended in the water
Pleistocene — the geologic period from two million years ago until ten thousand years ago
Plutonium — a rare chemical element, by-product of nuclear fission power plants, which is a dangerous carcinogen
Pollution — effects of unused wastes detrimental to the environment or to humans
Polar front — semi-permanent boundary between cold air moving south and warm air moving north
Pond — body of water smaller than a lake
Population — all the individuals of one species in a place at a time
Potable water — water pure enough for drinking
Potential energy — energy capable of doing work
Power — useful energy per time; e.g., electrical power can be measured in watts, or joules per second
Power plant — utility that produces electricity from fuel
ppm — parts per million
ppt — parts per thousand
Prairie — grassland ecosystem
Precipitation — rain, snow, sleet, etc.
Predator — animal that gets its food from consuming other animals
Prey — organisms eaten by a predator
Price — the amount of money to buy one unit of goods or services
Primary consumer — herbivore; the first consumer in a food chain
Producer — organism (green plant) that produces its own food from raw ingredients; a unit that combines inputs to form a new product
Production — the combination of two or more inputs to generate something new, as when plants produce food from inorganic materials or when industry generates a product from raw materials
Productivity — rate of production, as of plant growth
Program — a sequence of steps or events
Pulse — surge of storage and flow
Quantitative — dealing with numbers and measurements

Raptor — bird of prey, e.g., falcon, eagle
Reclamation — reorganization of lands and re-establishment of productive vegetative cover
Recycle — feedback of materials in a closed circle, as nutrients flow from decomposers back to plants
Red tide — bloom of a microscopic marine alga that produces fish-killing toxicity
Reef — mound of calcareous skeletons of animals and plants accumulated on shallow sea bottom
Respiration — process in an organism that releases energy from chemical storages; organic matter + $O_2 \rightarrow CO_2 + H_2O$ + nutrients + energy
Relative humidity — percent of water vapor in the air compared to saturation at that temperature
Renewable — a resource that can be replenished or renewed
Salinity — concentration of salt in water, measured in parts per thousand
Saltwater intrusion — movement of saltwater into groundwater that was fresh
Sand — loose granular crystals 0.05 to 2 mm in size, formed from disintegration of rocks and skeletons
Sand dune — hill built up from blowing sand
Sandstone — rock made of cemented sand
Sargassum — floating brown seaweed that forms large mats
Saturation — level where something is filled to capacity, as the amount of oxygen that can be dissolved in water at a certain temperature
Second energy law — the availability of energy to do work is used up in processes and in spontaneous dispersal of energy concentrations (depreciation)
Secondary consumer — animal that eats primary consumer
Sector — section, part; used as a part of an economy
Sediment — the matter that settles to the bottom of oceans, rivers, and lakes
Sedimentary cycle — formation, uplift, erosion, and redeposit of rock made of sand, shells, and clay
Seedling — young plant or tree grown from seed
Self-interactive drain — outflow in proportion to population interaction, a quadratic drain
Self-organization — process by which various parts of a system connect themselves to work together
Sewage — Fecal wastes and urine from people and animals; urban wastes
Shale — rock made of clay particles
Shell middens — shell piles accumulated over a long time
Silvaculture — the commercial production of trees
Simulation — prediction of what would happen as a result of a particular action or process, without the action or process actually occurring
Sinkhole — depression formed when soil and earth fall into a hole produced as limestone is dissolved
Smog — a polluting combination of smoke and fog

Softwater — water with few dissolved minerals; rainwater; occurs in areas without limestone; acidic
Soil — earthen part of a terrestrial ecosystem with mineral crystals, roots, animals, and microbes
Solar emergy — solar energy required directly and indirectly
Solar hot water heater — device to heat water using the sun
Solar technology — using technology to produce hot water and electricity from the sun
Solar transformity — the quantity of solar emergy required to make 1 joule of another type of energy
Solar voltaic cells — making electricity from the sun by using solar photons to separate electrons from matter
Solid waste — household trash, old cars, discarded machinery, etc.
Solution — liquid with solids dissolved within
Source — input to a system from outside
Species — organisms that naturally interbreed
Spreadsheet — page produced by an accounting program, such as Excel, that organizes data
Spring — source of water coming from the ground
Spring tide — largest tidal range, when sun and moon are in line
Standard of living — emergy per person; level of consumption of goods and services; sometimes measured as money income
Steady state — system where inflows equal outflows; dynamic equilibrium
Storage — a stock of matter, energy, or money
Storm — intensive weather with strong winds and rain or snow
Strand — broad area of vegetation with flowing water
Stratum — layer; strata (plural)
Stream — flowing water smaller than a river, larger than a brook
Strip mine — to dig deposits of minerals after removing the upper layers of earth
Sublime — to change from a solid to a gas state without becoming liquid
Subtropical evergreen forest — a complex high-diversity forest 30 degrees north or south of the equator with small thick leaves adapted to long dry periods
Succession — the process of self-organization of a system over time, resulting in a sequence of stages
Sun belt — southern and southwestern states of the U.S.
Supply, economic — stored amount available to purchase
Sustainable — capable of maintaining at a steady state
Swamp — wetland area covered mainly by trees (at least 35% tree cover)
Switch — turn on and off, as lightning turning on a fire
Symbiotic — relationship between two organisms that benefits both
Symbol — something that represents something else more complicated
System — set of items that operate together to form a unified whole
Technology — equipment and information for achieving a practical purpose

Temperate — at moderate temperatures, middle latitudes
Temperature — concentration of heat measured by the expansion of a liquid or gas
Tentacle — long, flexible extension of an animal, usually on head or mouth
Terrestrial — related to land
Thermodynamics — science of heat and energy
Thunder — echoes of lightning
Tide — large motions of water caused by changes in gravity due to the movement of the sun and moon
Topography — levels of landscape; mountains, valleys, and flat areas
Tornado — a storm with intense spiraling wind
Toxic — poisonous
Trade winds — zone of steady winds from the east in tropical latitudes
Transaction — an exchange, as in trade or sales
Transfer payments — payments by the federal government for social security, educational and medical programs, welfare, and other purposes
Transformation — a change from one form to another
Transformity — the ratio of energy of one type required to produce a unit of energy of another type
Transition — going from one area to another or from one time to another, as from a high-energy economy to a low-energy one
Transpiration — evaporation of water from pores in plant leaves
Trophic level — position within a food chain
Tropical — near the equator, below latitude 23 degrees
Truck farming — growing of vegetables transported by truck
Turbid — not clear, with suspended particles
Turbulence — mixing action of a body of water or air
Turnover time — time for things (water, organisms, systems) to be replaced
Ultraviolet light — light with wavelengths shorter than the visible spectrum
Understory — small trees and shrubs found above the groundcover and beneath the canopy of a forest
Upwelling — movement of deep, cold, nutrient-rich water to the surface of oceans
Uranium — radioactive chemical element used to produce electricity in present nuclear power plants
Urban ecosystem — a city and its environment
Utilities — enterprises that provide necessary energy and materials, such as electricity, water, gas
Viable — alive, successful
Waste — output that is not used
Wave energy — rhythmic motion of water in lakes and oceans caused by winds
Wealth — abundance of useful products
Weather — state of the atmosphere in regard to precipitation, winds, and storms

Weed — successional plant with high net production
Wetland — land area that is periodically covered with water and contains specialized wetland plants
Wildlife corridor — strip of vegetated land suitable for animal movements
Wind — motion of the air caused by differences in temperature
Work — the use of energy for a production process, an energy transformation
Yield — output resulting from a production process
Zooplankton — small animal life suspended in water

Suggested readings and references

Suggested readings

Systems

Odum, H.T., 1994, *Ecological and General Systems,* University of Colorado Press, Niwot (revision of Odum, H.T., *Systems Ecology,* John Wiley & Sons, New York, 1983.)
Odum, H.T. and E.C. Odum, 1981, *Energy Basis for Man and Nature,* 2nd ed., McGraw-Hill, New York.
Odum, H.T., 1996, *Environmental Accounting: Emergy and Environmental Decision Making,* John Wiley & Sons, New York.

Ecology and society

Milanich, J.T. and S. Proctor, Eds., 1994, *Tacachale: Essays on the Indians of Florida and Southeastern Georgia During the Historic Period,* University Press of Florida, Gainesville.
Odum, E.P., 1997, *Ecology — A Bridge Between Science and Society,* Sinauer Associates, Sunderland, MA.

Environmental systems of Florida

Bell, C.R. and B.J. Taylor, 1982, *Florida Wildflowers and Roadside Plants,* Laurel Hill Press, Chapel Hill, NC.
Headstrom, R., 1983, *Identifying Animal Tracks,* Dover, New York.
Kale, H.W. and D.S. Maehr, 1990, *Florida's Birds,* Pineapple Press, Sarasota, FL.
Lehr, P.E., R.W. Burnett, and H.S. Zim, 1987, *Weather,* Golden Press, New York.
Myers, R.L. and J.J. Ewel, Eds., 1990, *Ecosystems of Florida,* University of Central Florida Press, Orlando.
Reid, G.K., 1987, *Pond Life,* Golden Press, New York.
Soil Conservation Service, *Twenty-Six Ecological Communities of Florida,* U.S. Dept. of Agriculture, U.S. Government Printing Office, Washington, D.C.
West, E. and L.E. Arnold, 1956, *The Native Trees of Florida,* University Press of Florida, Gainesville.
Winsberg, M.D., 1990, *Florida Weather,* University of Central Florida Press, Orlando.
Zim, H.S., Ed., *Golden Guides: Birds, Reptiles, Fossils,* and others, Golden Press, New York.

Simulation

Odum, H.T. and E.C. Odum, 1991, *Computer Minimodels and Simulation Exercises for Science and Social Science*, Center for Wetlands and Water Resources Research, University of Florida, Gainesville.

Swartzman, G.L. and S.P. Kaluzny, 1987, *Ecological Simulation Primer*, Macmillan, New York.

Future

Brown, L.F., Ed., 1997, *State of the World 1997*, Norton, New York (published yearly).

Meadows, D.H. and J. Randers, 1992, *Beyond the Limits*, Chelsea Green, Post Mills, VT.

Naisbett, J., 1990, *Megatrends 2000*, Warner, New York.

Toffler, A. and H. Toffler, 1995, *Creating a New Civilization*, Turner Publishing, Atlanta, GA.

References

Allen, M., 1995–1996, *The Florida Handbook*, Peninsular Publishers, Tallahassee, FL.

Ballentine, T., 1976, A Net Energy Analysis of Surface Mined Coal from the Northern Great Plains, M.S. thesis, Environmental Engineering Sciences, University of Florida, Gainesville, 149 pp.

Browder, J., C. Littlejohn, and D. Young, 1977, *South Florida: Seeking a Balance of Man and Nature*, Bureau of Comprehensive Planning, Division of State Planning, Tallahassee, 120 pp.

Brown, M.T., 1976, *The South Florida Study — Lee County: An Area of Rapid Growth*, Center for Wetlands and Water Resources Research, University of Florida, Gainesville.

Brown, M.T., 1980, Energy Basis for Hierarchies in Urban and Regional Landscapes, Ph.D. dissertation, Environmental Engineering Sciences, University of Florida, Gainesville, 359 pp.

Brown, M.T. and C. Palmer, 1979, *How Things Work; Why Things Work; Systems and Symbols; Both Sides Now*, four multimedia film cassettes, National Park Service, U.S. Dept. of the Interior, U.S. Government Printing Office, Washington, D.C.

Brown, M.T., P.O. Green, A. Gonzalez, and J. Venegas, 1992, *Emergy Analysis Perspectives, Public Policy Options, and Development Guidelines for the Coastal Zone of Nayarit, Mexico.* Vol. 2. *Emergy Analysis and Public Policy Options*, report to The Cousteau Society and the Government of Nayarit, Mexico, Center for Wetlands and Water Resources Research, University of Florida, Gainesville.

Bureau of Mines, 1992, *Minerals Yearbook*, Vol. II. *Area Reports: Domestic*, U.S. Dept. of the Interior, U.S. Government Printing Office, Washington, D.C.

Burnett, M.S., 1978, Energy Analysis of Intermediate Technology Agricultural Systems, M.S. thesis, Environmental Engineering Sciences, University of Florida, Gainesville, 169 pp.

Campbell, D.E. et al., 1988, *Emergy Analysis of Maine* (unpublished report), Maine State Fisheries Laboratory, Boothbay Harbor.

Campbell, D.E. and H.T. Odum, 1997, *Calculation of the Solar Transformity of Tidal Energy Received and Tidal Energy Dissipated Globally* (manuscript).

Capehart, B.L., J.F. Alexander, and L.C. Capehart, 1982, *Florida's Electric Future: Building Plentiful Supplies on Conservation*, Center for Wetlands and Water Resources Research, University of Florida, Gainesville, 350 pp.

Cooke, C.W., 1945, *Geology of Florida*, The Florida Geological Survey, Tallahassee.

FCF, 1993, *Guide to Florida Environmental Issues and Information*, Florida Conservation Foundation, Winter Park, FL.

Fernald, E.A., Ed., 1981, *Atlas of Florida*, Florida State University Foundation, Tallahassee, FL.

Fernald, E.A. and E.D. Purdum, Eds., 1992, *Atlas of Florida*, University Press of Florida, Gainesville.

Florida Governor's Energy Office, 1981, *Forecasts of Energy Consumption in Florida, 1980–2000*, Executive Office of the Governor, Tallahassee, 120 pp.

Floyd, S.S., Ed., 1996, *Florida Statistical Abstract*, Bureau of Economic and Business Research, University of Florida, Gainesville.

Headstrom, R., 1983, *Identifying Animal Tracks*, Dover, New York.

Homer, M., 1977, in *Thermal Ecology II*, Energy Research and Development Administration (ERDA), U.S. Government Printing Office, Washington, D.C., pp. 259–267.

King, R.J. and J. Schmandt, 1991, *Ecological Economics of Alternative Transportation Fuels*, report to Texas State Department of Energy, Lyndon B. Johnson School of Public Affairs, University of Texas, Austin.

Lan, S., 1992, Emergy analysis of ecological-economic systems, in *Advances and Trends of Modern Ecology*, Jian, G., Ed., China Science and Technology Press, Beijing, pp. 166–286.

Lapp, C., 1991, Emergy Analysis of the Nuclear Power System in the United States (research paper for Mechanical Engineering degree), Environmental Engineering Sciences, University of Florida, Gainesville, 64 pp.

Ley, J., 1988, Emergy Analysis of Japan, student report, Environmental Engineering Sciences, University of Florida, Gainesville.

McPherson, B.F. et al., 1976, *The Environment of South Florida: A Summary Report*, Department of the Interior, U.S. Government Printing Office, Washington, D.C.

Meadows, D.H. and J. Randers, 1992, *Beyond the Limits*, Chelsea Green, Post Mills, VT, 300 pp.

McLachlan-Karr, J., 1995, *Evaluation of Eco-Tourism in Vallo de Vinales, Cuba*, draft report for Center for Latin American Studies, Johns Hopkins University and the Cuban Academy of Sciences, Baltimore, MD, 43 pp.

Miklanich, J.T. and C.H. Fairbanks, 1980, *Florida Archeology*, Academic Press, New York, 290 pp.

Myers, R.L and J.J. Ewel, Eds., 1990, *Ecosystems of Florida*, University of Central Florida Press, Orlando.

Odum, H.T., 1957, Trophic structure and productivity of Silver Springs, Florida, *Ecol. Monogr.*, 27, 55–112.

Odum, H.T. et al., 1976, Net energy analysis of alternatives for the United States, in *U.S. Energy Policy: Trends and Goals.* Part V. *Middle and Long-Term Energy Policies and Alternatives*, 94th Congress, 2nd Session, Committee Print, prepared for the Subcommittee on Energy and Power of the Committee on Interstate and Foreign Commerce of the U.S. House of Representatives, 66-723, U.S. Government Printing Office, Washington, DC., pp. 254–304.

Odum, H.T., 1983, *Systems Ecology*, John Wiley & Sons, New York, 644 pp.

Odum, H.T. and E.C. Odum, Eds., 1983, *Energy Analysis Overview of Nations*, working paper WP-83-82, International Institute for Applied Systems Analysis, Laxenburg, Austria, 469 pp.

Odum, H.T., M.J. Lavine, F.C. Wang, M.A. Miller, J.F. Alexander, Jr., and T. Butler, 1987, Energy analysis of environmental value, in *A Manual for Using Energy Analysis for Plant Siting with an Appendix on Energy Analysis of Environmental Value*, revised supplement from Final Report to the Nuclear Regulatory Commission, NUREG/CR-2443 FINB-6155, Energy Analysis Workshop, Center for Wetlands and Water Resources Research, University of Florida, Gainesville, 97 pp.

Odum, H.T. and E.C. Odum, 1991, *Computer Minimodels and Simulation Exercises for Science and Social Science*, Center for Wetlands and Water Resources Research, University of Florida, Gainesville.

Odum, H.T., E.C. Odum, and M. Blissett, 1987, *Ecology and Economy: "Emergy" Analysis and Public Policy in Texas*, Policy Research Project Report 78, Lyndon B. Johnson School of Public Affairs, University of Texas, Austin.

Odum H.T. and J.E. Arding, 1991, *Emergy Analysis of Shrimp Mariculture in Ecuador*, report to Coastal Studies Institute, University of Rhode Island, Narragansett, Center for Wetlands and Water Resources Research, University of Florida, Gainesville.

Odum, H.T., 1996, *Environmental Accounting: Emergy and Decision Making*, John Wiley & Sons, New York, 370 pp.

Pillet, G. and H.T. Odum, 1984, Energy externality and the economy of Switzerland, *Schweiz Zeitschrift for Volkswirtschaft and Statistik*, 120(3), 409–435.

Romitelli, S.M., 1993, *Emergy Analysis of Brazil* (unpublished course report), Environmental Engineering Sciences, University of Florida. Gainesville.

Scroggins, J.F. and A.C. Pierce, Eds., 1995, *The Economy of Florida*, University of Florida, Gainesville.

Sipe, N.G., 1978, A Historic and Current Energy Analysis of Florida, M.A. thesis, Department of Urban and Regional Planning, University of Florida, Gainesville.

Snedaker, S.C. and A.E. Lugo, 1972, *Ecology of the Ocala National Forest*, U.S. Forest Service, Atlanta, GA.

Swaney, D.P., 1978, Energy Analysis of Climatic Inputs to Agriculture, M.S. thesis, Environmental Engineering Sciences, University of Florida, Gainesville.

Tebeau, C.W., 1971, *A History of Florida*, University of Miami Press, Miami, FL.

Tootle, C.H., 1991, *Florida's Oil and Gas Reserves for 1991*, Open File Report No. 44, Florida Geological Survey, Tallahassee.

Ulgiati, S.H., H.T. Odum, and S. Bastianoni, 1994, Emergy use, environmental loading and sustainability: an emergy analysis of Italy, *Ecol. Model.*, 73, 215–268.

Wharton, C.H. et al., 1976, *Forested Wetlands of Florida — Their Management and Use*, Center for Wetlands, University of Florida, Gainesville.

Woithe, R.D., 1992, Emergy Analysis of the Exxon Valdez Oil Spill and Alternatives for Oil Spill Prevention, M.S. thesis, Environmental Engineering Sciences, University of Florida, Gainesville, 128 pp.

Yan, M., 1996, *Emergy Analysis of Tibet* (unpublished paper), Center for Environmental Policy, University of Florida, Gainesville.

Zucchetto, J., 1975a, Energy, economic theory and mathematical models for combining the systems of man and nature: case study of the urban region of Miami, *Ecol. Model.*, 1, 24–268.

Zucchetto, J., 1975b, Energy Basis for Miami, Florida, and Other Urban Systems, Ph.D. dissertation, Environmental Engineering Sciences, University of Florida, Gainesville.

Index

Index

A

Acid rain
- damage, 12
- definition, 405
- lakes, 353

Acidity, 352
Aesthetic benefit, 270
Agrarian
- city, 279
- definition, 405

Agriculture
- definition, 405
- intensive, 244
- self-sufficient, 250
- society, 305
- variety in Florida, 243

Agroecosystem
- definition, 405
- discussion, 243

Air circulation, 101, 105
Air pressure, 107
Air wastes
- acid rain, fluorine, 352
- greenhouse gases, 354
- ozone, 353
- sulfur dioxide, 13

Alafia River phosphorus, 167
Algae
- definition, 405
- phytoplankton, 141, 171

Alligators
- swamp, 179
- holes, 193, 195

Amplifier
- definition, 405
- reinforcing feedback, 70

Anaerobic
- definition, 405
- wetland soils, 186

Animal tracks, 36
Animals
- estuary, 159
- fish, 160

Annual rings
- definition, 405
- exercise, 252

Apalachee Indians, 305
Apalachicola Bay
- flood plain, 216
- oyster reefs and salinity, 168

Apatite, 262
Aquaculture
- definition, 405
- discussion, 259

Aquarium, 30
Archeology
- definition, 405
- Indian discoveries, 303

Army, 60

Assets
  definition, 405
  Florida, 312
Atmosphere, 101, 104
Audubon Sanctuary, 189

B

Balance of payments
  definition, 405
  Florida, 318
Barrier island
  definition, 405
  formation, 127
BASIC
  computer language, 78
  definition, 405
  simulation programs, 381
Beach
  definition, 405
  ecosystem, 127, 150
  erosion, 13
Beef cattle system, 248
Benthos
  bottom animals, 144
  definition, 405
Big Cypress, 187
Biodegradable
  definition, 405
  wastes, 348
Biodiversity
  definition, 405
  forests, 211
  gene pools, 239
  index, 154
  parks, 269
  soil restoration, 15
  speciation, 273
  succession, 65
Biomass
  definition, 405
  wildlife, 22
  yield ratio as a fuel, 337
Biomes
  definition, 405
  explanation, 114
Biosphere, definition, 406
Biosphere 2, 251
BioQUEST
  address, 382
  program, 241, 252
Birds
  wood storks, 12, 92
  woodpeckers, 66, 189
Bison, 67
Blue-water ecosystem
  definition, 406
  description, 141
Bog
  definition, 406
  wetland type, 402
Boil in a spring, definition, 406
Bottom land
  definition, 406
  hardwoods, 186
Box symbol, 23
Brazilian pepper, 195
Breeder reactor
  appraisal, 337
  definition, 413
Business
  finance, 227
  forest, 237
  stocks, 289
By-product
  definition, 406
  toxic decomposition, 164

C

Calcareous, definition, 406
Calorie, definition, 30, 406
Calusa Indians, 305
Canals, 193
Canopy, 208
  definition, 406

Cape Canaveral, 166
Carbon cycle, 38
Carbon dioxide
  carbonate equilibrium, 120
  definition, 406
  global climate, 354
Carnivore
  definition, 406
  ecosystems, 17
Carolina paraqueet, 189
Carrying capacity
  definition, 406
  of Florida's economy, 315
  population, 358
  wildlife, 48, 230
Cattails, 197
Cattle pasture system, 248
Cedar Keys
  hurricane, 113
  pencil industry, 307
Celsius temperature scale
  conversions, 399
  definition, 406
  explanation, 331
  other scales, 330
Central business district, 284
CFCs, *see* Chlorofluorocarbons
Channelization, 191
Chaos
  definition, 406
  prey-predator model, 96
Chaparral, 116
Charlotte harbor, 167
Chemical potential energy,
    definition, 406
Chlorine in waste disposal, 355
Chlorofluorocarbons (CFC)
  definition, 406
  depleting ozone, 13, 353
Chlorophyll, definition, 406
Chronological record of Florida, 304
Cities
  agrarian, 279
  fuel-based, 281
  spatial organization
  systems diagrams, 280, 282
Citrus
  groves, 44
  system diagram, 246
Clay
  definition, 406
  sediment, 121
Climate
  changes, 13, 354
  definition, 406
  Florida overview, 3
  Mediterranean, 116
  zones, 115
Climax
  definition, 406
  fire, 204
  stages, 63,69
Coal
  distribution, 325
  evaluation, 395
  yield ratio, 334
Coastline, length, 3,131
Coefficient, definition, 406
Cold front
  definition, 406
  in Florida, map and
      cross-section, 108
Colonial, definition, 406
Colonization, 5, 306
Commerce, definition, 406
Commodity, definition, 406
Community, definition, 406
Competition
  farmers and wild vegetation, 243
  fishermen, 257
  Indians and colonists, 306
  overfishing, 254
  simulation model, 309
  succession, 64
Computer simulations, 381
Conifers, definition, 406
Conservation
  definition, 407

energy, 27
matter, 38
Constant-flow source
computer model, 86
definition, 407
Constant-pressure source
computer model, 83
definition, 407
Consumers
definition, 407
economy, 312
systems, 17, 24
Continental shelf
definition, 407
ecosystem, 143
Florida, 132
Control burn
definition, 407
effects, 206
Convergence (spatial)
cities, 280
definition, 407
materials and energy, 347
Conversions, 399
Coquina clams, 150
Coral reef
definition, 407
ecosystems, 145,148
impacts, 273
Coriolis force
definition, 407
effect on winds, 107
Cracker house, 376
Crown fire
definition, 407
effects, 206
Crystal river estuary, 258
Culture
definition, 407
shared information, 291
Currents, 134
Custard apple swamp, 193
Cycles
broken, 346
carbon, 38
definition, 407
material, 37
nitrogen, 41
nutrients, 43
phosphorus, 39
sedimentary, 120
water, 37
Cypress domes, 187, 351

D

Deal log model, 87
Deciduous
definition, 407
forests, 206, 221
Decomposers
definition, 407
ecosystems, 17
Demand (economic)
definition, 407
supply and, 228
Denitrification, 42
definition, 407
Density
definition, 407
measurement, 136
Depreciation
definition, 407
ecosystems, 29
Descent scenario, policies, 373
Desert, 116
Detritus
continental shelf, 144
definition, 407
plankton-benthos, 161
streams, 178
Development
historical maps, south Florida, 232
principles, 229
Difference equation
definition, 407
simulation, 77
Dinoflagellates
blue water, 143

definition, 407
red tide, 145
Diversity
definition, 407
gene pools, 239
index, 154
role in protecting forests, 211
succession, 64
Dividend, 289
Drain
model, 87
program DRAIN, 385
Drainage ditches, 198
Dredge, definition, 407
Drift line
definition, 408
illustration, 151
Drill (snail), definition, 408
Dynamic equilibrium, definition, 408

E

Earth
cycle, 120
rotation, 103,134
Eclipse, 132
Ecological economics, 225
Ecological engineering
definition, 408
development, 198
planning, 233
transportation, 325
Ecology
definition, 408
suggested readings, 421
Economic use
development, 229
market, 228
system diagram, 227
Economics, 225
Economy
growth, 371
interface with environment, 225
sectors, 314
web, definition, 408
Economy of Florida
quantitative overview, 311
resource basis, 297
Ecosystem
definition, 408
pine forest example, 32
processes, 17
unlabeled diagram, 35
Eddy, definition, 408
Efficiency
concept, 52
definition, 408
food chain, 56
heat engines, 331
maximum power, 329
photosynthesis, 52
El Niño, 116
Electric cars
overall efficiency, 324
solar voltaic power, 340
Electric power
nuclear, 15, 336
state network, 325
Electricity yield ratios, 337
Embodied energy, *see* Emergy
Emdollar
definition, 408
explanation, 301
Emergy
definition, 408
evaluation, 226
evaluation of Florida, 395
explanation, 57
per person, 359
Emergy/dollar (emergy/$) ratio
calculation, 396
comparisons, 316
Florida diagram, 301
Emergy investment ratio
definition, 408
diagram, 226
parks, 269
values, 229

Emergy/money ratio
  calculations, 396
  comparisons, 316
  definition, 408
  diagram, 301
Emergy yield ratio
  definition, 408
  diagram, 332
Emjoules
  definition, 408
  emergy unit, 57
Endangered species
  definition, 408
  Florida, 274
  old growth forest, 212
Energy
  chain
  definition, 27, 408
  first law, 27, 409
  Florida, 325, 329
  forms, 28
  hierarchy, 59
  second law, 29, 415
  sources, 329
  transformation, 52, 57
Energy quality
  appropriate use, 342
  definition, 408
Environmental decision-making
  program, 381
Environmental problems, 11
Environmental systems, 421
Epidemics
  definition, 408
  Indians, 305
  pine bark beetle, 212
Epiphytes
  definition, 408
  hardwood hammock, 211
Equilibrium
  definition, 408
  dynamic, 86
Erosion
  definition, 408
  Florida rock, 123
Estuaries
  definition, 408
  description, 155
  food chain, 163
  generalized plan, 157
  nursery, 158
  species, 159
  systems diagram, 162
Ethanol
  definition, 408
  yield ratio, 334
Eutrophication, 139
Eutrophy, 175
  definition, 408
  freshwater, 174
  management problems, 176
Evaluation
  concepts, 57, 227
  energy sources, 329
  Florida economy, 311, 395
  Florida resources, 297
Evaporation
  definition, 408
  water budget, 123
Evapotranspiration
  definition, 408
  forest, 37
  water cycle, 39
Everglades
  definition, 408
  drainage effect, 49
  kite, 193
  map, 189
  park, 270
  problems, 12
  water flows, 192
Everglades National Park, 272
Evergreen
  definition, 409
  hardwood forest, 210, 217
Evolution
  biodiversity, 273
  cultural, 5
  gene flow, 239
  speciation, 274

Exotics
definition, 409
south Florida, 195
water table, 230
Experiments
definition, 409
"what if's", 79
Exponential
definition, 409
growth model, 83
program EXPO, 384
Exports
diagram, 315
evaluation, 395
EXTEND
address, 382
BioQUEST, 241, 252

F

Factors, limiting, 46
program FACTORS, 382
Fahrenheit
conversions, 399
definition, 409
explanation, 331
temperature scales, 330
Fakahatchee Strand
large cypress, 188
location, 219
Fecal pellets
definition, 409
plankton-benthos, 160
Feedback
concept, symbols, 21
control, 69
definition, 409
Fertility
definition, 409
soil, 243
Fertilizer
agriculture, 244
definition, 409
mining, 262
Filter feeder
benthos, 161
definition, 409
oyster reef, 164
Fire
model, 95
program FIRE, 338
switch symbol, 31
Fire climax
definition, 409
pine forests, 204
First energy law, 27
definition, 409
Fish
breeding at sea, 160
Crystal River food web, 258
spawn in freshwater, 161
Fish migrations
mullet and eels, 160
shad and sturgeon, 158
St. John's River, 166
Fisheries
definition, 409
diagram of competition for fish, 259
Florida, 253, 257
food web, 258
world fishery system, 155
Fishing
commercial, 257
sports, 257
yield and overfishing, 14, 253
Flatwoods, 206
Flood plain
definition, 409
forest, 191
Florida economy
emergy evaluation table, 395
emergy exchange, 317
emergy/money ratio, 301, 316
overview, 311
resource basis, 297, 312
simulation, 366
system diagram, 314
trade, 317

Florida environmental problems, 11
Florida divisions and networks
  cities, 321
  electric power, 325
  energy, 325, 329
  shipping, 323
  transportation, 321
Florida panther, 275
FLORIDA-WORLD program, 392
Fog, 106
Food chain
  definition, 409
  forest, 55
Food web
  Crystal River estuary, 258
  definition, 409
  evergreen hardwood forest, 210
  forest system, 53
Forest
  comparing management, 239
  conservation, 211
  ecosystem, 22
  food chain, 55
  maritime, 128
  rainfall, 38
  succession, 201
  walk, 35
Forestry, 235
Fossil fuel
  definition, 409
  Florida, 266
Fossils
  Florida, 121
  large foraminifera, 124
Free trade, 247
Freeze
  mangroves, 191
  oranges, 244
  wetlands, 115, 197
Freshwater
  ecosystems, 171
  rivers entering the sea, 137
Fronts (weather), 109
Fuels
  net emergy evaluation, 332
  sources for Florida, 299
  use in Florida, 395
  yield ratios, 334
Full moon, 132
Future
  readings, 422
  scenarios, 371
  simulation, 363

G

Gene pool
  definition, 409
  forests, 239
Geology, 119
Geology of Florida
  cycle evaluation, 395
  definition, 409
  periods, 122
  uplift, 299
Geothermal
  power, definition, 409
  yield ratio, 337, 341
Ghost town, 89
Global
  definition, 409
  mini-model, 364
  program WORLD, 391
Glossary, 405
Gram
  conversions, 399
  definition, 409
Grass flats
  definition, 409
  estuaries, 161
  marine ecosystem, 145
Grasslands
  biome, 116
  definition, 409
Green Swamp, 215
Greenhouse effect
  causes, 13
  definition, 409
  gases, 354

Greenways
definition, 410
wildlife, 275
Gross production
definition, 410
photosynthetic, 45
succession, 65
Gross state product
definition, 410
explanation, 312
Florida, 300
historical trend, 359
Groundwater
beach, 128
definition, 410
hydrological cycle, 123
pine forests, 207
wetlands, 188
Growth models, 83
Growth policies, 371
Guano, 346
Gulf Stream
definition, 410
Florida, 135
Gypsum, 264
Gyre
definition, 410
Florida continental shelf, 137

H

Habitat
definition, 410
species, 273
Hammocks
subtropical, 209
tropical, 211
Hardwater
definition, 410
wet prairie, 193
Hardwood forests
deciduous, 208
definition, 410
ecosystem, 210
subtropical evergreen, 209
tropical evergreen, 211
types, 206
Hare, 94
Heat engines
efficiency, 331
power, 329
weather, 101
Heat sink
definition, 410
symbol, 21
Heat
definition, 410
explanation, 27
Herbaceous, definition, 410
Herbivore
definition, 410
ecosystems, 17
Hibernate, definition, 410
Hierarchy
army, 60
definition, 410
energy, 61
population centers, 321
High pressure (weather)
definition, 410
global pattern, 105
High-quality energy, definition, 410
Highways
fill and underpasses, 193, 196
replacement, 322
system diagram, 323
Home
appropriate landscape, 276
system, 286
Homestead, hurricane damage, 114
Housing, 283
Humidity
definition, 410
water vapor in air, 106
Humus
definition410
soil, 209

Hunting, 276
Hunting and gathering
  definition, 410
  society in Florida, 303
Hurricanes, 112
  definition, 410
  evaluation, 395
Hydroelectric power
  contribution to Florida, 395
  in Florida, 341
  yield ratio, 337
Hydrogen
  fuel, 336
  ions, 352
  yield from natural gas, 334
Hydrologic cycle
  definition, 410
  diagram, 120
  Florida, 123
Hydrology
  definition, 410
  in Florida, 119
Hydroperiod
  definition, 410
  lake borders, 171
  wetlands, 186
Hydroponics
  definition, 410
  example, Epcot Center, 48

I

Ice ages
  definition, 410
  water levels, 125
Icon
  definition, 410
  EXTEND, 241
  symbols, 19
Illegal drugs evaluation, 395
Image
  definition, 410
  Florida system, 315
  tourists, 15
Immigration, in Florida, 300
  definition, 410
Imports
  evaluation, 396
  Florida summary, 315
  overview, 15
Indian River estuary, 166
Indians
  cultures, history, 303
  overview, 5
  system diagrams, 306
  territories, 307
Industry
  definition, 410
  system, 289
Information
  definition, 411
  emergy, 300
  system in Florida, 290
Infrastructure
  definition, 411
  Florida, 322
Intensive agriculture
  definition, 411
  Florida, 243
Intertropical convergence
  definition, 411
  world circulation, 105
Interaction
  definition, 411
  symbol, 20
Interdependence, 218
International
  emergy per person, 360
  trade, 318, 325
Intertidal zone
  breaking waves, 150
  definition, 411
  oyster reefs, 164
Inversion
  definition, 411
  microclimate, 114
Invertebrates
  definition, 411
  marine animals, 158

Ion, 138
  definition, 411
Isobars
  definition, 411
  global pattern, 105
  winds, 107
Iteration
  definition, 411
  simulation program, 79
Ivory-billed woodpecker, 189

J

Joule
  conversions, 399
  definition, 411
  energy flow, 30
Juncus, 191

K

Kelvin temperatures, 330
Key (geographical), definition, 411
Kilocalorie, 30
  definition, 411
Kissimmee River
  channelization, 192
  Lake Okeechobee, 189
  role in south Florida, 4

L

Lake Apopka
  eutrophy, 177
  wetland, 188
Lake Okeechobee
  emergency discharge, 167
  management history, 177
  overflow to Everglades, 193
  overview, 4
  water flows, 189
  water reservoir, 177
Lakes, 171
Land rotation
  definition, 411
  sustainable landscape, 276
Landfill
  definition 411
  problem, 355
Landscape
  definition, 411
  ecosystem mosaic, 215
  Florida overview, 6
Larvae
  definition, 411
  estuarine nursery, 158
Latitude, 102
  definition, 411
Law of diminishing returns
  definition, 411
  graph, 47
Leach
  definition, 411
  wetland, 262
Lead filtration by wetland, 351
Leaves, decomposition, 34
Legume
  definition, 411
  nitrogen-fixing, 48
Lichens
  definition, 411
  tree limbs, 211
Light, 103
  definition, 411
  night light from satellite, 293
Lightning, 111
Lignin, 352
Lignite fuel, 337
  definition, 411
Limestone
  definition, 411
  Florida mass, 122
  mining, 266
Limiting factors
  concept, 46
  definition, 412
  program FACTORS, 383

Litter
  definition, 412
  forest, 33
  model of decay, 80
Logistic
  growth, 84
  growth, definition, 412
  program LOGISTIC, 384
Long-leaf pine, 207
Longshore current, 135
  definition, 412
Loop current, 136
Low-energy agriculture, definition of, 412
Low pressure (weather)
  definition, 412
  global pattern, 105
Loxahatchee National Wildlife Refuge
  location, 219
  river of grass, 196
Luminescence, 141
Lynx, 44

M

Macroeconomics, 225
Maize
  definition, 412
  Indian agriculture, 305
Mangroves
  definition, 412
  dwarf, 193
  ecosystem, 191
  estuaries, 164
Manure, 346
Maps
  cities in Florida, 322
  city, 284
  electric power network, 326
  rain, 125
  regional, 215
  rivers, 5
  satellite view, 9
Marine ecosystems, 141
Market
  supply and demand, 228
  value, definition, 412
Marshes
  definition, 412
  filters, 197
  freshwater, 185, 190
  map, 186
  saltwater, 164, 191
Mature forest, 239
Maximum power principle, 67
  definition, 412
Maximum sustainable yield, 254
Mediterranean climate, 116
Megafauna, definition, 412
Megajoule, definition, 412
Melaleuca
  transpiration, 195
  water table, 69
  wetland strategy, 231
Mercury, 356
Metabolism
  definition, 412
  ecosystem, 29
Methane, 354
Metric conversions, 399
Microbes, *see* Microorganisms
Microclimate
  definition, 412
  inversions, freezing, 114
  protecting orange groves, 244
Microcosms
  definition, 412
  examples, instructions, 71
Microeconomics
  definition, 412
  resource use, 225
Microorganisms (microbes)
  activated sludge, 350
  definition, 412
  pine forest, 33
Migrant labor
  definition, 412
  Florida agriculture, 250

Migration
definition, 412
land birds, 209
estuary, 158
Mining, 261
Model
competition
definition, 20, 412
drain, 87
exponential growth, 83
fire, 95
logistic growth, 84
nonrenewable, 87
prey-predator, 91
pulse, 94
renewable, 86
tank, 75, 80
two-source, 88
Money, 226
Money transaction
definition, 412
sales and costs, 226
symbol, 23
Monoculture
definition, 412
forests, 237
Moon and tide, 103
Mosaic
definition, 412
Florida ecosystems, 215
Mycorrhizae, 239
Mosquitoes, 165, 271

N

National forests, 271
Native, definition, 412
Natural gas
electricity yield ratio, 337
evaluation, 395
yield ratio of compressed gas, 334
Neap tide
definition, 413
explanation, 132
Nematodes, 245
Net ban, 14, 158
Net benefit
mining, 266
emergy evaluation, 229
Net emergy
concept, 332
definition, 413
yield of sources, 329
Net production, 45
definition, 413
Networks, 321
New moon, 132
Niche
definition, 413
species, 65
Nitrates
groundwater pollution, 351
nitrogen cycle, 42
Nitrogen
cycle, 41
definition, 413
Nitrogen fixation, 42
definition, 413
soil, 48
pasture, 248
Nitrous oxide, 354
Nonrenewable resources
definition, 413
model, 87
program NONRENEW, 386
source, 261
sources evaluation, 396
Northern Florida, 3
Nuclear power
breeder reactor, definition, 413
contribution in Florida, 395
cooling, Turkey Point, 167
fission power plant, definition, 413
fusion power, definition, 413
net yield, 336
Nursery
definition, 413
estuary, 158

Nutrients
cycle, 43
definition, 413
release in ecosystems, 17

O

Ocala National Forest
dunes and water table, 207
location, 217
sand scrub ecosystem, 272
Ocean thermal electrical conversion (OTEC)
definition, 413
Florida, 341
yield ratio, 337
Oceanography
definition, 413
Florida, 131
Oil yield ratio, 334
Okefenokee Swamp, 178, 185, 215
Oklawaha River
Rodman dam, 166
Silver Springs, 180
Oligotrophic lakes
definition, 413
Florida, 174
Optimum efficiency, 329
Organic consumption, equation, 29
Organic matter
definition, 413
ecosystem, 17
Oscillation
chaotic, 96
definition, 413
ecosystems, 67
program PREYPRED, 387
program PULSE, 389
systems, 91
Overfishing
definition, 413
Florida, 253
Oxidize, definition, 413
Oxygen
definition, 413
diurnal changes in waters, 175
Silver Springs, 183
Oyster reefs
ecosystem, 162
system diagram, 165
Ozone
definition, 413
formation and distribution, 353
pollution, 12

P

Panther, 275
Paper mill wastes, 352
Parks
aquatic, 272
Florida, 270
Pathway, 24
Peace River, phosphorus, 167
Peat
definition, 413
evaluation, 395
explanation, 188
oxidation, 196
reserves and mining, 265
yield ratio, 334
Peatland, 147
Pennekamp National Park, 150
People model
program PEOPLE, 390
simulation, 361
Percolate, definition, 413
Petroleum contribution to Florida, 395
pH, 352
Phosphorus
calcareous sands, 212
cycle, 39
definition, 414
evaluation, 395
mining, 218, 261

Photons
  definition, 414
  spectrum, 103
Photosynthesis
  definition, 414
  ecosystems, 17
  equation, 29
Phytoplankton
  definition, 414
  estuaries, 163
  freshwaters, 171
  oceanic, 141
  oscillating, 91
Pine flatwoods, 206
Pine forests
  distribution, 221
  fire, 204
  food web, 53
  groundwater, 207
  system diagram, 32
  types, 206
Pine plantation
  solid waste dispersal, 355
  system, 235
Pioneer
  definition, 414
  species, 64
Plankton
  definition, 414
  estuaries, 163
  oceanic waters, 143
Plankton-benthos (ecosystem)
  components, 153, 160
  definition, 414
  in estuaries, 155
Plastic waste, 355
Pleistocene
  archeological site, 303
  definition, 414
  geological record, 122
Plot, two populations, 93
Plutonium
  definition, 414
  nuclear power, 337
Polar front
  definition, 414
  global position, 105
  role in Florida, 106
Polar ice, 116
Policies for prosperity, 371
Pollen records, 304
Pollution, *see* Wastes
  definition, 414
  dilution, 347
Pond cypress, 188
Ponds
  components, 171
  cypress, 187
  definition, 414
  simple diagram to complete, 34
  simulation, 361
  system diagram, 173
Population
  carrying capacity, 357
  contribution to Florida, 395
  crowding, 85
  definition, 414
  increase in Florida, 15, 358
  problems, 14
  resource limits, 360
  simulation, 361
Potable water, definition, 414
Potassium, as nutrient, 39
Potential energy
  definition, 414
  ecosystem, 27
  from the ocean, 139
Power
  definition, 329
  maximum, 67
  optimum efficiency, 329
Power plants, 137, 414
ppm, definition, 414
ppt, definition, 414
Prairie, definition, 414
Precipitation, definition, 414
Predator, 51
  definition, 414

Prey-predator model, 91
oscillation, 93
program PREYPRED, 387
prey, definition, 414
Prices
market supply, 227
seafood, 158
Primary consumer
definition, 414
food chain, 56
Primary treatment, 349
Producers
definition, 414
ecosystem, 17
Production
definition, 414
ecosystem, 45
Productivity, definition, 414
Programs (computer), definition, 414
2SOURCE, 386
DRAIN, 385
FACTORS, 383
FIRE, 388
FLORIDA, 392
LOGISTIC, 384
NONRENEW, 386
PEOPLE, 390
PREYPRED, 387
PULSE, 389
RENEW, 385
TANK, 383
WORLD, 391
Pulse model, 94
definition, 414
program PULSE, 389

R

Radioactive fallout, 347
Rain
evaluation in Florida, 395
forest, 116
formation, 106
map, 125
Reclamation
definition, 415
mining, 264
Recycle
definition, 415
ecosystems, 22,37
wastes of human society, 345
Red tide
definition, 415
shelf ecosystem, 145
Reefs
coral, 145
definition, 415
oyster, 162
stress, 14
Reflecting infrared light, 188
Reinforcing feedback, 69, 254
Relative humidity, 106, 415
Renewable resources
definition, 415
model 86, 385
program RENEW, 385
sources, 337, 396
Reorganization, 67
Reset, 68
Residential neighborhood
landscape, 276
system, 285
Resources
purchased resources, 299
renewable, 298
stored reserves, 300, 398
summary table, 312
Respiration
definition, 415
equation, 29
Restoration
phosphate mining, 264
refuges for reseeding, 276
Reuse, 345
Riptide (rip current), 135
Rivers
entering sea, 137
evaluation, 395
Florida map, 5

Rockland pines, 207
Rodman pool
dam obstruction to fish, 166
effect on Silver Springs, 180
Rootlets, 41

S

Salinity
definition, 415
density, 136
Florida marine waters, 155
Saltwater intrusion
coastal pumping, 125
definition, 415
Sand, 265
definition, 415
Sand-dune ecosystems, 152
Sand pines, 207, 272
Sandstone, 121
definition, 415
Santa Fe river, 188
Sargassum
accumulated radioactivity, 347
definition, 415
ecosystem, 143
Satellite
night lights, 293
view of Florida, 9
Saturation
definition, 415
oxygen in water, 175
water vapor in air, 106,
Savannah, 116
Sawgrass, 197
Scenarios, future, 371
Sea level
ice ages, 126
rise, 14, 129
Seawater
chemistry, 138
composition, 138
Second energy law, 29
definition, 415
Secondary consumer
definition, 415
food chain, 56
Secondary treatment, 349
Sector
definition, 415
economy consumers, 313
Sedimentary cycle
definition, 415
diagram, 120
Sedimentary rock, 121
Sediments
definition, 415
rock formation, 121
Seedling, definition, 415
Self-interactive drain
definition, 415
logistic model, 85
Self-organization
definition, 415
interface wetlands, 197
succession, 63
waste recycle, 349
Self-sufficient agriculture, 250
Seminole Indians, 305
Sewage
definition, 415
municipal wastes, 349
Shale
definition, 415
kinds of sedimentary rock, 121
Shell middens
definition, 415
Florida Indians, 304
Silver Springs, 180
Simulation
definition, 415
future, FLORIDA model, 365
future, PEOPLE model, 390
future, WORLD model, 363
hand, 77
procedure, TANK model, 75
readings, 422
Sinkhole
definition, 415

formation, 13
limestone solution, 121
Slash pine, 235
Sludge, 350
Smoky Bear, 206
Softwater
definition, 416
lakes, 174
Soil
definition, 416
formation, layers, 203
geologic action, 119
limiting factors, 48
part of ecosystem, 18
types, 219
Solar emergy
definition, 416
explanation, 57
Solar technology
definition, 416
hotwater heaters, 340
solar voltaic cells, 339
Solar transformities
definition, 416
table, 59
Solar voltaic cells
definition, 416
efficiency, net yield, 339
pine plantation, 355
Solar water heaters, 340
Solid waste
content, processing, 354
definition, 416
Source
definition, 416
renewable, 86
symbol, 19
Southern Florida, 3
Space, 294
Spanish colony, 306
Spanish moss, 211
Spartina, 191
Spatial hierarchy, 284
Species, 274
definition, 416
Spreadsheet
definition, 416
program calibration, 79
Spring tide, 132
definition, 416
Springs, 180
definition, 416
Squall line
hurricane, 113
prefrontal, 108
St. Marys River, 178, 187
St. Johns River
bird's-eye view, 4, 9
estuary, 166
Standard of living
emergy measure, 357
definition, 416
Steady-state scenario, 374
climax concept, 63
definition, 416
Storage
definition, 416
symbol, 20
Storms
definition, 416
effect on ocean, 138
kinds, 101
thunderstorms, tornadoes, 111
tropical, hurricanes, 112
Storm-water retention ponds, 198
Strand
definition, 416
wetland type, 401
Streams
definition, 416
ecosystems, 178
predominant animals, 179
Strip mine
definition, 416
limestone, fuels, 266
peat, sands, 265
phosphate, 261
Subtropical evergreen forest
definition, 416
hardwoods, 209

Succession
characteristics of stages, 69
definition, 416
early, 64
ecosystem self-organization, 63
growth model, 86
upland forests, 201
Sugarcane, 247
Sulfur dioxide, 13
Summer weather, 110
Sun belt
definition, 416
image, 315
Sun, contribution to Florida, 395
Sunlight, 297
Superphosphate, 262
Supply
definition, 228, 416
Surfbeat, 135
Sustainable
definition, 416
growth, 371
management, 212
scenario, 372
society policies, 372
Suwannee River
diversion south, 231
nitrate wastes, 351
Okefenokee Swamp source, 178, 185
Swamps
definition, 416
kinds and distribution, 185
stream, 179
Switch
definition, 416
symbol, 31
Switching model, 93
Symbiosis
definition, 416
self-organization, 63
Symbols for systems diagrams
definition, 416
explanation, 18
summary, 24
Systems
concepts, 17
definition, 416
readings, 421

T

Taiga, 116
Tamiami Trail, 196
Tampa Bay, 167
Tank model
explanation, 75
program TANK, 79, 383
simulations, examples, 79
Taxes, 15, 290
Taylor self-sufficient farm, 251
Technology
definition, 416
environmental, 349
industry, 289
Television, 292
Temperature
definition, 417
measurement, 27, 331
scales, 330
Tentacle, definition, 417
Tequesta Indians, 305
Tertiary treatment, 349
Thermodynamics
definition, 417
energy, 27–36, 329–343
Thousand Islands
Everglades flow, 193
location, 189
Thunder
definition, 417
storms, 111
Tide
contribution to Florida, 395
definition, 417
explanation, 131
yield ratio, 341
Timucua Indians, 305
Titanium mineral sands, 265

Tornadoes
  definition 417
  severe thunderstorms 111
Tourists, 269
Toxic
  definition, 417
  metals in sludge, 349
Tracks of animals, 36
Trade
  Florida, 317
  free, 247
  simulation, 366
Trade winds
  climatic zones, 115
  definition, 417
  ocean currents, 135
Transaction
  definition, 417
  symbol, 23
Transfer payments
  cities, 283
  definition, 417
Transformation
  definition, 417
  energy, 27
Transformity
  definition, 417
  Florida inputs, 395
  table, 59
Transpiration
  definition, 417
  explanation, 37
  wetland, 187
Transportation
  alternatives, 324
  ecological engineering, 325
  fuels, 335
  network, 321
Transporting fuels, 333
Treatment of wastes, 349
  wetland, 197
Tree islands, 219
Truck farming
  definition, 417
  south Florida, 245
Tundra, 116
Turbidity
  definition, 417
  streams, 178
Turbulence
  definition, 417
  marine ecosystems, 141
Turkey oak, 207
Turkey Point nuclear plant, 167
Turnover time, definition, 417
Turtle grass
  estuaries, 161
  south Florida illustration, 46
Turtles, 152
Two-population plot, 92
Two-source model
  description, 88
  program 2SOURCE, 386

U

Ultraviolet light
  definition, 417
  ozone, 353
Understory
  definition, 417
  pine forest, 235
Universities, 291
Upland forests, 201
Upwelling
  definition, 417
  storm nutrients, 138
Uranium
  definition, 417
  nuclear power, 337
Urban system
  definition, 417
  diagrams, 279
Utilities, definition, 417

V

Vegetarian diet, 250

W

Warm front, 105, 109
Warm Salt Springs, 183
Wastes
alternatives, 347
city, 286
definition, 417
history, 345
overview diagram, 348
wastewaters, 349
Water
budget, 126
cycle, 37
hard, 120
vapor, 106
Water pollution, 11
Waterways, 323
Waves
breaking zone, 150
energy, definition, 417
evaluation, 395
yield ratio, 341
Wealth, definition, 417
Weather, 101
definition, 417
Weathering, 120
Weeds
definition, 418
succession, 64
Wetlands
constructed, 197
definition, 418
distribution, 85
filters, 197
management, 197
system diagram, 190
types, 401
waste treatment, 351
Wet prairie, 193
"What if" experiments, 79
Wilderness, 270
Wildlife corridor
definition, 418
forest ecosystem, 22
greenways, 275
Wildlife habitat, 230
Wind
air pressure, 107
contribution to Florida, 395
definition, 418
waves, 134
yield ratio, 340
Winter
vegetables, 245
weather, 109
Withlacoochee River, 185
Wood storks, 12, 92
Wood stoves, 92
Wood yield ratio, 334
Work
definition, 418
concept, 27
World simulation
model, diagram, 363
program WORLD, 391

Y

Yield
definition, 418
fisheries, 253
systems, 224

Z

Zooplankton
definition, 418
lakes, 171
oscillation, 91
vertical migration, 141
Zooxanthellae, 145